INTERNATIONAL TECHNOLOGICAL UNIVERSITY
This Book is Donated by:
Mr. Martin Dost
Date: 02/10/95

MODELING AND ANALYSIS
OF
LINEAR PHYSICAL SYSTEMS

MODELING AND ANALYSIS OF LINEAR PHYSICAL SYSTEMS

J.F. LINDSAY
V. RAMACHANDRAN

Department of Electrical and Computer Engineering
Concordia University
Montreal, PQ
CANADA

THE LATE SILAS KATZ

Department of Mechanical Engineering
Concordia University
Montreal, PQ
CANADA

©Copyright, Weber Systems, Inc. 1990

All rights reserved. No part of this book may be reproduced or utilized in any form or by any means, electronic or mechanical, including photocopying, recording or by any information storage and retrieval system, without written permission from the publisher.

Weber Systems, Inc.
8437 Mayfield Rd.
Chesterland, OH 44026
Phone: (216) 729-2858
FAX: (216) 729-3203

Production: Jean Knox and Andy Jager
Printing: Cushing-Malloy

ISBN: 0-929704-19-3

CONTENTS

PREFACE	xiii
INTRODUCTION	**1**
1.1 The Circuit as a System	1
1.2 The Basic Physical Components	3
1.3 Input and Output Signals	8
1.4 Modeling	15
1.5 Signal Variables	17
SINGULARITY FUNCTIONS	**2**
2.1 The Unit Step Function	24
2.2 Functions Derived by Integration of Unit Step	32
2.2.1 The Ramp Function	33

	2.2.2 The Parabolic Function	37
2.3	Functions Derived by Differentiation of Unit Step	41
	2.3.1 The Unit Impulse	41
	2.3.2 The Unit Doublet	45
2.4	Exponential Functions	46
2.5	Composite Functions	55
2.6	Mathematical Operations on Functions	61
	2.6.1 Time Integration	61
	2.6.2 Time Differentiation	67
	2.6.3 Proportional Plus Integral	69
	2.6.4 Proportional Plus Derivative	74
	2.6.5 Proportional Plus Derivative Plus Integral	78
2.7	Composite Functions by Graphical Approach	84
2.8	Summary and Discussions	86
2.9	Problems	87

CIRCUIT COMPONENTS 3

3.1	Introduction to Circuit Components	101
3.2	Electric Circuit Components	106
	3.2.1 Electric Signal Variables	106
	3.2.2 Electrical Resistance	108
	3.2.3 Electrical Inductance	111
	3.2.4 Electrical Capacitance	114
3.3	Fluid Circuit Components	117
	3.3.1 Fluid Signal Variables	118
	3.3.2 Fluid Resistance	121
	3.3.3 Fluid Inertance	125
	3.3.4 Fluid Capacitance	128
3.4	Mechanical Circuit Components	137
	3.4.1 Mechanical Signal Variables	138

3.4.2 Damping (Mechanical Resistance)	141
3.4.3 Spring (Mechanical Inductance)	145
3.4.4 Mass (Mechanical Capacitance)	150
3.5 Thermal Circuit Components	152
3.5.1 Thermal Signal Variables	153
3.5.2 Thermal Resistance	154
3.5.3 Thermal Inertance	156
3.5.4 Thermal Capacitance (Heat Storage)	157
3.5.5 General Thermal Components	160
3.6 Sources and Loads	161
3.6.1 Across Variable Sources	162
3.6.2 Through Variable Sources	166
3.6.3 Loads	170
3.6.4 Modeling of Coulomb Friction	172
3.6.5 Modeling of Weight	174
3.7 Elements as Mathematical Operators	175
3.8 Summary and Discussions	181
3.9 Problems	183

BASIC NETWORK MODELS 4

4.1 Formulation of Circuits	196
4.2 Basic Circuit Laws	208
4.3 Properties of Storage Elements	216
4.3.1 Initial Conditions of T-type and A-type Elements	217
4.3.2 Final Conditions with Step Inputs	224
4.4 The Mathematical Approach to Initial Conditions	234
4.5 Circuit Models for Initial Energy Storage	243
4.5.1 A-type Storage	243
4.5.2 T-type Storage	246

4.6 Realization of Mathematical Operations on
Singularity Functions by the Circuit Models   249
4.6.1 Realization of Proportional plus Integral   250
4.6.2 Realization of Proportional plus Derivative   253
4.6.3 Realization of Proportional plus Derivative
plus Integral   256
4.7 Summary and Discussions   260
4.8 Problems   260

RESPONSE OF FIRST ORDER CIRCUITS   5

5.1 Step Response   273
5.2 Ramp, Parabolic and Impulse Responses   292
5.3 Response to Composite Functions   303
5.4 Response with Initial Storage   314
5.5 Summary and Discussions   326
5.6 Problems   328

RESPONSE OF SECOND ORDER CIRCUITS   6

6.1 Formulation of Differential Equations   341
6.2 Natural Response of Second Order Circuits   346
6.3 Step Response of Second Order Circuits   355
6.4 Response of Second Order Circuits to Other Inputs   371
6.5 Response of Second Order Circuits with Initial Storage   381
6.6 Summary and Discussions   393
6.7 Problems   394

GENERALIZED IMPEDANCES AND SYSTEM FUNCTIONS   7

7.1 The General Exponential Input Function   406

7.2	Response to Exponential Function	408
7.3	Impedance of Basic Components	417
7.3.1	Impedances in Combination	421
7.3.2	Driving Point Impedance and Admittance	427
7.4	System Functions	431
7.5	General Formulation of System Equations	440
7.5.1	Determination of Analogous Impedance Circuits	442
7.5.2	Node-to-datum Equation Formulation	444
7.5.3	Loop and Mesh Equations	456
7.6	Some Properties of Linear Systems	467
7.6.1	Superposition Theorem	471
7.6.2	Reciprocity Theorem	473
7.6.3	Thevenin's Theorem	474
7.6.4	Norton's Theorem	480
7.7	Summary and Discussions	482
7.8	Problems	482

SINUSOIDAL RESPONSE 8

8.1	Response to Sine and Cosine Functions	497
8.1.1	First Order Circuits	497
8.1.2	Second Order Circuits	508
8.2	Sinusoidal Steady State	518
8.3	Power in the Sinusoidal State	526
8.3.1	Maximum Power Transfer	531
8.4	Signal Filters	536
8.4.1	Low-Pass Filters	536
8.4.2	High-Pass Filters	543
8.4.3	Band-Pass and Notch Filters	549
8.5	Resonance	554
8.5.1	Series Resonance	555

8.5.2 Parallel Resonance	559
8.6 Bode Plot	561
8.6.1 Magnitude Plot	563
8.6.2 Phase Plot	574
8.7 Modeling Based on Sinusoidal Response	580
8.8 Summary and Discussions	587
8.9 Problems	588

LAPLACE TRANSFORMS 9

9.1 Definition of Laplace Transform	593
9.2 Inverse Laplace Transform	596
9.3 Properties of Laplace Transform	597
9.4 Solutions of Differential Equations	611
9.5 Applications in the Analysis of Networks	622
9.6 Models of Basic Components in the Transform Domain and Some of Their Uses	625
9.7 Summary and Discussions	634
9.8 Problems	635

TWO-PORT NETWORKS AND MUTUALLY COUPLED COMPONENTS 10

10.1 Two-Terminal Pairs	640
10.1.1 Open-Circuit Impedance Parameters	640
10.1.2 Short-Circuit Admittance Parameters	643
10.1.3 Hybrid Parameters	652
10.2 Transformers	654
10.2.1 Mutually Coupled Coils (Electrical Transformers)	654
10.2.2 Mutually Coupled Mass (Mechanical Transformer)	660
10.3 Ideal Couplers	666
10.3.1 Levers	667

10.3.2 Pulleys	670
10.3.3 Gears	674
10.3.4 Differential Piston	678
10.4 Equivalent Network for Ideal Couplers	679
10.5 Selection of Coupler for Maximum Power Transfer	689
10.6 Summary and Discussions	691
10.7 Problems	692

NUMERICAL SOLUTIONS 11

11.1 Introduction	703
11.2 Block Diagram of a Differential Equation	705
11.3 State Equations	709
11.4 Numerical Integration	714
11.5 Runge-Kutta Methods	723
11.6 Summary and Discussions	735
11.7 Problems	737

BIBLIOGRAPHY	741
ANSWERS TO SELECTED PROBLEMS	743
INDEX	753

PREFACE

To many of the students entering engineering or technology programs, this is an instantaneous world. For them, transients either do not exist or are not very important. They proceed from a wish to the initiation of an action and then the desired result occurs zero seconds later. The neglect of transients is not restricted to students alone. It is common to all of us, occasionally. For example, a package is to be moved from one place to another and it gets there - somehow. The velocity during transit usually does not concern us. Of course, if the package contains a block of ice and it is to be delivered across town on a hot day, we do give more thought to the transit time. Similarly, when a television is switched on, we pay little attention to the warm-up time. Has anyone ever seen anybody measuring the duration of the warm-up? Yet, if important news was about to be broadcast and we feared missing it, we would notice the warm-up period and be irritated by it.

Since so much of our previous training and environment emphasize instantaneous action (that is, quick food, overnight success), we feel that it is necessary to call attention to dynamic effects as early as possible in a student's education. Accordingly, this text considers transients and is designed for use in the first or second year of an engineering or technology program. In this way, students are made aware at the beginning of their program that there are no instantaneous processes in real life and that sometimes processes require a long time to complete. The precise dynamics of a process depend on geometric and material properties which students normally encounter in subsequent years when they study fluid mechanics, electronics, heat transfer and vibrations. However, from a few basic ideas presented here, students will be able to estimate and appreciate the dynamic performance of certain processes before they are exposed to the more detailed studies.

This text is a revised version of the previous book on the same topic retaining the basic principle of presenting the system dynamics in the simplest possible way. Additional material has been introduced in order to clarify the concepts to students. We do believe that the stated objective is best accomplished through the use of circuits and circuit components. Most students have been exposed briefly to only electrical circuits in their high school physics courses. Thus the circuit concept offers an excellent framework with which to bridge the gap from static to dynamics analysis. More important perhaps is that circuit theory lends itself very well to analogy. Through the use of circuit methods, we may show that it is possible to treat problems in fluid, thermal and mechanical systems as well as in electrical systems. For example, the similarity between charging a capacitor and filling a bucket of water is made obvious by comparison of the mathematical descriptions of these processes. Of course, we might have recognized this similarity intuitively. However, we find that these processes are also analogous to the change in velocity of a mass when a force is applied. This is not quite so obvious. Thus the circuit concept proves extremely useful in reducing physical situations in different physical media to a common form from which a common type of solution may be obtained.

We have tried to present the material in a logical order. However, it seemed that there were always several things that had to be introduced simultaneously as in the earlier book. Chapter I contains some basic ideas on system response and attempts to introduce the technical concepts that follow. In Chapter II, we deal with time functions. Although the emphasis is on specification of certain functions as system inputs, the purpose is rather broader. It is to make the student feel comfortable in the mathematical formulations and operations with time dependent signals. Chapter III describes the circuit components in the different physical media and stresses the analogy between them. It also reinforces the concepts of time dependent signals by applying these signals to the components. The combination of components into simple circuits and the basic circuit laws are introduced in Chapter IV.

The initial and final properties of storage elements and the effect of initial storage are also presented here. The reason for this is that these properties are best described by referring to their effects in circuits. Chapters V and VI present the solution of the equilibrium circuit equations for first and second order circuits respectively. The generalized impedance concept is reserved for Chapter VII, since it facilitates the formulation of equations in more complex circuits. This material is presented without recourse to Laplace transforms but is based rather on the concept of exponential inputs. This has the advantage that students work in the time domain and thereby develop a sound appreciation of the dynamic behavior of physical situations. Chapter VIII considers the frequency response of circuits and this follows quite properly from the impedance concepts. This chapter also contains two additional sections namely Bode plots and modeling based on sinusoidal response. Chapter IX discusses Laplace transforms and their applications in the analysis of such systems. Chapter X considers mutually coupled devices based on two-terminal pair circuits. Finally, Chapter XI deals with numerical solutions based on state-variable representations.

We acknowledge and appreciate the excellent suggestions made by Dr. Rama Bhat and Mr. N. Suresh, based on their experiences in teaching this course. We thank Ms. Maureen Kennedy for her help with the word-processing unit. We thank our respective wives Judith Lindsay and Kamala Ramachandran and our children for their patience and understanding during the preparation of the book. We also wish to place on record our appreciation of Ms. Estelle Katz for consenting to use her husband's name, the Late Silas Katz, as an honorary author, since he was a coauthor of the previous edition.

After tossing a coin, it was decided that the authors be listed strictly in the alphabetical order.

J.F. Lindsay
V. Ramachandran

Montreal, Canada
January 1991

CHAPTER 1

INTRODUCTION

1.1 THE CIRCUIT AS A SYSTEM

Most engineering endeavors are concerned with physical devices or groups of devices for which the word **system** is generally applied. By definition, a system is a set of things or parts forming a whole and serving a common purpose. The components or devices that combine to form a physical system are based on phenomena that engineering students normally encounter in a first course in physics. Our purpose here is to introduce methods for describing and analyzing physical systems.

The concept of a physical system is very broad. We attempt to be more specific by restricting our consideration to physical circuits. Basically, a circuit is a closed path. However, we use the term **circuit** to indicate a special type of system in which the components represent only one physical area. Thus we will deal with electrical mechanical fluid or thermal circuits. The advantage of the circuit approach is that the techniques used for the analysis of electric circuits are highly developed. These same techniques may be applied by analogy to problems in mechanical, fluid and thermal circuits. In this way, the similarities between problems, in different physical media, are made obvious. As a result, solutions in one physical medium are transferable, by analogy, to the other media.

The description of a circuit may be qualitative or quantitative. If the latter is the case, there is often the necessity for taking measurements, either to provide basic data or to confirm that a design does what it is supposed to do. The process of describing a physical situation is often referred to as **modeling**. To determine a suitable model requires the ability to observe the physical situation and to describe it first in qualitative terms. For example, when an automobile passes over a bump in the road, there is an up and down motion which takes place immediately afterwards. The manufacturer wants to know before building a million cars whether the motion will be small or large. It is necessary, therefore, to think in terms of a model that can give quantitative information on the amplitude of the motion. To obtain such a model, however, requires a more detailed qualitative description of the physical phenomena involved. In this case, it requires an understanding of the main parts of the suspension and the ability to observe exactly what happens at the point of contact between the wheels and the road. It is only after the qualitative description is complete that the quantitative description may be obtained.

There is, of course, a temptation to believe that once the quantitative description has been obtained (that is, the formula), the problem is solved. The danger in this attitude is that physical situations may change. Springs may break or shock absorbers may wear out. The only sure defense against this situation is to observe continually the physical situation.

1.2 THE BASIC PHYSICAL COMPONENTS

One of our primary objectives is the determination of suitable models for physical situations. Specifically, we want mathematical models based on established natural laws. However, we limit our consideration to physical situations where the mathematical modeling can be accommodated by means of linear constant coefficient equations. Mathematical models will be developed only for the most basic physical components. More complicated situations can then be described with combinations of basic components.

For example, let us consider one of the basic fluid system components - the fluid line. Fig. 1.1(a)

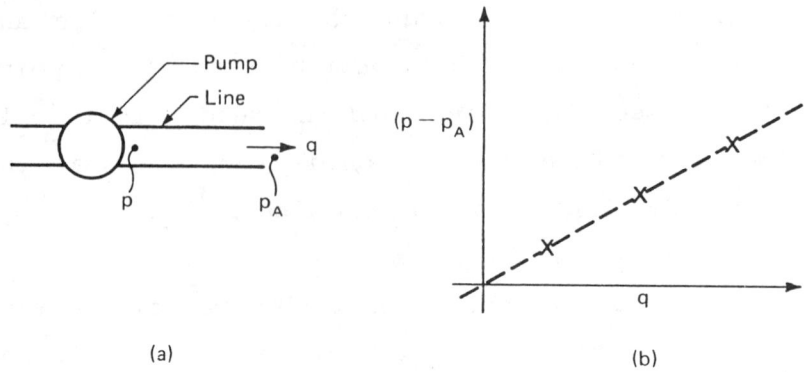

Figure 1.1 Fluid line component

shows a pump supplying fluid to a line. How are we to describe this component mathematically? To answer this question, we must first decide what we wish to know about the passage of the fluid in the line. One thing we want to know is the rate at which liquid is passing through the line. We call this the flow and for a fluid system, it is designated as q. Another thing we might need to know is the size of the pump required to produce the flow. Instead of physical size, the capability of a pump is often indicated by the pressure p, it can develop. If we were able to relate the pressure, p, to the flow, q, we would have a type of mathematical model of the line. But can we do this? We cannot because the flow through the line depends also on the pressure, p_A, at the end of the line. If p_A were equal to p, there would be no flow through the line. Thus our mathematical model must relate the flow, q, to the pressure difference, $p - p_A$. These quantities, flow and pressure difference, are the signal variables for fluid systems. Flow is the through variable (TV) and pressure difference is the across variable difference (AV). To determine the mathematical model of the line. we could perform some experiments on the line in which the pressure difference was changed and the resulting flow measured. Fig. 1.1(b) shows some typical results for three data points (indicated by x). As pressure difference increases, the flow increases. The mathematical model, of course, must hold for an infinite number of pressure difference - flow combinations - not just three data points. To expand the data so that it will prove useful for the other cases, we use the data to draw a straight line between the points, shown dashed on Fig . 1.1(b). This dashed line is based on our limited observations and is assumed to represent the fluid component at values of the signal variables where we have made no measurements. The mathematical description of the dashed line is our model of the fluid component.

It is possible to describe most of the basic physical situations by means of two-terminal components, but there are situations which require a model having two pairs of terminals. The common two-terminal-pair components are described in Chapter X. However, at this stage, we consider only the two terminal components. The general two terminal component is shown schematically in Fig. 1.2. The across variable difference represents a difference in a physical

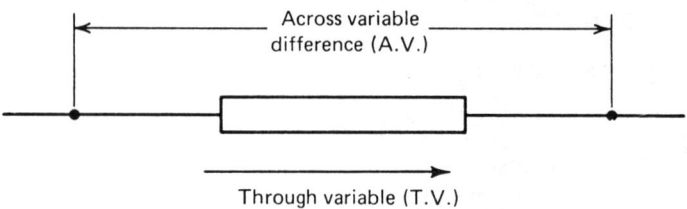

Figure 1.2 Two terminal component

condition on either side of the component and the through variable represents the quantity transmitted through the component. The value of the two-terminal representation is that it permits components to be combined together into circuits that can be analyzed by a well developed circuit theory. In this way, we may use a common method of analysis for all physical circuits.

In general, there are three basic components in each of the physical media:
  (a) one dissipative component in which energy is lost, and
  (b) two storage components in which energy is retained.
Fig. 1.3 shows the circuit symbol for each component performance. The symbol for the dissipative component, Fig. 1.3(a), usually represents electrical resistance. However, we use this symbol to represent a component in any of the media in which the across variable difference and through variable are directly proportional to each other. This is the situation that fits into our

Introduction

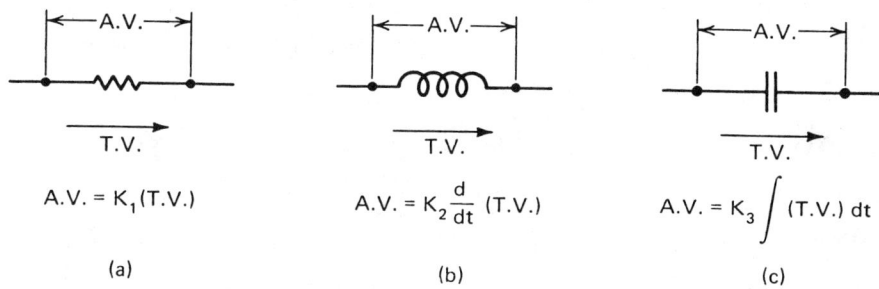

Figure 1.3 Basic components

observation of the fluid line. Thus the fluid line may be considered as a fluid resistance. In a similar way, the storage element symbols shown in Figs. 1.3(b) and (c), usually indicate electrical inductance and capacitance. For our purposes, we use these same symbols for components in other media that produce the same relationship between the through and the across variables. We shall find, for example, that mechanical springs produce the same signal variable relation as an inductance and that a mass produces a capacitive type relation. These components are discussed, in detail, in Chapter III. For the present, however, it will be sufficient to observe that the relations between the variables are either proportional, differential, or integral. The differential and integral components perform time dependent operations on the signal variables. These mathematical operations are discussed fully in Chapter II.

1.3 INPUT AND OUTPUT SIGNALS

Initially the circuits that we deal with consist of particular combinations of two terminal components. These combinations act together to produce a more complicated mathematical function than is produced by a single component. However, in Chapter X, we introduce mutually coupled components which have more than two terminals. A typical circuit with two terminal components is shown in Fig. 1.4. The input represents a signal variable (either across or through type) that is applied to the circuit. The input is basically a signal. The circuit operates on the input signal according to the circuit mathematical function and produces an output signal. The output signal is sometimes called the response of the circuit. It may also be either of the across or through type.

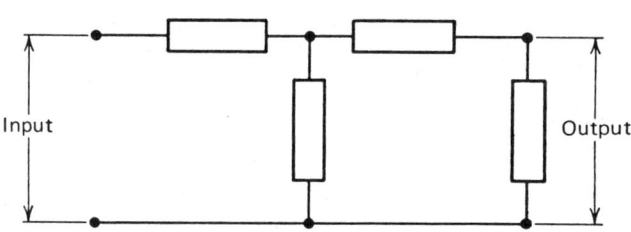

Figure 1.4 Typical circuit

In normal use, a system or circuit produces some results, hopefully those desired by the person using the device. The usual term used for such results is **output**. That is, the **output** of an automobile could be its velocity. However, in order to get an output, there must be some process of giving information or energy (or both) to the system. This is the **input**. Inputs to an automobile may include the energy stored in the gasoline, position of the steering wheel, road surface, accelerator position, and brake pedal position. These concepts are usually shown graphically by means of a block diagram such as shown in Fig. 1.5.

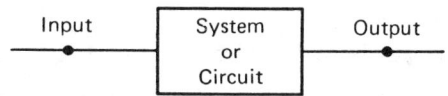

Figure 1.5 Block diagram representation of circuit

For an automobile, if efficiency is of no concern, there is no need to consider the output power as one of the outputs nor is there any need to consider the chemical energy of the fuel as one of the inputs. In this case, the output torque of the engine would be an input to the automobile transmission system.

The meaning of the word **system** therefore includes the concept of a boundary within which all the parts constitute the system. In this example of the automobile, it must be noted that it only the energy conversion process which would not be included in the system; the mass of the engine, both its case and moving parts, would still be included in the system. To a certain extent, the choice of the boundary for a particular problem is an art, the mastery of which is one of the attributes of the best engineers.

The terms **input**, **output** and **system** may therefore seem to be rather nebulous. Nevertheless it is essential to be fully aware of what initiates the action in a system (input) and what constitutes the results (output). Usually there is little difficulty in determining the inputs and outputs of a system, if it is possible to give a clear qualitative description of it.

The prediction of the behavior of a physical system involves determination of the response (output) to certain hypothetical stimuli (inputs). Although many systems eventually operate under steady conditions (for example, constant velocity for an automobile), there is the problem of getting them into such a steady state. Systems generally pass through a dynamic or transient state in which numerical values for their outputs are changing. Sometimes this dynamic state is the critical attribute in determining the acceptability of the performance, and therefore it is the dynamic response which must be determined. Since the numerical values of the outputs change with time, the simplest way to describe dynamic response is by means of a mathematical function of time. In general, the inputs also change with time and are described by mathematical time functions. However, there is the inherent understanding that the input function can be specified. It is the cause. The output function, on the other hand, is the effect and it depends on both the input and the circuit.

Fig. 1.6 shows two typical input functions, the

Introduction

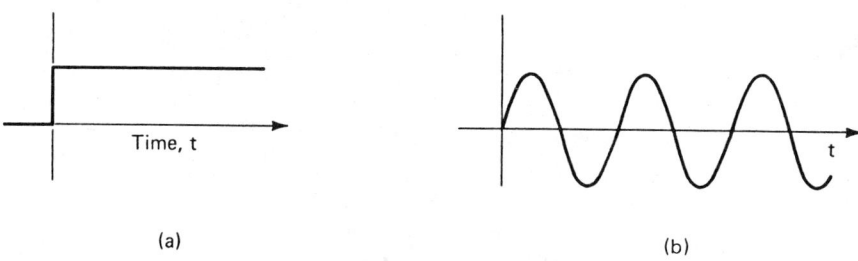

Figure 1.6 Typical inputs

step input [Fig. 1.6(a)] and the sine wave input [Fig. 1.6(b)]. In an actual system, the input may not be known in advance. Some drivers step down hard on the gas while others drive smoothly. However, the input functions shown are really test inputs. They enable us to compare the performance of systems and circuits. The output that results from a step input is termed the step response. The output from a sine wave input is the sinusoidal response. It consists of a transient portion and a sinusoidal **steady state** portion. The sine wave portion of the output has the same frequency, but, in general, a different amplitude and phase than the sine wave input. The manner in which the amplitude and phase of the outputs change for input sine wave over a range of different frequencies is called the **frequency response**.

An examination of the physical nature of some typical responses should prove useful. Consider a simple pendulum having a mass at the end of a string, Fig. 1.7(a). If a constant force is applied

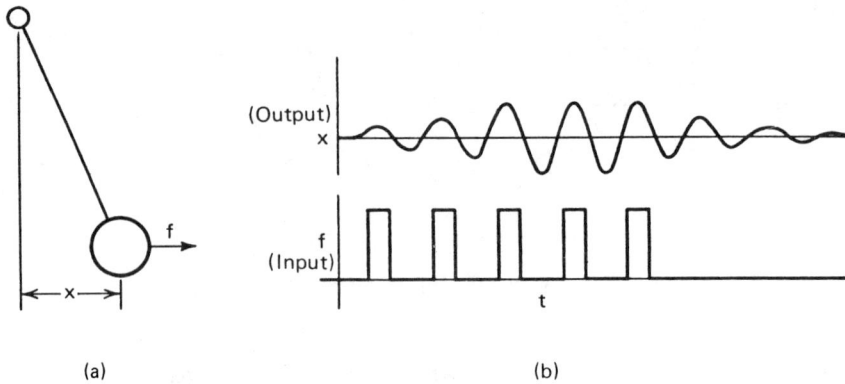

Figure 1.7 Response of pendulum

periodically, the pendulum can be made to swing with a constant amplitude. If the force is no longer applied, the pendulum will continue to swing, although its amplitude will decrease due to the air resistance [Fig. 1.7(b)]. The main point to note is that there is motion (output) even when there is no applied force (input). The form this motion takes is a property of the pendulum and is called the **natural response** or **force-free response**. The force is applied periodically to take advantage of this fundamental property of the system, the difference between the natural response and the actual response being the component of the response directly attributable to the applied force. It is usually called the **forced response**.

An example in which the forced response is more easily seen is that of a bicycle being pedaled along a horizontal road on a calm day. If the rider applies a constant force to the pedals, the speed will increase until a steady value is reached [Fig. 1.8]. This speed is directly related to the applied force and is the forced response of the system. After the rider stops pedaling, the bicycle still moves, although at a decreasing speed. Again, there is a response without any stimulus. This is the natural response of the system which can also be identified as being the difference between the forced response and the actual response during acceleration. Other examples in which the forced and natural responses may readily be identified by a similar argument are the heating and cooling of an oven and the switching on and off of a large fan.

The mathematical model of any system must be capable of giving solutions which adequately describe the physical situations considered above. The model of a dynamic system often is put into the form of one or more differential equations, the individual terms of which describe each individual or elemental effect within the system. The solution of a differential equation has two parts: the particular-integral solution and the complementary solution. The **particular-integral** solution is identified with the forced response and the **complementary solution** with the natural response (also sometimes called the **homogeneous solution**). Indeed, in an engineering context, it is usual to refer to these solutions by their physical descriptions, that is, the forced and natural responses.

14 Modeling and Analysis of Linear Physical Systems

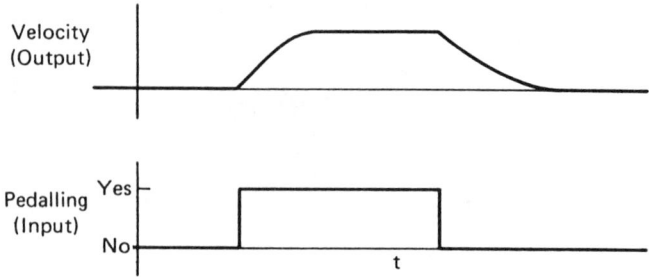

Figure 1.8 Response of bicycle

For linear systems, (that is, systems in which each elemental part has the property that its response is directly proportional to the input or its time derivative or time integral), the two component solutions are quite independent of each other. The forced response (both physically and mathematically) has exactly the same form as the input or driving function. The natural response has a form which is completely independent of the driving function, although the magnitude does depend on the input and also on the initial stored energy. Unfortunately, for nonlinear systems do not have this simplifying property and solutions are significantly more difficult and outside the scope of this text. The solutions of linear differential equations that are often encountered in physical circuits are considered in Chapters V, VI and IX..

1.4 MODELING

The elemental effects within a system are physical phenomena which may be concentrated at one point in space, but more generally are distributed over a region. As might be expected, it is more difficult to describe a distributed phenomenon in detail than to describe the situation at a point. However, in many cases, it is possible to consider a limited number of points and to produce a model which considers the action to take place at these discrete points, yet still gives a sufficiently accurate description of the interaction with the remainder of the system. For example, a coil of wire produces a magnetic field throughout its length; the conductor also has resistance which is distributed similarly. Despite the fact that in a real coil these two effects exist simultaneously, it is in a great many cases acceptable to consider the resistive effect entirely separately from the magnetic inductive effect, and the usual model consists of two elements: one is a conductor which has resistance but produces no magnetic field and the other is a conductor which produces a magnetic field but has no resistance. This situation is analogous to the performance of a fluid line where we may also distinguish two effects. One effect is the frictional dissipation that we have considered in Fig. 1.1. Another effect, that we had neglected, is the one due to the inertia of the fluid. As a result, the fluid line may also be considered as two components in series.

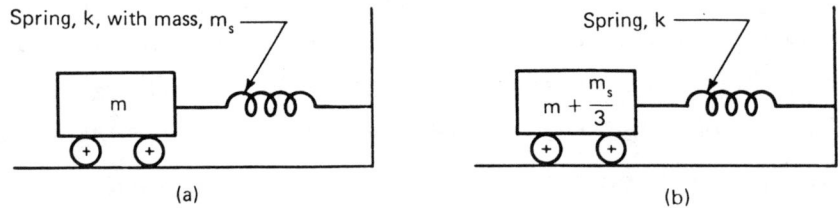

Figure 1.9 Modelling a spring having significant mass

Another example is that of a spring, [Fig. 1.9]. The material of which the spring is made of inevitably has mass and therefore there must be an inertia force involved in any change in velocity of the spring [Fig. 1.9(a)]. However, this mass may be negligible in comparison with the mass of the object to which spring is connected. In these circumstances, the spring is considered to exhibit only the characteristic of the stiffness. If this is not sufficiently accurate, part of the mass of the spring is sometimes considered as being situated at one end [Fig. 1.9(b)]. That is, the actual spring is represented by an ideal spring and a mass connected to an end. It is fairly common to describe practical situations which may reasonably be represented by an ideal spring as **spring-like**. Similarly, if only the mass of an object is significant, it is **mass-like**.

This process of lumping each phenomenon into a simple elemental situation produces a **lumped** model as opposed to a **distributed** model. Since distributed

Introduction

models usually require the solution of sets of partial differential equations and lumped models involves ordinary differential equations, there is a definite preference to use lumped models whenever possible. Inevitably, there are doubtful situations in which it is not clear that the conditions are such that a lumped model is valid. In such cases, it may be necessary to obtain both lumped and distributed models and to compare solutions. Alternatively, the lumped model may be compared with experimental data. Either approach is necessary to provide the experience which is needed for the development of reliable models.

1.5 SIGNAL VARIABLES

We have touched on the signal variables briefly in Section 1.2 and they will be elaborated upon in Chapter III. However, they are so basic to physical circuit analysis that we discuss them very generally again now.

Both lumped and distributed models involve the use of mathematical variables to represent physical quantities. Although the subject matter of this course is limited to lumped models, the basic nature of the variables is also applicable to distributed models. This should be expected since an electric current, for example, may equally be called upon to flow along the axis of a cylindrical conductor or along the surface of a sheet of conducting material. Basically, there are only two types of physical quantities: those which flow along or through a constrained path, such as water flowing through a pipe, and those which imply a measurement which must be made between two points such as the distance between the points. In connecting a measuring instrument between two points, it must be placed across everything that lies between them, and such quantities are described as **across variables**. In connecting an instrument to measure the flow along a path, it is necessary to cut the path and insert the instrument so that the flow is constrained to pass through it, and therefore such quantities are described as **through variables**.

The classification of all physical quantities being modeled as either **through** or **across** provides a powerful tool for the exploitation of analogous situations. Historically, these analogies become evident only after the mathematical model, in the form of a set of equilibrium equations, has been obtained. That is, it is possible for a particular mechanical system, fluid system, thermal system and electric circuit to have equilibrium equations which have identical mathematical form. Having obtained a solution for any one of them, the solutions for the others have effectively been obtained. The benefits resulting from these analogies can be realized more quickly, if they can be recognized early in the modeling process. The concept of the through and across variables is the key to the formation of the analogous circuit diagram which effectively forms a bridge between the physical model and the detailed mathematical model. This diagram has the appearance of an electric circuit. The main advantage of the analogous circuit diagram is that each different type of system is modeled by a diagram having the same general appearance. As a result, the process of obtaining a set of equilibrium equation for any physical circuit is a straightforward technique which is very simple to apply, and has the effect of providing a unified approach to the analysis of electrical, mechanical, fluid and thermal systems.

In the succeeding chapters, the details of this approach are developed. The through and across variables are identified for systems involving fluid flow, electric circuits, mechanical translation, mechanical rotation, and heating problems (Chapter III). After the process of obtaining the analogous circuit has been discussed (Chapter IV), some very basic physical situations are considered. Thereafter, the analytical techniques (Chapters V,VI and IX) are developed with examples drawn from any or all of these types of systems so that by the end of the course a student should have equal skill and confidence when faced with problems of modeling and analysis of any of the systems included in this study.

CHAPTER 2

SINGULARITY FUNCTIONS

In general, a real system is subject to a wide variety of random input functions. The system operates on these inputs to produce the output response. However, in the design and analysis of systems, it is usually impossible to know in advance the exact input function that the system will experience in service. Several examples can be quoted. One such example is to consider an automobile (Fig. 2.1) as it approaches a bump on the road. Let us assume that the automobile has a constant forward velocity, v_x. After the front wheels reach the bump, a transverse velocity, v_y, is imparted to the body through the springs and shock absorbers. This transverse velocity is an input function to the car suspension. The magnitude of such an input depends on the size and shape of the bump and also on the forward speed of the car. Designers of cars cannot know exactly what kind of roads will be encountered or how fast each driver will choose to drive.

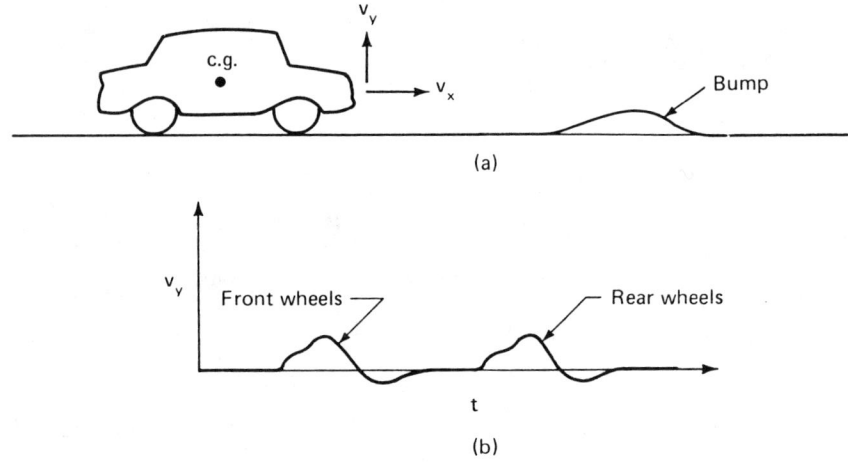

Figure 2.1 Transverse velocity, v_y, of car going over bump

Another function which is commonly encountered in practice is human speech. Human speech differs from one individual to another. Each individual possesses a distinct and unique speech waveform which is different from any other human speech waveform. Nevertheless, all speech waveforms have certain common features. In spite of these common features, standardization of human speech is impossible. However, we can represent a typical speech waveform [Fig. 2.2(a)] noting that it can vary considerably. Fig. 2.2(b) shows the same speech waveform represented by a sequence of impulse functions. (Impulse functions are discussed in Section 2.3). Such

Singularity Functions   23

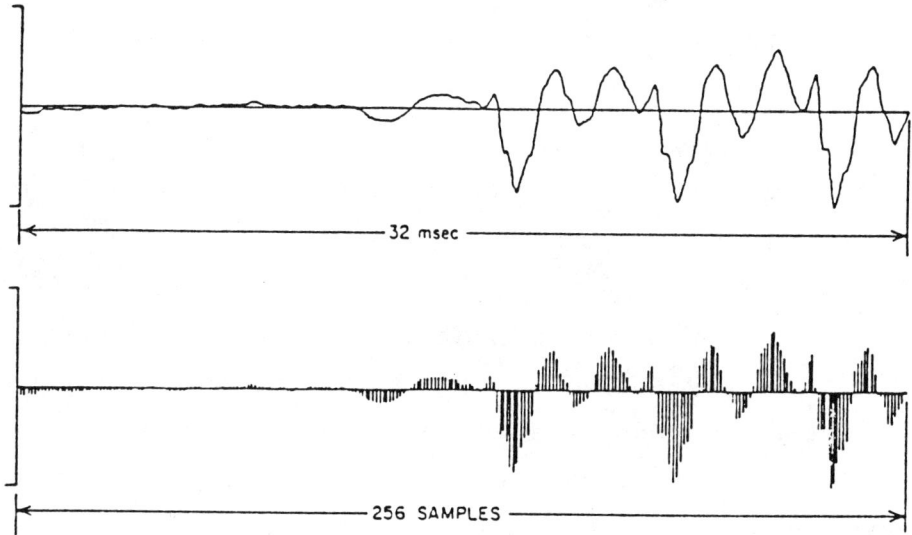

Figure 2.2 Representations of a speech signal.

(L. R. Rabiner/R. W. Schafer, DIGITAL PROCESSING OF SPEECH SIGNALS, © 1978, p. 11. Adapted by permission of Prentice Hall, Inc., Englewood Cliffs, New Jersey.)

a wide variety of waveforms constitutes input to a human ear or to any sound reproducing system.

From the foregoing discussion, it is imperative that input functions have to be studied in a formal fashion, so that systems meant to process these inputs have to be designed to meet these requirements. For this reason, it is convenient to use a set of idealized functions. Thus, if two different systems are designed, we may determine which one gives the better performance for a prescribed idealistic input function. Furthermore, we may formulate more realistic input functions from linear combinations of the idealistic input functions.

In this chapter, we will introduce most of the standard singularity functions which are used to express these input functions in mathematical form.

Basically, a singularity function is used to describe any sudden change in the amplitude, slope or any other higher order derivative of a function. The discontinuity at a sudden change in amplitude is the most evident and is regarded as a **step** as shown in the next section. A sudden change in slope may not be as easily seen but nevertheless it is simply modeled by a ramp which will be shown to be the integral of a step. Thus the singularity functions are related to each other by either differentiation or integration.

2.1 THE UNIT STEP FUNCTION

The most common singularity function is the unit step function. This type of function is often approximated in real life when we switch on the lights or turn on a water tap in a house. The step represents a sudden change in the input from one level to another. We may get a better understanding of a step change by referring to Fig.2.3. Initially [Fig. 2.3(a)], a butterfly valve blocks the exit line from a water tank. With the valve closed, there is no flow, q, from the tank. Fig. 2.3(b) shows the butterfly valve in its fully opened position. The valve, however, is a

Singularity Functions   25

mechanical part and cannot move from the closed to the open position instantaneously. This time restriction on sudden changes applies to all mechanical systems. The time required to open the valve is reflected in the shape of the actual flow, q(t), from the tank [(Fig.2.3(c)]. Note also that in this example the flow decreases slightly after the valve is opened. After sufficient time, the tank will be empty and the flow is reduced to zero. However, if the supply is from a large tank where the water level is maintained constant (as in a city water distribution system), the flow of water

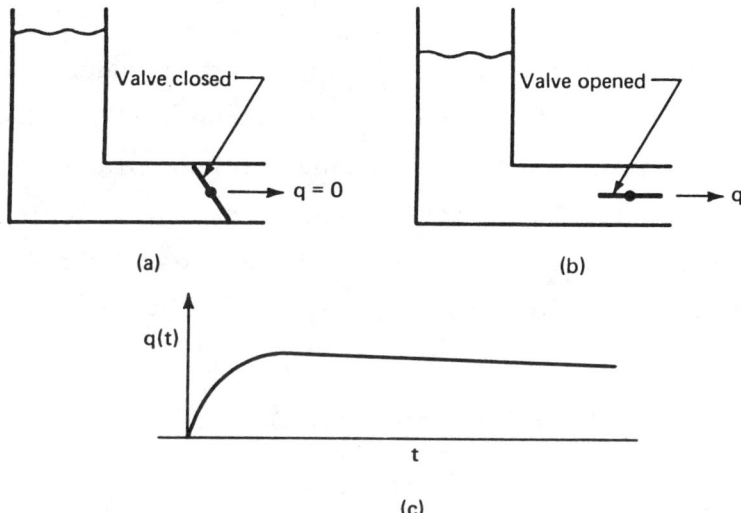

Figure 2.3 Flow from a tank through a quick-opening valve

will be steady. Also the time elapsed between the instant of opening the tap and the instant the water begins to flow can be considered small enough to be negligible.

Another example of a unit step is an ON\OFF electric or electronic switch. What is in common use is a mechanical switch which connects the electric supply to an electric light bulb or an appliance. However, there are electronic circuits where the electronic components act like switches. In either case, from the instant of time the contacts of the switch are closed to the instant of time the electric current begins to flow, the time elapsed is negligibly small, since the speed of electric current flow is quite high. In this way, this operation can be considered as instantaneous and hence can be represented by a step function.

The function we use to approximate sudden changes is the unit step function, $U_s(t)$, shown in Fig. 2.4. In this function, an abrupt change takes place in zero time; that is, the function goes from the **zero** level to the **one** level instantaneously. We express this function in mathematical form as

$$U_s(t) = 1, \text{ for } t > 0$$
$$= 0, \text{ for } t < 0$$

..(2.1)

The argument of a particular step function is the quantity in parenthesis. In (2.1), the argument is the time, t. However, the function holds for other arguments. In other words, the unit step function is a function that has a value of unity when the argument is greater than zero, and has a value of zero when the

Singularity Functions

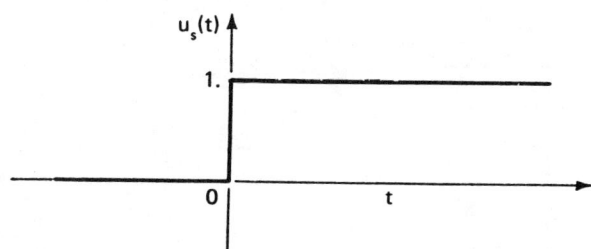

Figure 2.4 The unit step function

argument is less than zero. The function is not defined when the argument is exactly equal to zero.

As a consequence of the definition given above, we may shift the unit step along the time axis by changing the argument. Fig. 2.5 shows, for example, the advance and delay of the unit step function. It is to be remembered that the change always occurs when the argument is zero. Therefore, to advance the step function T time units, we use the argument, t + T, with the result that:

$$U_s(t + T) = 1, \text{ for } t + T > 0 \text{ or } t > -T$$
$$= 0, \text{ for } t + T < 0 \text{ or } t < -T$$

..(2.2)

In a similar way, we may delay the step function T time units by using the argument, t - T. This results in:

$$U_s(t - T) = 1, \text{ for } t - T > 0 \text{ or } t > T$$
$$= 0, \text{ for } t - T < 0 \text{ or } t < T$$

..(2.3)

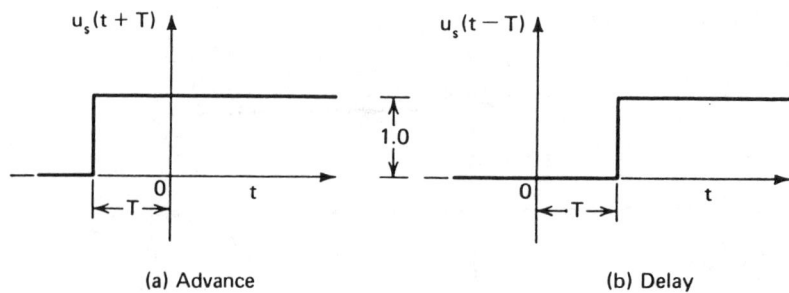

Figure 2.5 Advancing and delaying the unit step function

Eqs. (2.2) and (2.3) can be combined together by expressing the shifted unit step function as

$$U_s(t + T) \qquad \qquad ..(2.4)$$

where T may be positive, zero or negative.

If T is positive, (2.4) corresponds to (2.2), and if T is negative, (2.4) corresponds to (2.3). However, if T = 0, (2.4) corresponds to (2.1). Hence, (2.4) can be considered as the general version of the unit step function, so that the discontinuity may occur anywhere on the time axis.

In order to avoid any confusion which may arise when a unit step function is used in analysis, we shall write (i) T^- as the instant of time just before the discontinuity occurs, and (ii) T^+ as the instant of time just after the discontinuity occurs. As will be seen later, these will prove to be highly consistent and useful.

Singularity Functions

The unit step function has only two possible values, zero and unity. However, we may represent step changes of other amplitudes A in the form $AU_s(t)$. This means that:

$$AU_s(t) = A, \text{ for } t > 0$$
$$= 0, \text{ for } t < 0 \qquad \qquad ..(2.5)$$

It is also possible to combine step functions algebraically to attain composite functions of greater complexity. Two types of problems are of particular interest. In one, we decompose a function presented graphically into step function parts and then express the function in mathematical form by summing the parts. In the other, we may need to interpret a mathematical formulation by plotting the function. We will consider one problem of each type in the following two examples. The examples provide the various steps to be followed in solving such problems.

Example 2.1

We are given the rectangular pulse shown in Fig. 2.6 and asked to determine, g(t), in mathematical form.

Our first step must be to decompose the given pulse into individual parts whose mathematical form we know. From the given function, we may observe that there is a sudden change from 0 to 2 at t = 3 seconds. That is, at $t = 3^-$ seconds, g(t) is zero and at $t = 3^+$ seconds, g(t) is 2. This sudden change may be

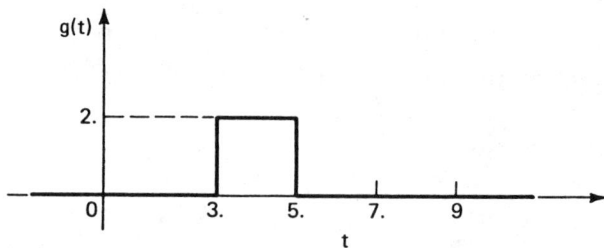

Figure 2.6   Rectangular square pulse (example 2.1)

represented by the step function, $2U_s(t - 3)$, which is shown in Fig. 2.7. This function remains at the level of two for all times greater than 3 seconds. The given function, on the other hand, is equal to zero for all times greater than 5 seconds. We can obtain this result by adding a step of -2 units at 5 seconds. This sudden decrease may be written mathematically as $-2U_s(t - 5)$. The given function is seen to be simply the sum of the component parts shown in Fig. 2.7. Thus the function given in Eq. (2.5) is:

$$g(t) = 2U_s(t - 3) - 2U_s(t - 5) \qquad ..(2.6)$$

Example 2.2

Show the graphical representation of the function, $g(t)$, where

$$g(t) = 3U_s(t + 1) - 5U_s(t - 1) + 4U_s(t - 3) \qquad ..(2.7)$$

In this type of problem, we must sketch each segment separately and then add the ordinates algebraically at each value of time. Fig. 2.8 shows the

Singularity Functions

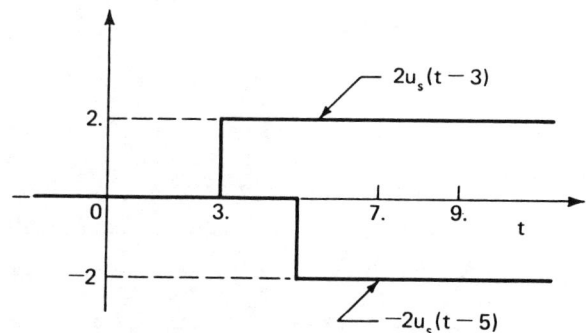

Figure 2.7 Decomposition of square pulse (example 2.1)

graphical representation of the three individual segments that comprise the given function.

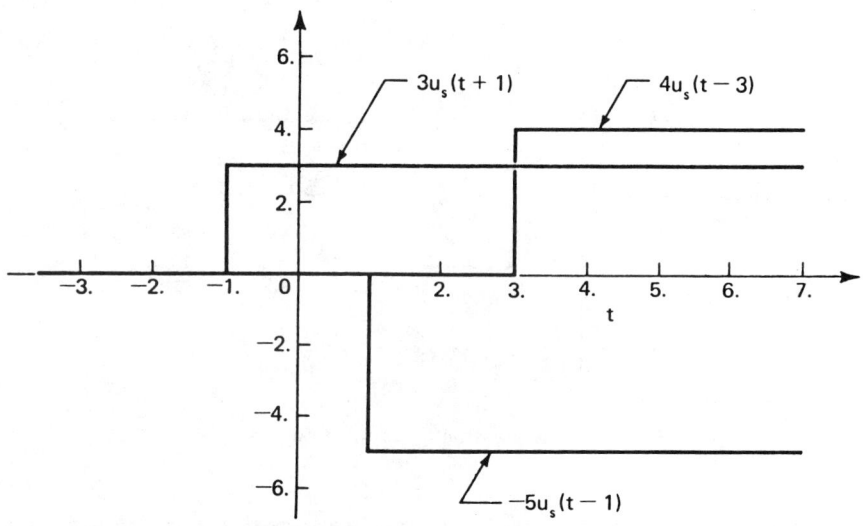

Figure 2.8 Segments of function (example 2.2)

The individual segments shall be added in order to obtain g(t). To avoid any confusion, it is a good policy to do the summation immediately before and after each sudden change. Thus, at t = 1^- second, we are adding amplitudes of 3, 0 and 0. At t = 1^+ second, we are adding amplitudes of 3, -5 and 0. Thus the level changes from +3 to -2 at t = 1 second. This is shown in the reconstruction in Fig. 2.9. There is another sudden change at 3 seconds. When t = 3^+ seconds, we must add amplitudes 3, -5 and 4. As a result, the level changes from -2 to +2 at a t= 3 seconds. Remember that we must always add the same number of segments as there are terms in the given function. In this example there are three terms and this has produced three quantities to add. The complete function is shown in Fig. 2.9.

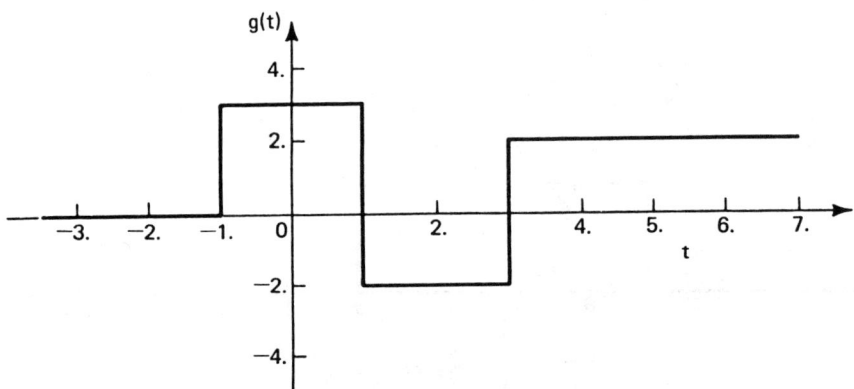

Figure 2.9   Reconstruction of function (example 2.2)

2.2 FUNCTIONS DERIVED BY INTEGRATION OF UNIT STEP

In this section, we shall consider functions which can be obtained by the integration of unit step functions. Specifically, two functions namely (i) the Ramp function and (ii) the Parabolic function will be

Singularity Functions

discussed in detail.

2.2.1 The Ramp Function

A single integration of the unit step function with respect to time results in the unit ramp function, $U_r(t)$. That is,

$$U_r(t) = \int_{-\infty}^{t} U_s(t') \, dt' \qquad ..(2.8)$$

The auxiliary variable, t', is introduced, since the upper limit of integration is to be the original independent variable, t.

If we substitute the unit step function of (2.1) into (2.8) and perform the integration, we may express the unit ramp function as:

$$\begin{aligned} U_r(t) &= t, \quad \text{for } t \geq 0 \\ &= 0, \quad \text{for } t \geq 0 \end{aligned} \qquad ..(2.9)$$

Note that $U_r(t)$ is defined correctly for $t = 0$ by either condition.

The unit ramp function is shown plotted in Fig. 2.10.

The argument of the unit ramp may be replaced in a manner similar to the unit step to advance or delay the function. The amplitude of the ramp may also be other

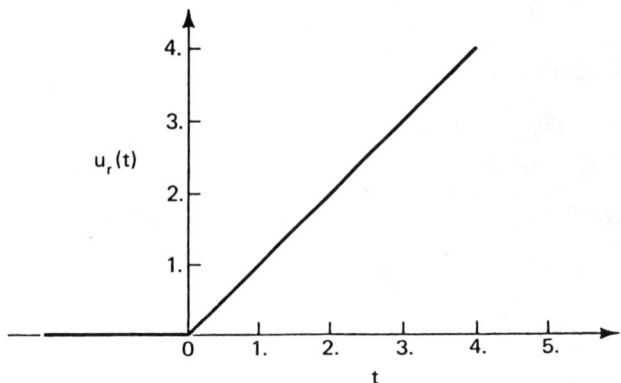

Figure 2.10 The unit ramp function

than unity. This changes the slope of the ramp. The function $AU_r(t + T)$ is defined as

$$AU_r(t + T) = A(t + T), \quad \text{for } t + T \geq 0$$
$$= 0, \quad \text{for } t + T \leq 0 \quad ..(2.10)$$

where T may be positive, zero or negative and these have the same meanings as explained before.

Fig. 2.11 is a plot of the function described in (2.10) with a time advance of 1 second (T = 1) and an amplitude of A = 1/2. In mathematical form, we may write

$$(1/2)U_r(t + 1) = (t + 1)/2, \quad \text{for } t \geq -1$$
$$= 0, \quad \text{for } t \leq -1 \quad ..(2.11)$$

Some composite functions may be separated into a

Singularity Functions   35

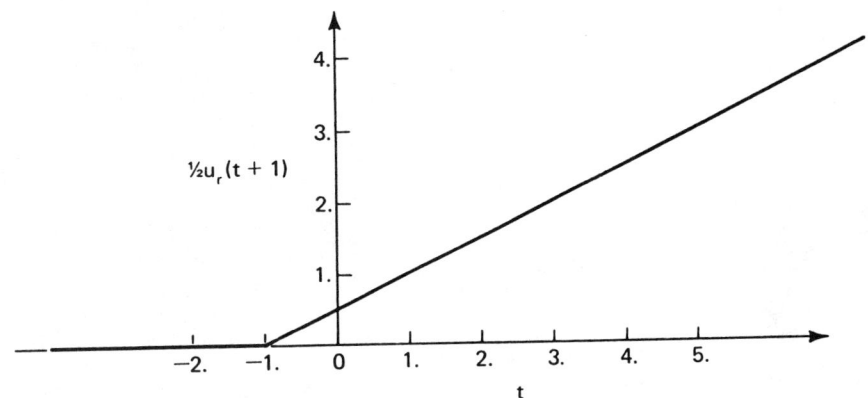

Figure 2.11 The ramp function advanced

sum of ramp functions with different arguments. Here again, we must use decomposition to identify the basic segments. This may be done by noting that the addition of two straight lines of the form $y_1 = mx_1 + b_1$ and $y_2 = m_2x + b_2$ yields a straight line of the form $y = (m_1 + m_2)x + (b_1 + b_2)$. The gradient of the summed relation is the sum of the gradients. Thus, when adding or subtracting ramp functions, we merely add or subtract their amplitudes. The following example will clarify the procedure.

Example 2.3

Fig. 2.12(a) shows a function which begins like a ramp delayed and then levels off. This is a useful input function since switches and valves do not change levels instantaneously (refer to Fig. 2.3). The problem is to describe this function in terms of the known basic functions.

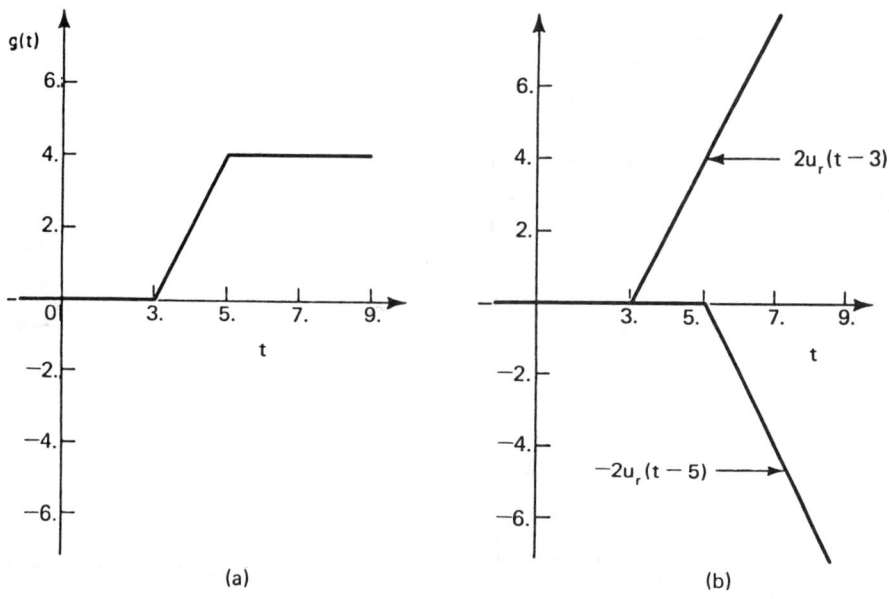

Figure 2.12 Ramp decomposition (example 2.3)

To decompose the given function, the initial ramp portion suggests a ramp function delayed. Fig. 2.12(b) shows this portion identified as $2U_r(t - 3)$. When t = 5 seconds, this portion has a magnitude of 4 and this is the correct leveling off magnitude. However, the function $2U_r(t - 3)$ continues to rise. At t = 7 seconds, it has a magnitude of 8. Thus the function $2U_r(t - 3)$ only describes the given function for t ≤ 5 seconds. We must, therefore, subtract a second function beginning at t = 5 seconds. This function must have the same slope as the initial ramp portion and be opposite in sign. The composite function will then have zero slope. This second portion is therefore $-2U_r(t - 5)$. The given composite function may be described as:

Singularity Functions

$$g(t) = 2U_r(t - 3) - 2U_r(t - 5) \qquad ..(2.12)$$

If this is really the function given in Fig. 2.12(a), it must be correct for all times. We may check by evaluating the function at a few specific times and comparing the result with Fig. 2.12(a). Thus

$$g(3) = 0 - 0 = 0$$
$$g(4) = 2 - 0 = 2$$
$$g(5) = 4 - 0 = 4$$
$$g(n) = 10 - 6 = 4, \, n \geq 5$$

The functional form yields the same values as Fig. 2.12(a) for all times.

2.2.2 The Parabolic Function

A double integration of the unit step function with respect to time results in the unit parabolic function $U_p(t)$:

$$U_p(t) = \int_{-\infty}^{t} \int_{-\infty}^{t'} U_s(t'') \, dt'' \, dt' \qquad ..(2.13)$$

A unit parabolic function may also be defined in terms of the unit ramp function as:

$$U_p(t) = \int_{-\infty}^{t} U_r(t') \, dt' \qquad ..(2.14)$$

If we substitute the functional forms for the unit ramp [given in (2.9) into (2.10)], the unit parabolic function becomes:

$$U_p(t) = t^2/2, \quad \text{for } t \geq 0$$
$$= 0, \quad \text{for } t \leq 0 \qquad ..(2.15)$$

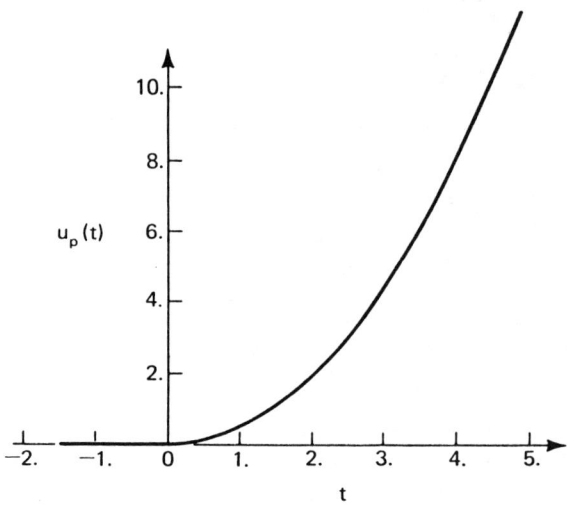

Figure 2.13 Unit parabolic function

The unit parabolic function is shown in Fig. 2.13. As in the cases of the step and the ramp functions, we may modify the amplitude and the argument of the parabolic function. The function $AU_p(t + T)$ is defined as

$$AU_p(t + T) = A(t + T)^2/2, \quad \text{for } t + T \geq 0$$
$$= 0, \quad \text{for } t + T \leq 0 \qquad ..(2.16)$$

Singularity Functions

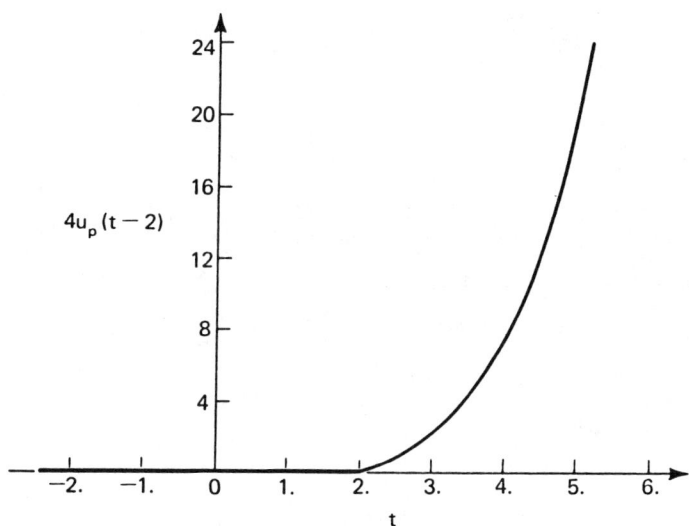

Figure 2.14 Parabolic function delayed

where T may be positive, zero or negative and these have the same meanings as before. Fig. 2.14 shows the parabolic function $4U_p(t - 2)$.

The addition or subtraction of parabolic functions requires somewhat more care than with ramp or step functions. The operation may proceed directly from the definition of the parabolic functions. Another approach, however, is to differentiate the composite parabolic function, which will be discussed later (See Section 2.7). The resulting ramp functions may be treated as previously indicated and then converted back to the parabolic function by integration. The following example demonstrates the first procedure.

Example 2.4

Plot the function described by

$$g(t) = U_p(t) - 2U_p(t-1) + U_p(t-2) \qquad ..(2.17)$$

We merely plot each component of the given function separately and then add. Fig. 2.15(a) shows the three segments of the given function plotted separately. To compute the function rather than depend on graphical addition, it is convenient to rewrite the given function as

$$g(t) = U_p(t) = t^2/2 + 0 + 0, \text{ for } 0 \le t \le 1 \qquad ..(2.18a)$$

$$g(t) = U_p(t) - 2U_p(t-1) + 0 = \frac{t^2}{2} - (t-1) \qquad ..(2.18b)$$

$$\text{for } 1 \le t \le 2$$

$$g(t) = U_p(t) - 2U_p(t-1) + U_p(t-2)$$

$$= \frac{t^2}{2} - (t-1)^2 + \frac{(t-2)^2}{2}$$

$$= 1, \text{ for } t \ge 2 \qquad ..(2.18c)$$

From this formulation, we may compute the value of the function at any time.

Thus, $\qquad g(1.5) = 1.125 - 0.250 = 0.875.$

Fig. 2.15(b) shows the required function.

Singularity Functions

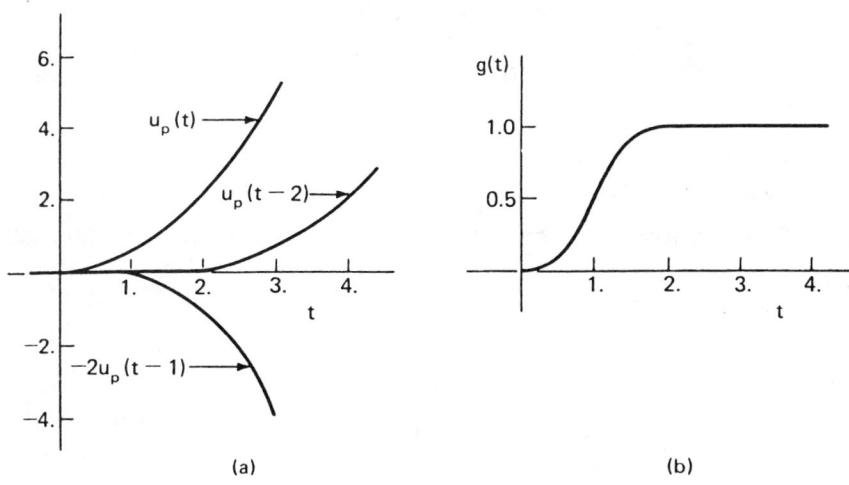

Figure 2.15 Development of function for example 2.4

2.3 FUNCTIONS DERIVED BY DIFFERENTIATION OF UNIT STEP

In this section, we shall discuss functions obtained by the operation of differentiation of the unit step function.

2.3.1 The Unit Impulse

The time derivative of the unit step function is the unit impulse function, $U_i(t)$. In mathematical form, this will be

$$U_i(t) \equiv \frac{d}{dt}\left[U_s(t)\right] \qquad ..(2.19)$$

We may evaluate the unit impulse function by substituting the values of the unit step [as in (2.1)] into (2.19). The result is:

$$U_i(t) = 0, \text{ for } t > 0$$
$$= 0, \text{ for } t < 0 \quad \quad ..(2.20)$$
$$= \infty, \text{ for } t = 0$$

Fig. 2.16 shows a graphical representation of the unit impulse function.

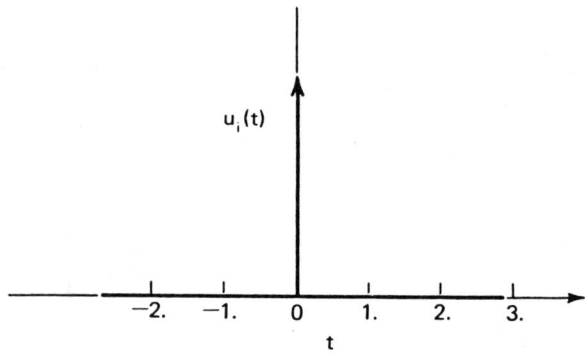

Figure 2.16 Graphical representation of unit impulse function

A useful property of the impulse function can be obtained by integrating both sides of (2.19) with respect to time so that

$$\int_{-\infty}^{+\infty} U_i(t) \, dt = U_s(t) \Big|_{-\infty}^{+\infty} = 1 \quad \quad ..(2.21)$$

Thus the area bounded by the unit impulse function and the time axis is always equal to unity. An alternate approach to the impulse function is through a limiting process. Consider the function g(t) shown in Fig. 2.17(a).

Singularity Functions

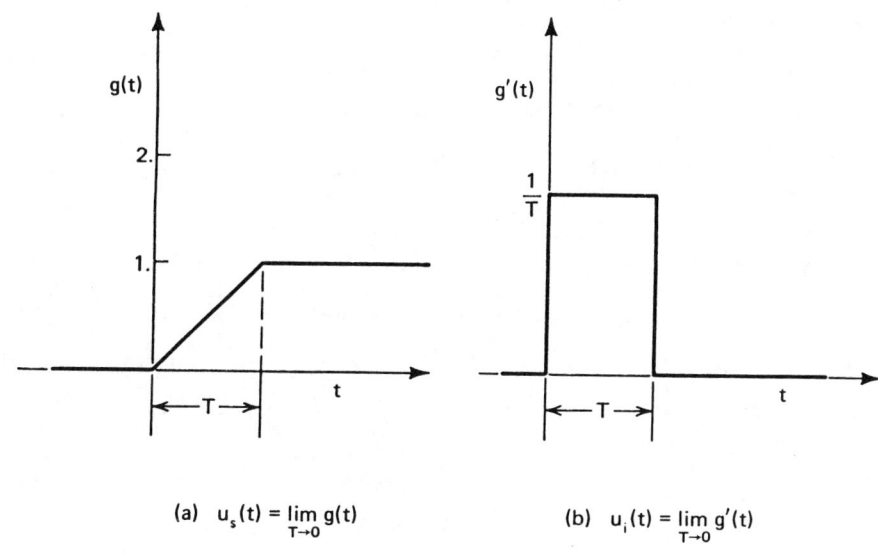

(a) $u_s(t) = \lim_{T \to 0} g(t)$ (b) $u_i(t) = \lim_{T \to 0} g'(t)$

Figure 2.17 The impulse function as a limit

We may express this function in terms of ramp functions as:

$$g(t) = \frac{1}{T}\left[U_r(t) - U_r(t - T)\right] \qquad ..(2.22)$$

Now note that as T becomes smaller, g(t) tends towards the unit step function. That is:

$$U_s(t) = \lim_{T \to 0} \frac{1}{T}\left[U_r(t) - U_r(t - T)\right] \qquad ..(2.23)$$

The advantage of this formulation is that we may perform the differentiation of (2.23) which yields the unit impulse function as:

$$U_i(t) = \lim_{T \to 0} \frac{1}{T}\left[U_s(t) - U_s(t - T)\right] \quad ..(2.24)$$

Fig. 2.17(b) shows the function described in (2.24). As time approaches zero, the amplitude approaches infinity. However, the area under the impulse still remains unity. If we wish to designate functions with other than unit area, we may do so by multiplying the impulse by the appropriate constant [that is, $AU_i(t)$]. Thus the function

$$g(t) = 5U_i(t - 1) + 2U_i(t - 3) + 3U_i(t - 4) \quad ..(2.25)$$

which represents a particular series of impulse functions, may be represented graphically as shown in Fig. 2.18.

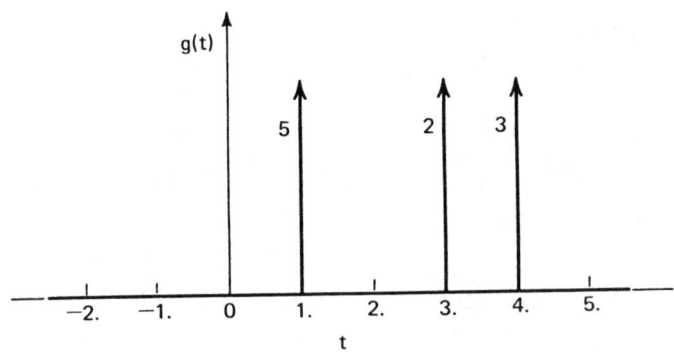

Figure 2.18 Graphical representation of a series of impulse functions

The **strength** of each impulse is indicated by the number at the side and refers to the area under the impulse.

Singularity Functions

2.3.2 The Unit Doublet

The second time derivative of the unit step function is the unit doublet function $U_d(t)$ given by

$$U_d(t) \equiv \frac{d^2}{dt^2}\left[U_s(t)\right] = \frac{d}{dt}\left[U_i(t)\right] \qquad ..(2.26)$$

Let us use the limiting approach to determine values for the doublet function. As a first step, we construct a function which approaches the impulse function in the limit. The desired function must also enclose unit area. In the construction, we use segments that are readily differentiated. Fig. 2.19(a) shows the function $g_1(t)$ which may be expressed as

$$g_1(t) = \frac{1}{T^2}\left[U_r(t+T) - 2U_r(t) + U_r(t-T)\right] \qquad ..(2.27)$$

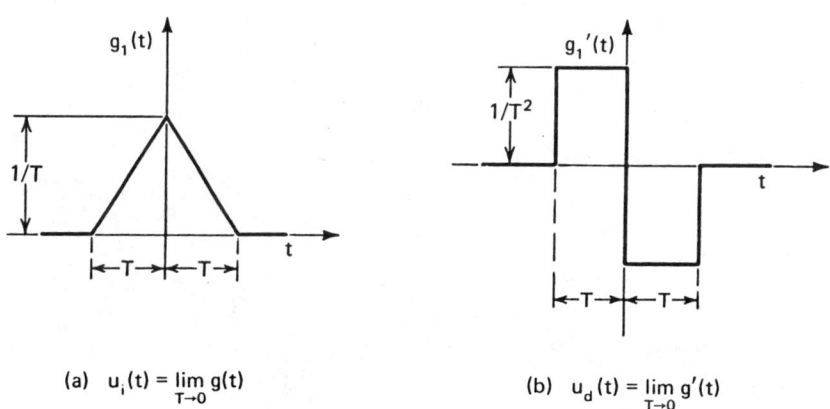

Figure 2.19 The unit doublet function as a limit

To verify that (2.27) is the mathematical expression for Fig. 2.19 (a), the student should apply

(2.9). Now note that the unit impulse function is:

$$U_i(t) = \lim_{T \to 0} g_1(t) \quad \quad ..(2.28)$$

From (2.26), (2.27) and (2.28), the unit doublet function may be written as

$$U_d(t) = \lim_{T \to 0} g_1'(t)$$

$$= \lim_{T \to 0} \frac{1}{T^2} \left[U_s(t + T) - 2U_s(t) + U_s(t - T) \right]$$

$$..(2.29)$$

Fig. 2.19(b) shows the limiting function that represents the unit doublet in (2.29). As T approaches zero, the unit doublet may be evaluated as:

$$\begin{aligned} U_d(t) &= 0, \quad \text{for } t < 0^- \\ &= 0, \quad \text{for } t > 0^+ \\ &= +\infty, \quad \text{for } t = 0^- \\ &= -\infty, \quad \text{for } t = 0^+ \end{aligned} \quad ..(2.30)$$

The graphical representation of the unit doublet function is shown in Fig. 2.20.

2.4 EXPONENTIAL FUNCTIONS

The exponential functions that we are about to consider are not singularity functions. However, they

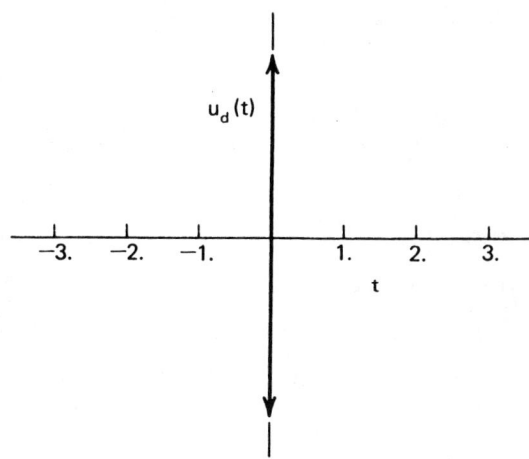

Figure 2.20 Graphical representation of unit doublet function

are often combined with singularity functions to describe system inputs. It is therefore convenient to include them at this point.

The most general form of the exponential function is:

$$g(t) = \varepsilon^{st} \qquad ..(2.31)$$

where s may be a complex number $\sigma + j\omega$. We will consider the following cases:

i) s is purely real ($\omega = 0$)

ii) s is purely imaginary ($\sigma = 0$)

iii) s is complex .

When s is only a real number, (2.31) becomes

$$g(t) = \varepsilon^{\sigma t} \qquad ..(2.32)$$

Fig. 2.21 is a plot of (2.32) in which σ has the values +0.3 and -0.3. When the exponent, σ, is negative, the function is known as the exponential decay. A positive exponent results in the exponential rise function. For the special case where $\sigma = 0$, the function has the value unity for all times.

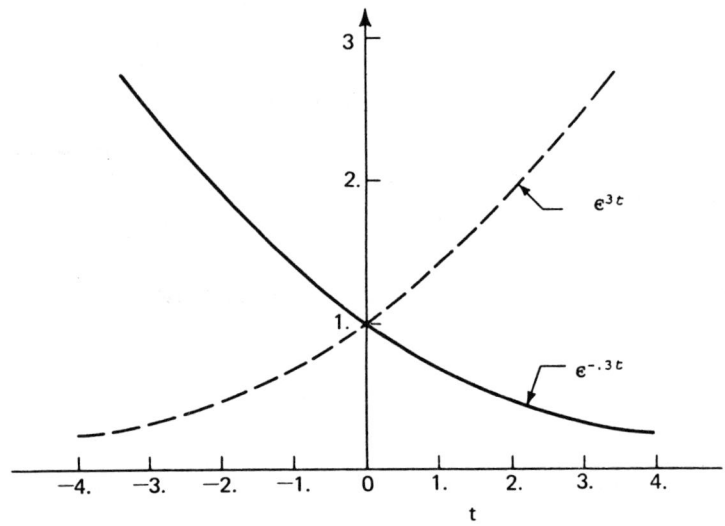

Figure 2.21 Exponential decay and rise functions

When s is a purely imaginary number, (2.31) is written as

$$g(t) = \varepsilon^{j\omega t} \qquad ..(2.33)$$

Singularity Functions   49

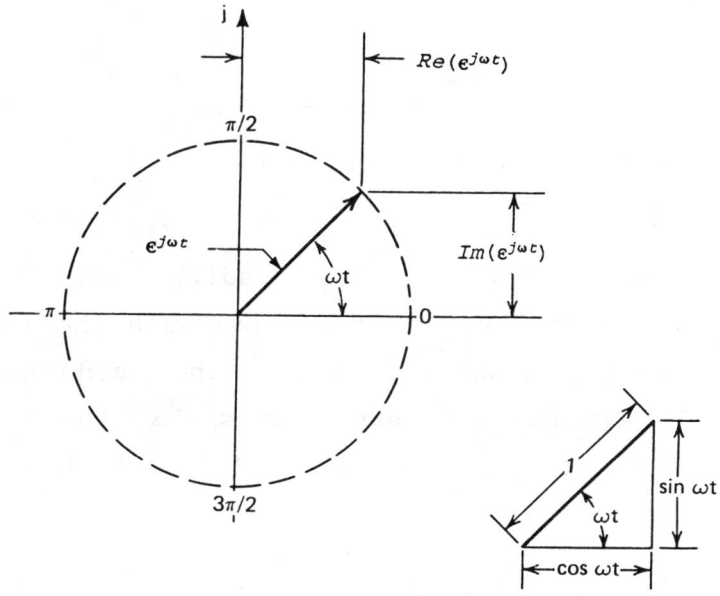

Figure 2.22 Representation of exponential as rotating vector

Fig. 2.22 shows the representation of the exponential function in (2.33) as a rotating unit vector. The angular frequency of the rotation depends on the magnitude of ω. The vector rotates counterclockwise as time increases. According to Euler's formula, the imaginary exponential may be expressed as

$$\varepsilon^{j\omega t} = \cos \omega t + j \sin \omega t \qquad ..(2.34)$$

If we use the symbols Re [ ] and Im [ ] to indicate the real and the imaginary parts of the bracketed quantity, then (2.34) may be defined the circular functions, cos ωt and sin ωt.

$$\text{Re}\,(\varepsilon^{j\omega t}) = \cos \omega t$$
$$\text{Im}\,(\varepsilon^{j\omega t}) = \sin \omega t$$
..(2.35)

Eq. (2.35) may be more readily visualized by comparing the unit vector components with the inserted triangle on Fig. 2.22. If we plot the functions given in (2.35) against a linear time scale, the familiar cosine and sine shapes appear. Fig. 2.23 shows this plot.

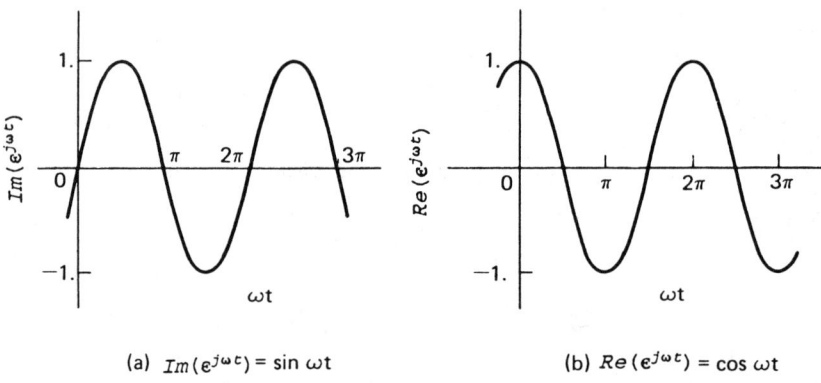

(a) $Im(e^{j\omega t}) = \sin \omega t$ (b) $Re(e^{j\omega t}) = \cos \omega t$

Figure 2.23 Real and imaginary parts of exponential function

When s is a complex number, (2.31) is written as

$$g(t) = \varepsilon^{\sigma t} \cdot \varepsilon^{j\omega t}$$
..(2.36)

As explained before, $\varepsilon^{j\omega t}$ represents a rotating

Singularity Functions

unit vector. Therefore, $\varepsilon^{\sigma t}$ determines the amplitude of the rotating vector. Two possibilities arise, namely (i) $\sigma > 0$ and (ii) $\sigma < 0$. When σ is positive, the amplitude increases with time. Fig. 2.24(a) shows the representation of (2.36) as a rotating vector with increasing amplitude. Fig. 2.24(b) shows the real and imaginary parts of such a function, which are recognized as the growing sinusoids.

When σ is negative, the amplitude decreases with time. Fig. 2.25(a) shows the representation of (2.36) as a rotating vector with decreasing amplitude. Fig. 2.25(b) shows the real and imaginary parts of such a function, which are recognized as a damped sinusoids.

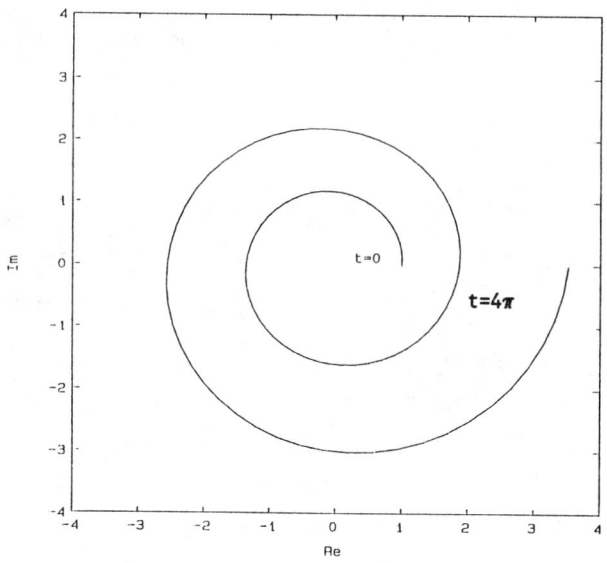

(a)

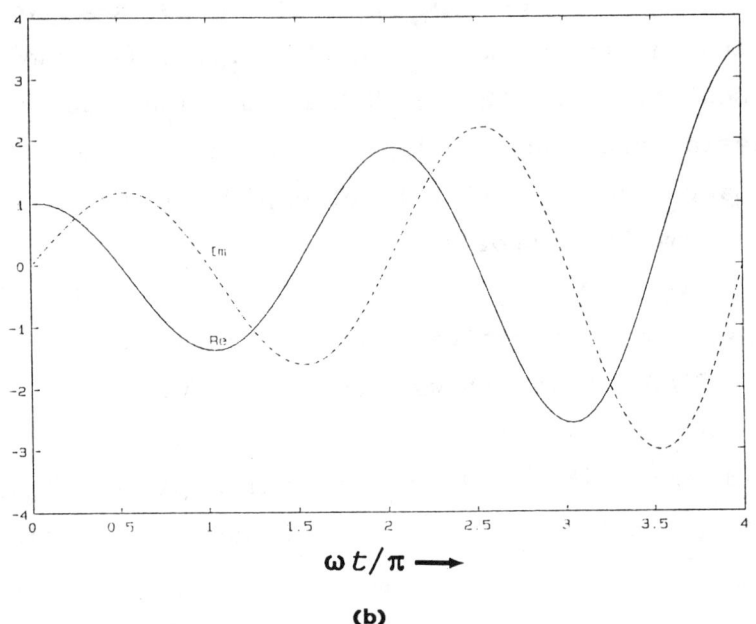

Figure 2.24 (a) Rotating vector with increasing amplitude
(b) The real and imaginary parts of figure 2.24a

Example 2.5

In the solution of problems, we often obtain functions that are the sums or the differences of sine and cosine terms of the same frequency. We may always express such functions in terms of sine or cosine functions alone. Suppose

$$g(t) = A \sin \omega t + B \cos \omega t \qquad ..(2.37)$$

It is not easy to interpret this function directly and

Singularity Functions

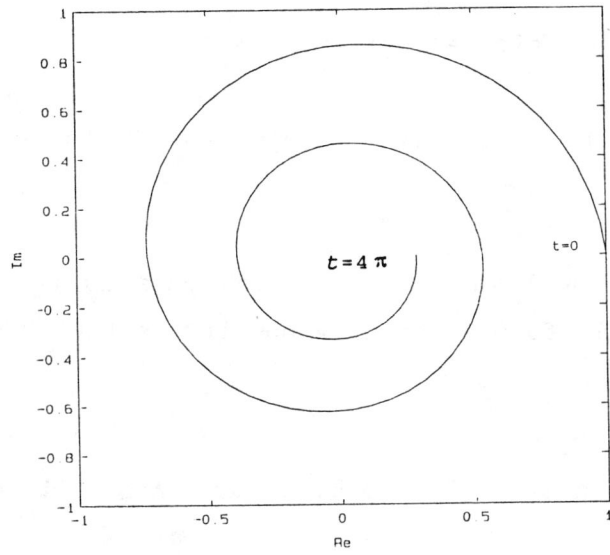

(a)

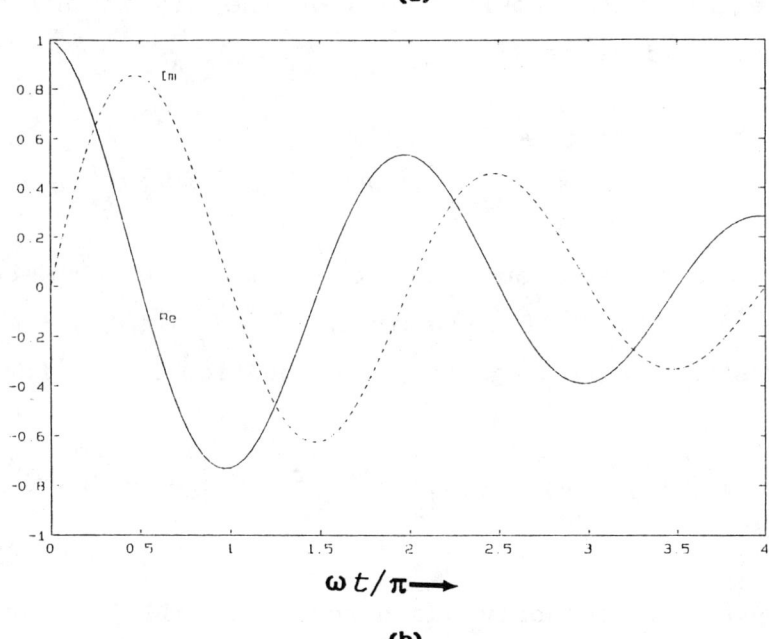

(b)

Figure 2.25 (a) Rotating vector with decreasing amplitude
(b) The real and imaginary parts of figure 2.25b

therefore, it usually is essential to convert it to the form

$$g(t) = D \sin(\omega t + \varphi_1) \quad \quad ..(2.38)$$

where φ_1 is a constant angle and D is a constant amplitude.

The usual way to accomplish this is to apply the double angle formula of trigonometry so that $D \sin(\omega t + \varphi_1)$ is expanded to

$$(D \cos\varphi_1) \sin \omega t + (D \sin\varphi_1) \cos \omega t = A \sin \omega t + B \cos \omega t$$
$$..(2.39)$$

By equating the coefficients of the sin ωt and cos ωt terms, we find that:

$$D = \sqrt{A^2 + B^2}, \quad \quad \varphi_1 = \text{Tan}^{-1}(B/A) \quad \quad ..(2.40)$$

We obtain the same result by working with complex numbers. Thus, if we desire $g(t) = D \sin(\omega t + \varphi_1)$, we may start by putting the given equation in the form:

$$\text{Im}\left[D \, \text{Exp}\{j(\omega t + \varphi_1)\}\right] = \text{Im}\left[A \, \varepsilon^{j\omega t}\right] + \text{Re}\left[B \, \varepsilon^{j\omega t}\right]$$
$$..(2.41)$$

However, we can only add directly if all the terms are either Im or Re. This can be changed to that form by rotating the last terms on the right so that

Singularity Functions

$$\text{Im}\left[D \exp\{j(\omega t + \varphi_1)\}\right] = \text{Im}\left[A\, \varepsilon^{j\omega t}\right] + \text{Im}\left[jB\, \varepsilon^{j\omega t}\right]$$

..(2.42)

Now, since all the terms are Im, we may drop this designation and

$$D\, \varepsilon^{j\varphi_1}\, \varepsilon^{j\omega t} = A\, \varepsilon^{j\omega t} + jB\, \varepsilon^{j\omega t} \qquad ..(2.43)$$

In polar form, this reduces to:

$$D\, \varepsilon^{j\varphi_1} = \sqrt{A^2 + B^2}\; \exp(j\, \text{Tan}^{-1}\, B/A) \qquad ..(2.44)$$

from which

$$D = \sqrt{A^2 + B^2} \quad \text{and} \quad \varphi_1 = \text{Tan}^{-1}\, B/A \qquad ..(2.45)$$

The procedure to place the expression in the form $g(t) = D \cos(\omega t - \varphi_2)$ is similar and is left as an exercise for the student.

2.5 COMPOSITE FUNCTIONS

The basic singularity functions may be combined to describe a wide variety of time functions. These composite functions are usually formed by adding or multiplying basic functions. In the case of multiplication, the unit step and the unit step shifted are most often used. We must be able to find both the functional representation from the graphical time plot and the inverse process. Examples 2.6 and 2.7 will demonstrate some composite functions and their formulation.

Example 2.6

We are given the function shown in Fig. 2.26 and asked to find an expression which represents it.

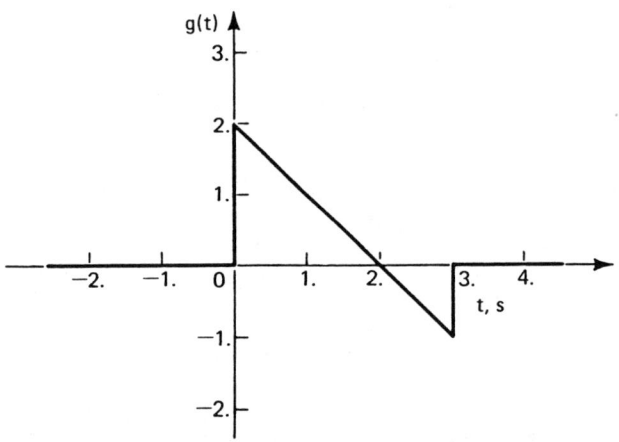

Figure 2.26 Function for example 2.6

The procedure to follow in this type of problem is to start at $t = -\infty$ and traverse along the time axis till we reach a point when the function is not zero and see what has to be done at this time. From the function, we note that at $t = 0$, there is a change from 0 to 2. This means immediately that one of the components of the required function is $2U_s(t)$, because in no other way we

Singularity Functions

can duplicate the sudden change at $t = 0$. Now, the function magnitude decreases from 2 at $t = 0^+$ to -1 at $t = 3$. The decrease, however, begins at $t = 0^+$ and for this reason, we must begin subtracting from the step function at this time. The slope of the given function between $t = 0$ and $t = 3$ is -1 and this suggests we subtract a unit ramp, $U_r(t)$. If we combine these two functions, we obtain the function $2U_s(t) - U_r(t)$. To see what we have done so far, refer to Fig. 2.27(a).

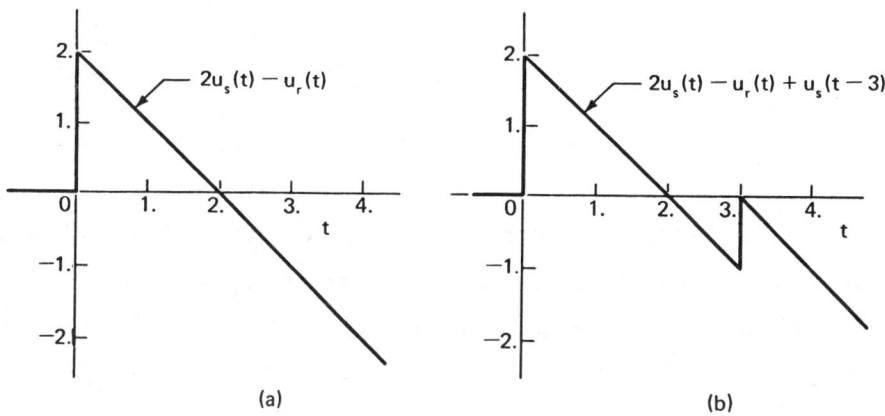

Figure 2.27 Stages in building up function for example 2.6

Comparison of Fig. 2.27(a) with the desired function shows that they are the same until $t = 3^-$. However, at this time the given function again changes suddenly from -1 to 0. To accomplish this change, we must add a unit step shifted by 3 seconds. Remember that the functions formed up to 3 seconds remain in effect for all time. Thus by adding the step delayed by 3 seconds, we have accumulated a function in the form

$[2U_s(t) - U_r(t) + U_s(t-3)]$. This function is shown in Fig. 2.27(b). If we again compare the function with Fig. 2.26, we see that our combined function decreases for times greater than 3 seconds, whereas the desired function remains at zero. This means that we must add another function beginning at t= 3 to counteract the decrease of the unit ramp. The required component is a unit ramp which starts increasing at t = 3. Thus the complete function which describes the function in Fig. 2.26 is:

$$g(t) = 2U_s(t) - U_r(t) + U_s(t - 3) + U_r(t - 3)$$

..(2.46)

We limit our consideration of the multiplication of two time functions to cases where one of the functions is a combination of unit step functions. Thus, in effect, we will be multiplying a given function, g(t), by 0 or 1 in specific time intervals. The resulting functional product will then represent the given function g(t) with sections deleted. Suppose, we wish to express an exponential decay that holds only when t > 0. To do this, we would use the product form $\varepsilon^{\sigma t} U_s(t)$. Fig. 2.28(a) shows the result.

The step function effectively blocks out the function for t < 0. To shift this exact function in time, we need only to replace all values of t by (t ±

Singularity Functions

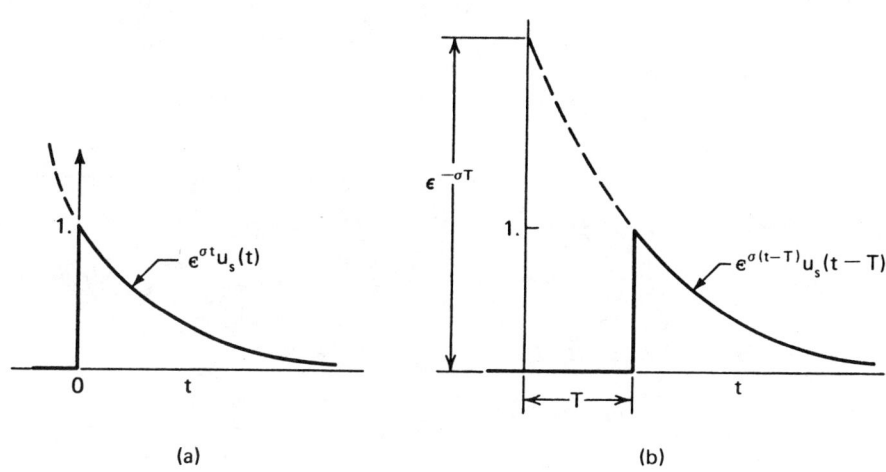

Figure 2.28 Effect of multiplication by unit step function

T). Thus to delay $\varepsilon^{\sigma t} U_s(t)$ by T, we would obtain $\varepsilon^{\sigma(t-T)} U_s(t - T)$ as the shifted function [Fig. 2.28(b)]. Another way of interpreting this function is as $(\varepsilon^{-\sigma T}) \varepsilon^{\sigma t} U_s(t - T)$. Instead of shifting the function of Fig. 2.28 (a), we could have blocked out an exponential with amplitude, $\varepsilon^{-\sigma t}$, for all times less than T.

Example 2.7

The function shown in Fig. 2.29(a) is one half cycle of a sinusoid. Find the functional representation.

The required function h(t) will be the product of two time functions. One of the functions must obviously be the sine function. Thus we may write

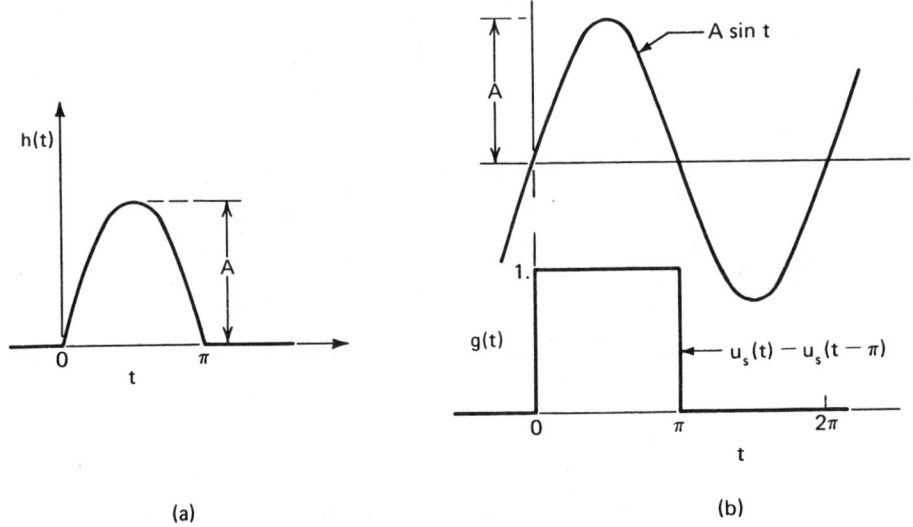

Figure 2.29 Multiplication of functions (example 2.7)

$$h(t) = g(t) \sin t \qquad ..(2.47)$$

We must determine the other time function g(t), to produce the given function, The characteristics of g(t) must therefore be

$$g(t) = 1, \text{ for } 0 < t < \pi$$
$$= 0, \text{ for all other values of } t \qquad ..(2.48)$$

Fig. 2.29(b) is a graphical representation of g(t). To obtain this function, we must add a unit step and a negative unit step delayed by π. The result is g(t) = $U_s(t) - U_s(t - \pi)$ and the required function is

$$h(t) = \left[U_s(t) - U_s(t - \pi) \right] \sin t \qquad ..(2.49)$$

Singularity Functions

2.6 MATHEMATICAL OPERATIONS ON FUNCTIONS

In the previous sections, we have considered a broad class of input functions, $g_i(t)$. We now turn our attention to the production of output functions, $g_o(t)$, by mathematical operations on the input functions. Fig. 2.30 is a block diagram representation of this process in which an input function is operated on to produce an output function. We deal here with five common operations that are performed by physical components and systems. For the time being, however, we will treat the operations as mathematical exercises. Their physical implementations will be discussed later, (See Chapter IV). The operations are: (1) Time Integration; (2) Time Differentiation; (3) Proportional plus Integral; (4) Proportional plus Derivative; and (5) Proportional plus Derivative plus Integral.

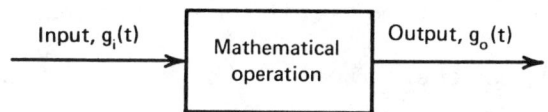

Figure 2.30 Block diagram representation of input-output relation

2.6.1 Time Integration

When the mathematical operation is integration, the output function is related to the input function by:

$$g_o(t) = \int g_i(t)\, dt \qquad \ldots (2.50a)$$

Recall that with an indefinite integral, there is a

constant of integration that must be evaluated. However, integrations performed by real physical components are always definite and reflect the entire history of the components. Thus, in this context, the operation of integration should be expressed as:

$$g_o(t) = \int_{-\infty}^{t} g_i(t') \, dt' \qquad ..(2.50b)$$

where t' is the dummy variable. Eq. (2.50b) may be conveniently separated into two component integrals:

$$g_o(t) = \int_{-\infty}^{t_b} g_i(t') \, dt' + \int_{t_b}^{t} g_i(t') \, dt' \qquad ..(2.50c)$$

or

$$g_o(t) = g_o(t_b) + \int_{t_b}^{t} g_i(t') \, dt' \qquad ..(2.50d)$$

where t_b is the time at the beginning of the operation and $g_o(t_b)$ is the value of the output function at $t = t_b$. Eq (2.50d) is therefore the basic form of integration which is appropriate for physical systems.

An important feature of the integration operation is that the output function always depends on both the integration and the previous value of the output. Thus integrating devices have memory. To demonstrate this, let us return to the operation indicated in (2.50d). Suppose that the input function in a particular interval has a zero value, the output function then retains its

Singularity Functions   63

last value and thus reflects the effects of all previous inputs.

To visualize the mathematical operation, it is often advantageous to employ graphical methods. Thus, in the case of integration, the output function at any time will equal the area under the input function up to the time under consideration.

The following example will make the integration process clear.

Example 2.8

The input function shown in Fig. 2.31(a) is fed into an integrator. Determine the output function, if $g_o(0) = 0$.

There are several ways of finding the output function. We may use either a mathematical or graphical approach. In this example, we will demonstrate both methods.

(a) <u>Mathematical Integration</u>:

The most direct way to perform mathematical integration is to break the input function into an appropriate number of intervals in which the functional form remains the same. In this example, they will be

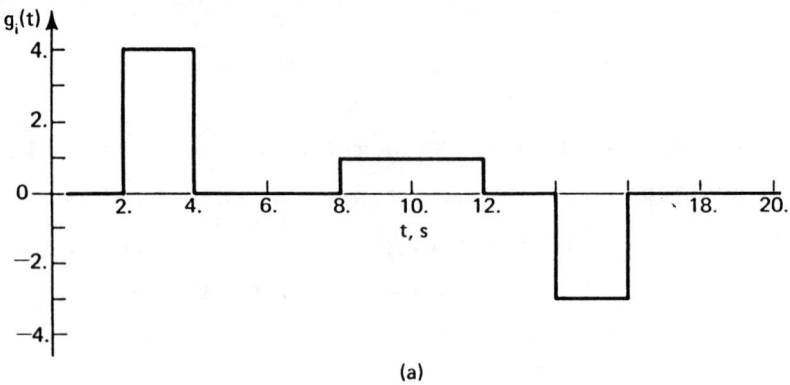

(a)

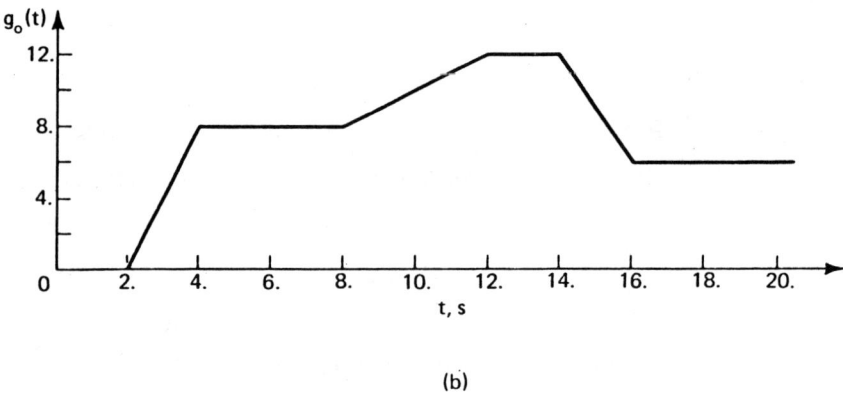

(b)

Figure 2.31 Integration of time functions (example 2.8)

$$
\begin{aligned}
&1^{\text{st}} \text{ interval} \longrightarrow g_i(t) = 4, && 2 < t < 4 \\
&2^{\text{nd}} \text{ interval} \longrightarrow g_i(t) = 0, && 4 < t < 8 \\
&3^{\text{rd}} \text{ interval} \longrightarrow g_i(t) = 1, && 8 < t < 12 \\
&4^{\text{th}} \text{ interval} \longrightarrow g_i(t) = 0, && 12 < t < 14 \\
&5^{\text{th}} \text{ interval} \longrightarrow g_i(t) = -3, && 14 < t < 16
\end{aligned}
$$

$$\dots (2.51)$$

Singularity Functions

We now proceed by applying (2.50d) in each interval in turn beginning with the first interval. Thus,

$$g_o(t) = \int_2^t 4\, dt + 0 = 4t - 8, \quad 2 \le t \le 4 \quad ..(2.52)$$

We may use the equation to plot the function in the first interval. Fig. 2.31(b) shows the output. At the end of the first interval, $t = 4$ so that $g_o(4) = 8$. In the second interval, the input function is zero. Thus the output throughout this interval remains at $g_o(4) = g_o(8) = 8$. In the third interval, the output has the form

$$g_o(t) = \int_8^t (1)\, dt + g_o(8) = t, \quad 8 \le t \le 12 \quad ..(2.53)$$

We find values of the output in the third interval from this relation. At the end of the third interval, the output $g_o(12) = 12$. Finally, the last interval has an output of:

$$g_o(t) = \int_{14}^t (-3)\, dt + g_o(14)$$

$$= 54 - 3t, \quad 14 \le t \le 16 \quad ..(2.54)$$

This relation holds for the output in the fifth interval. The output then stays at $g_o(16) = 6$ until another input is applied.

Another way of solving this problem is to express the entire input function in mathematical form. Thus:

$$g_i(t) = 4U_s(t-2) - 4U_s(t-4) + U_s(t-8) - U_s(t-12)$$
$$- 3U_s(t-14) + 3U_s(t-16) \qquad ..(2.55)$$

If $g_i(t)$ is placed into (2.50d), the output $g_o(t)$ is:

$$g_o(t) = 4U_r(t-2) - 4U_r(t-4) + U_r(t-8) - U_r(t-12)$$
$$- 3U_r(t-14) + 3U_r(t-16) + g(0) \qquad ..(2.56)$$

A plot of the above function also yields the output shown in Fig. 2.31(b).

(b) Graphical Integration:

In this approach, the output $g_o(t)$ is equal to the area under the $g_i(t)$ curve from 0 to t. For example, to find $g_o(4)$, we need the area under the input function from t = 0 to t = 4. This area is $4(4-2) = 8$. Thus $g_o(4) = 8$.

To find $g_o(16)$, we must sum all the area under $g_i(t)$ from t = 0 to t = 16. Negative areas are subtracted from the total. Thus:

$$g_o(16) = 4(4-2) + 1(12-8) - 3(16-14)$$
$$= 6 \qquad ..(2.57)$$

Singularity Functions   67

In those regions where the input has no value (that is, for t between 4 and 8, or t between 12 and 14), the output function remains constant at its previous value.

2.6.2 Time Differentiation

The relation between the output function and the input function when the mathematical operation is differentiation is:

$$g_o(t) = \frac{dg_i(t)}{dt} \qquad ..(2.58)$$

Differentiation emphasizes a change in the input function. The output does not depend on any previous inputs. Whenever, when the input function is not changing in value, the output is zero. From a graphical viewpoint, the output function is the slope of the input function at each instant of time.

Example 2.9

Fig. 2.32(a) shows the input function supplied to a differentiator. Find and plot the corresponding output function.

(a) Mathematical Differentiation:

We will use two mathematical methods to find the output function. In the first method, we express the given function in terms of time in specific intervals. For the given input function, we may write:

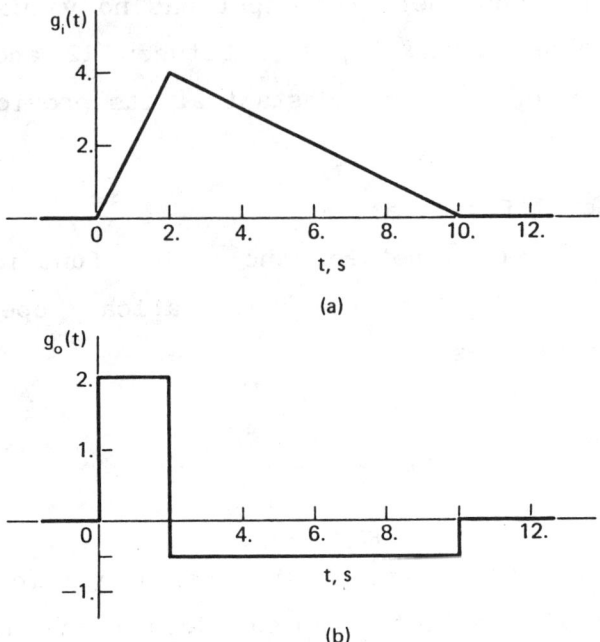

Figure 2.32 Differentation of time functions (example 2.9)

$$g_i(t) = 2t, \quad 0 < t < 2$$
$$= 5 - \frac{t}{2}, \quad 2 < t < 10 \quad \quad ..(2.59)$$

If these inputs are substituted into (2.58), the output function is:

$$g_o(t) = 2, \quad 0 < t < 2$$
$$= -1/2, \quad 2 < t < 10 \quad \quad ..(2.60)$$

The result is shown in Fig. 2.32(b).

In the second method, we write the input function in terms of time functions so that for this example:

Singularity Functions

$$g_i(t) = 2U_r(t) - (5/2)U_r(t-2) + (1/2)U_r(t-10)$$

..(2.61)

Now, the application of (2.58) leads immediately to:

$$g_o(t) = 2U_s(t) - (5/2)U_s(t-2) + (1/2)U_s(t-10)$$

..(2.62)

A plot of this function yields the same result shown in Fig. 2.32(b).

(b) <u>Graphical Differentiation</u>:

The output function is the slope of the input function at each value of time. The slope from $t = 0$ to $t < 2$ is constant and equal to $4/2 = 2$. Thus $g_o(1) = 2$. At $t = 2$, the slope changes discontinuously from positive to negative. The value of the output is constant between $t > 2$ and $t < 10$, because the slope is constant in this region. At $t = 6$, for example, $g(6) = 4/(2-10) = -1/2$. Any horizontal segment of the input function results in a zero output function, since the horizontal line has zero slope. Thus for times greater than 10 seconds, $g_o(t) = 0$.

2.6.3 Proportional Plus Integral

In this case, the output function is the sum of two operations on the input function. The output due to the proportional operation is designated as $g_{op}(t)$ and the one due to the integration as $g_{oi}(t)$. Thus the operation performed can be represented mathematically by:

$$g_{op}(t) = A_1 g_i(t) \qquad \qquad ..(2.63a)$$

$$g_{oi}(t) = A_2 \int_{t_b}^{t} g_i(t')\,dt' + g_{oi}(t_b) \qquad ..(2.63b)$$

$$g_o(t) = g_{op}(t) + g_{oi}(t) \qquad \qquad ..(2.63c)$$

where A_1 and A_2 are constants. The output from the proportional plus integral operation differs from the integration operation. The output of the proportional plus integral operation retains only the portion due to integration, when the input function goes to zero.

Example 2.10

The input function of Fig. 2.33(a) is subjected to proportional plus integral action according to the formula

$$g_o(t) = 2g_i(t) + \int g_i(t)\,dt \qquad ..(2.64)$$

Plot the output function, if the initial condition is zero.

The input function may be separated into two intervals as:

Singularity Functions

$$g_i(t) = (2t/3) + 1, \quad 0 < t < 3$$
$$= 0, \quad t > 3 \quad \quad ..(2.65)$$

The proportional portion of the output may be obtained directly as:

$$g_{op}(t) = (4t/3) + 2, \quad 0 < t < 3$$
$$= 0, \quad t > 3 \quad \quad ..(2.66)$$

This portion of the output is shown in Fig. 2.33(c). The integral portion of the output is:

$$g_{oi}(t) = \int_0^t [(2/3)t' + 1]dt'$$

$$= (1/3)t^2 + t, \quad 0 < t < 3$$
$$= 6, \quad t > 3 \quad \quad ..(2.67)$$

The integral operation on the input is shown in Fig. 2.33(b). Now we may add to obtain the complete output from both the operations as

$$g_o(t) = (1/3)t^2 + (7/3)t + 2, \quad 0 < t < 3$$
$$= 6, \quad t > 3 \quad \quad ..(2.68)$$

Fig. 2.33(d) shows the required output for proportional plus integral operation.

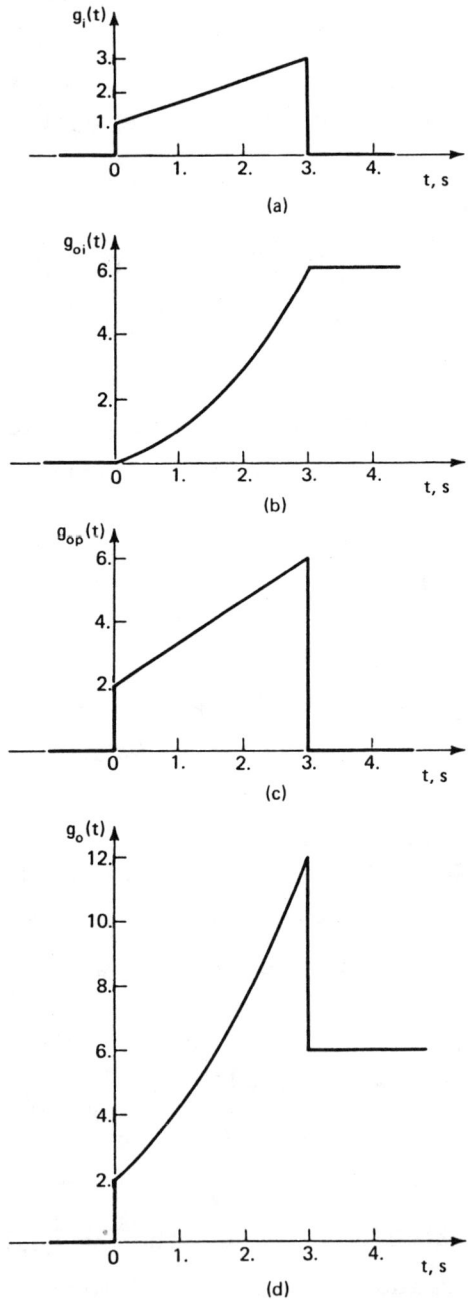

Figure 2.33 Proportional plus integral operation on time functions (example 2.10)

Singularity Functions

The same operations may be dealt with strictly in terms of time functions. Thus we begin with

$$g_i(t) = U_s(t) + (\tfrac{2}{3})U_r(t) - 3U_s(t-3) - (\tfrac{2}{3})U_r(t-3)$$

..(2.69)

The output due to proportional operation is:

$$g_{op}(t) = 2U_s(t) + (\tfrac{4}{3})U_r(t) - 6U_s(t-3) - (\tfrac{4}{3})U_r(t-3)$$

..(2.70)

and that due to integral operation is:

$$g_{oi}(t) = U_r(t) + (\tfrac{2}{3})U_p(t) - 3U_r(t-3) - (\tfrac{2}{3})U_p(t-3)$$

..(2.71)

The sum of these functions gives the same answer as shown in Fig. 2.30 (d). However, the plot is more difficult to obtain, because eight separate time functions must be considered.

The problem can also be solved using a strictly graphical approach described in Section 2.7. This has advantages particularly when parabolic functions are considered. However, when the composite function contains $U_s(t)$ and $U_r(t)$ only, either method can be used to advantage.

2.6.4 Proportional Plus Derivative

This operation yields an output that depends on the sum of a proportional part, $g_{op}(t)$, and a derivative part, $g_{od}(t)$. The complete output may be expressed as

$$g_o(t) = A_1 g_i(t) + A_2 \frac{dg_i(t)}{dt} \quad ..(2.72)$$

The operation produces an output that depends on the rate of change of the input to a lesser extent than in the differentiation operation alone. Thus the output is somewhat less sensitive to input changes.

Example 2.11

The input function given in Fig. 2.34(a) is applied at the input of a proportional plus derivative controller. Determine the output function, if the mathematical operation is:

$$g_o(t) = 1.5\, g_i(t) + \frac{dg_i(t)}{dt} \quad ..(2.73)$$

The input function is separated into five time intervals as shown:

Singularity Functions

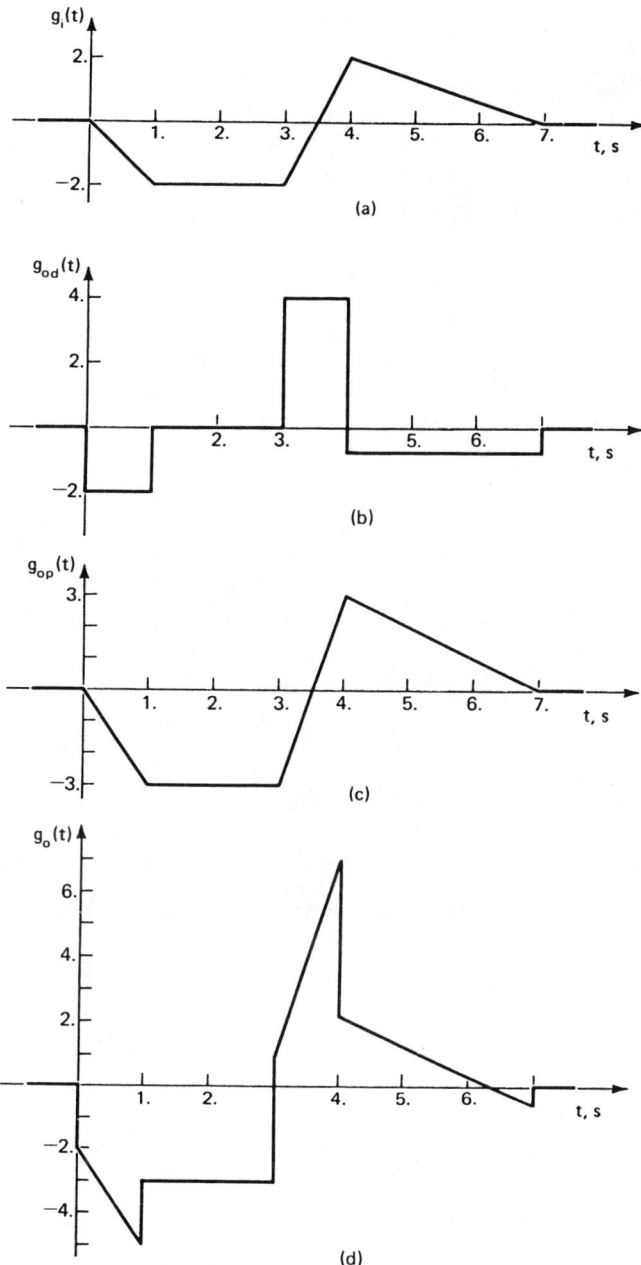

Figure 2.34 Proportional plus derivative operation on time functions (example 2.11)

$1^{\underline{st}}$ interval $\longrightarrow$ $g_i(t) = -2t,$ $\qquad$ $0 < t < 1$

$2^{\underline{nd}}$ interval $\longrightarrow$ $g_i(t) = -2,$ $\qquad$ $1 < t < 3$

$3^{\underline{rd}}$ interval $\longrightarrow$ $g_i(t) = 4t - 14,$ $\qquad$ $3 < t < 4$

$4^{\underline{th}}$ interval $\longrightarrow$ $g_i(t) = (-2/3)t + 14/3,$ $\qquad$ $4 < t < 7$

$5^{\underline{th}}$ interval $\longrightarrow$ $g_i(t) = 0,$ $\qquad$ $7 < t$

$\qquad\qquad\qquad\qquad\qquad\qquad\qquad\qquad\qquad\qquad$ .. (2.74a)

Thus the output due to the derivative portion is:

$\qquad g_{od}(t) = -2,$ $\qquad 0 < t < 1$

$\qquad g_{od}(t) = 0,$ $\qquad 1 < t < 3$

$\qquad g_{od}(t) = 4,$ $\qquad 3 < t < 4$ $\qquad$ .. (2.74b)

$\qquad g_{od}(t) = -2/3,$ $\qquad 4 < t < 7$

$\qquad g_{od}(t) = 0,$ $\qquad 7 < t$

This output portion is shown in Fig. 2.34(b). The proportional portion is merely the original input multiplied by 1.5 and is:

$\qquad g_{op}(t) = -3t,$ $\qquad 0 < t < 1$

$\qquad g_{op}(t) = -3,$ $\qquad 1 < t < 3$

$\qquad g_{op}(t) = 6t - 21,$ $\qquad 3 < t < 4$ $\qquad$ .. (2.74c)

$\qquad g_{op}(t) = -t + 7,$ $\qquad 4 < t < 7$

$\qquad g_{op}(t) = 0,$ $\qquad 7 < t$

Singularity Functions

Fig. 2.34(c) shows the proportional part of the output. The complete output is the sum of the proportional and derivative parts and is:

$$g_o(t) = -3t - 2, \quad 0 < t < 1$$
$$g_o(t) = -3, \quad 1 < t < 3$$
$$g_o(t) = 6t - 17, \quad 3 < t < 4 \quad \quad ..(2.74d)$$
$$g_o(t) = -t + 19/3, \quad 4 < t < 7$$
$$g_o(t) = 0, \quad 7 < t$$

The required output function is shown in Fig. 2.31(d).

In this case also, we may use the time functions:

$$g_i(t) = -2U_r(t) + 2U_r(t-1) + 4U_r(t-3) - \frac{14}{3} U_r(t-4)$$
$$+ \frac{2}{3} U_r(t-7) \quad \quad ..(2.75a)$$

The output due to proportional operation is:

$$g_i(t) = -3U_r(t) + 3U_r(t-1) + 6U_r(t-3) - 7U_r(t-4)$$
$$+ U_r(t-7) \quad \quad ..(2.75b)$$

When (2.75a) is differentiated, we get

$$g_{oi}(t) = -2U_s(t) + 2U_s(t-1) + 4U_s(t-3) - \frac{14}{3} U_s(t-4)$$

$$+ \frac{2}{3} U_s(t-7) \qquad \qquad ..(2.75c)$$

The sum of (2.75b) and (2.75c) gives the same answer as shown in Fig. 2.31(d).

From the above, it can be seen that the result in terms of time functions will contain ten separate functions. Thus, it may become time-consuming and laborious to obtain a sketch of the output. For the mathematical operations of integration and differentiation, the time function approach may sometimes be more convenient. However, the summation operations of proportional plus integral and proportional plus derivative appear to be easier by using the piecewise approach.

Here again, in this example, it is possible to treat the operation by graphical means alone and is left as an exercise to the student.

2.6.5 Proportional Plus Derivative Plus Integral

This operation yields an output that depends on the sum of a proportional part $g_{op}(t)$, a derivative part $g_{od}(t)$ and an integral part $g_{oi}(t)$. The complete output is expressed as

Singularity Functions

$$g_o(t) = A_1 g_i(t) + A_2 \frac{dg_i(t)}{dt} + A_3 \int g_i(t) \, dt \qquad ..(2.76)$$

In fact, this may be considered as a combination of the two previously discussed cases, namely proportional plus integral and proportional plus derivative.

Example 2.12

The input function of Fig. 2.35(a) is subjected to Proportional plus Derivative plus Integral action according to the formula

$$g_o(t) = 2g_i(t) + 3 \frac{dg_i(t)}{dt} + \int g_i(t) \, dt \qquad ..(2.77)$$

Sketch the input function, if the initial condition is zero.

The input function may be separated into four time segments as follows:

1st interval $\quad g_i(t) = t, \qquad 0 \le t \le 1$

2nd interval $\quad g_i(t) = -t + 2, \qquad 1 \le t \le 3$

3rd interval $\quad g_i(t) = t - 4, \qquad 3 \le t \le 4$

4th interval $\quad g_i(t) = 0, \qquad 4 \le t$

$\qquad\qquad\qquad\qquad\qquad\qquad\qquad ..(2.78a)$

The proportional portion is twice the original input and is:

$$g_{op}(t) = 2t, \quad 0 \leq t \leq 1$$
$$g_{op}(t) = -2t + 4, \quad 1 \leq t \leq 3$$
$$g_{op}(t) = 2t - 8, \quad 3 \leq t \leq 4 \quad \quad ..(2.78b)$$
$$g_{op}(t) = 0, \quad 4 \leq t$$

This is as shown in Fig. 2.35(b).

The output due to the derivative portion is:

$$g_{od}(t) = 3, \quad 0 \leq t \leq 1$$
$$g_{od}(t) = -3, \quad 1 \leq t \leq 3$$
$$g_{od}(t) = 3, \quad 3 \leq t \leq 4 \quad \quad ..(2.78c)$$
$$g_{od}(t) = 0, \quad 4 \leq t$$

This is as shown in Fig. 2.35(c).

The output portion due to the integral operation is:

$$g_{oi}(t) = \frac{t^2}{2}, \quad 0 \leq t \leq 1$$
$$g_{oi}(t) = \frac{-t^2}{2} + 2t - 1, \quad 1 \leq t \leq 3$$
$$\quad\quad ..(2.78d)$$
$$g_{oi}(t) = \frac{t^2}{2} - 4t + 8, \quad 3 \leq t \leq 4$$
$$g_{oi}(t) = 0, \quad 4 \leq t$$

Singularity Functions

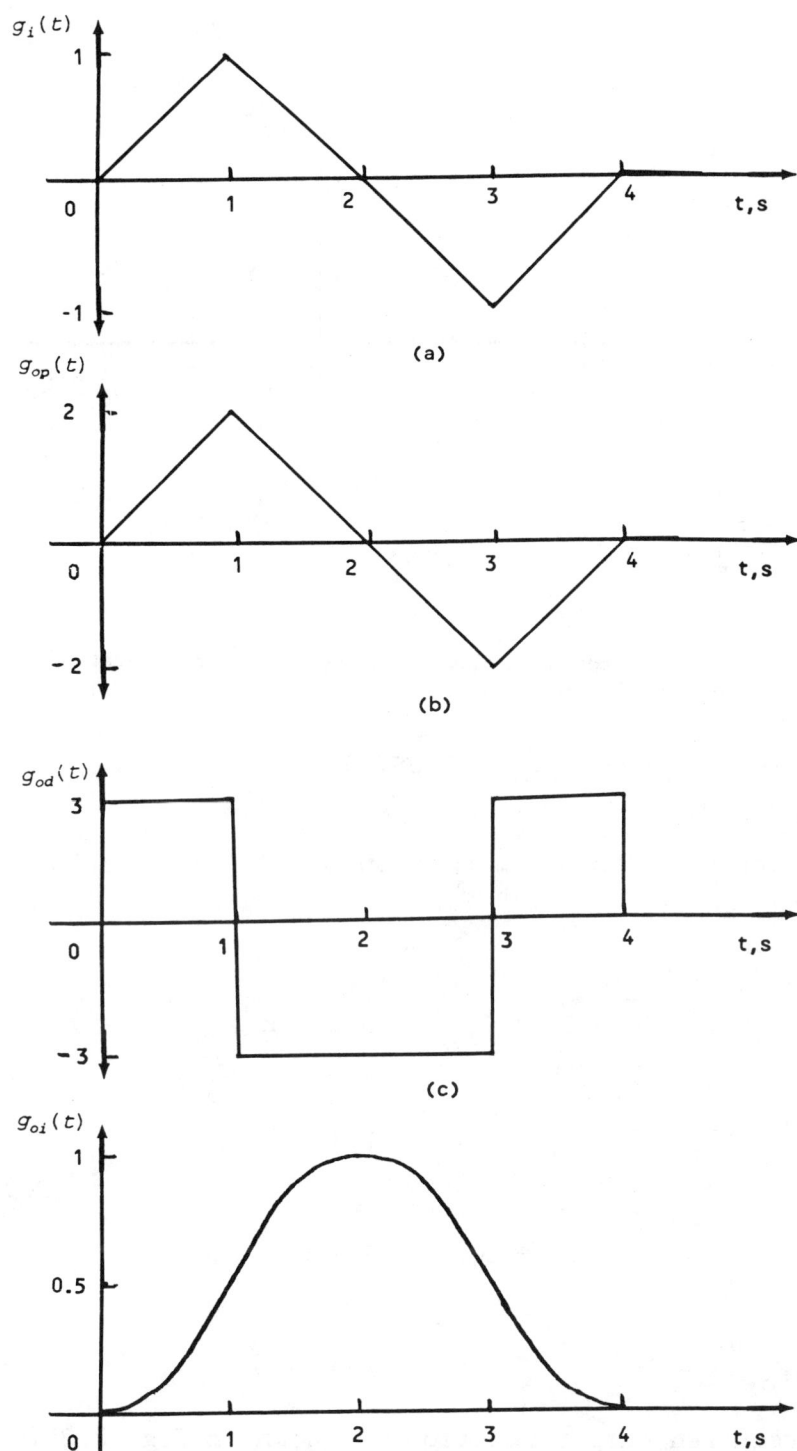

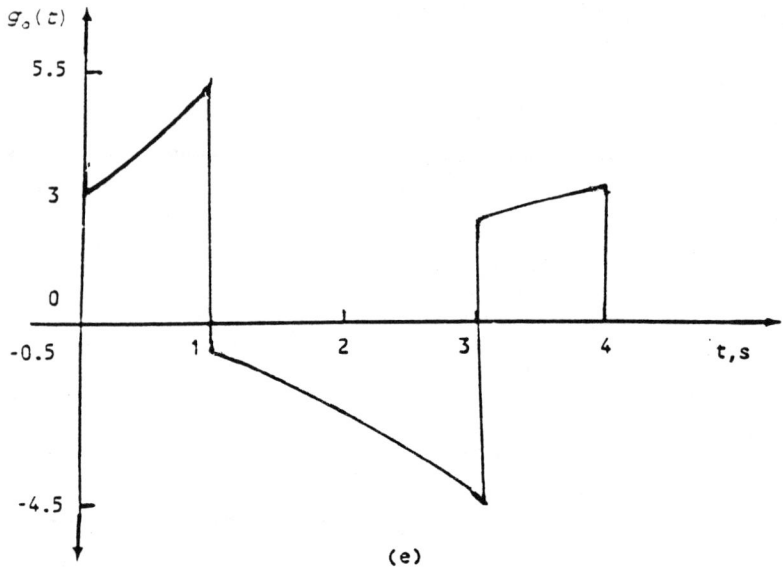

Figure 2.35. Proportional plus derivative plus itegral operation on time functions (example 2.12).

This is as shown in Fig. 2.35(d).

The complete output is the sum of (2.78b), (2.78c) and (2.78d) and is

$$g_o(t) = \frac{t^2}{2} + 2t + 3, \qquad 0 \le t \le 1$$

$$g_o(t) = \frac{-t^2}{2}, \qquad 1 \le t \le 3$$

$$g_o(t) = \frac{t^2}{2} - 2t - 3, \qquad 3 \le t \le 4$$

$$g_o(t) = 0, \qquad 4 \le t$$

..(2.78e)

The required output function is shown in Fig. 2.35(e).

Singularity Functions

The above problem is solved by the use of time functions as follows:

$$g_i(t) = U_r(t) - 2U_r(t-1) + 2U_r(t-3)$$
$$- U_r(t-4) \quad ..(2.79a)$$

The output due to proportional operation is

$$g_{op}(t) = 2U_r(t) - 4U_r(t-1) + 4U_r(t-3)$$
$$- 2U_r(t-4) \quad ..(2.79b)$$

The output due to the derivative operation is

$$g_{od}(t) = 3U_s(t) - 6U_s(t-1) + 6U_s(t-3)$$
$$- 3U_s(t-4) \quad ..(2.79c)$$

The output due to the integral operation is

$$g_{oi}(t) = U_p(t) - 2U_p(t-1) + 2U_p(t-3)$$
$$- U_p(t-4) \quad ..(2.79d)$$

The sum of (2.79b), (2.79c) and (2.79d) gives the same answer as shown in Fig. 2.35(e). It is observed that twelve different time functions need be considered in this case.

It is also possible to treat the operation by graphical means also and is left as an exercise for the student.

2.7 COMPOSITE FUNCTIONS BY GRAPHICAL APPROACH

In the previous section, it has been stated that graphical approach can be used to obtain the waveform of a composite function. Also in Section 2.2, the functions derived by integration have been discussed. In many of these cases, plotting these functions may become laborious and difficult. This is particularly so, when parabolic functions are involved. Under such circumstances, it is highly preferable to differentiate the given function resulting in a combination of U_s and U_r functions which can be plotted in an easier manner. The original function can then be determined by graphical integration which has been discussed earlier. An example will illustrate this method.

Example 2.13

Sketch the function described by

$$g(t) = U_p(t) - 2U_p(t-1) + U_p(t-2) \qquad (2.80)$$

It is noted that this is the same function considered in Example 2.4.

The first step is to differentiate $g(t)$ so that the resulting function $g'(t)$ contains only $U_s(t)$ and $U_r(t)$ only. In this case, it will be

$$g'(t) = U_r(t) - 2U_r(t-1) + U_r(t-2) \qquad ..(2.81)$$

Singularity Functions

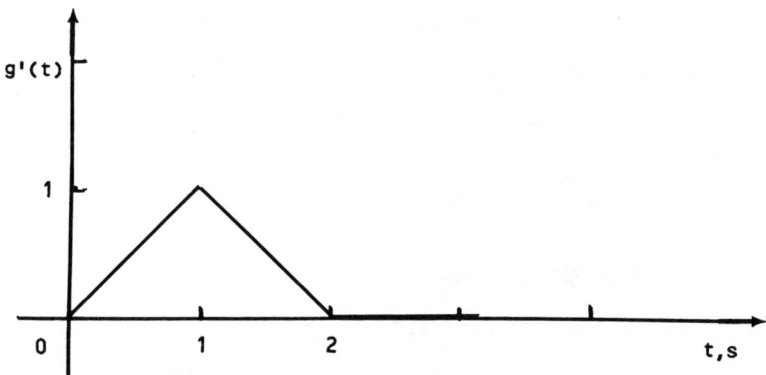

Figure 2.36. Sketch of g'(t) of (2.81)

This is a triangular pulse shown in Fig. 2.36.

Now, we can perform graphical integration on the function given in Fig. 2.36 and obtain g(t). For example, if we wanted to know g(1), we would find the area under g'(t) from time t = 0 to t = 1. The result would be g(1) = 0.5. If, on the other hand, we wanted g(3), we would need the area from t = 0 to t = 3. In this case, it is only the area under the triangular function from t = 0 to t = 2. There is no additional area from t = 2 to t = 3. Thus g(2) = g(3) = 1.0. This process can be continued to obtain g(t) graphically. The final result will be the same as shown in Fig. 2.15(b).

In the above approach, it is possible that impulse functions may appear in g'(t). This can be taken care of easily, because the integral is the impulse function is a unit step. This means that wherever a unit impulse appears in g'(t), we add a unit step in g(t) by graphical means.

2.8 SUMMARY AND DISCUSSIONS

In this chapter, a unit step function, $U_s(t)$, is first defined. Starting from this function, two singularity functions are obtained by successive integrations and they are the unit ramp and the unit parabolic functions. Also, starting from the unit step function, two more functions are obtained by successive differentiations and they are the unit impulse and the unit doublet functions. The time shifting and magnitude scaling properties of all these functions have been discussed. In particular, it is demonstrated that the graphical approach is highly convenient to determine the integral or the derivative of such functions. Though exponential functions do not belong to this category, they are defined and some of their properties are given, as they prove useful in later chapters.

Starting from these functions, composite functions are obtained. These composite functions are time-shifted and weighted combinations of $U_s(t)$, $U_r(t)$ and $U_p(t)$. These are constructed in two ways, either mathematically or graphically. The method of

Singularity Functions

decomposition of a composite signal into its constituent parts has been given. Examples have been included to illustrate these concepts.

2.9 PROBLEMS

2.1 Obtain the graphical representation of the functions given below using the mathematical approach:

(a) $g_1(t) = U_s(t + 1) - 2U_s(t) - 3U_s(t - 2)$

(b) $g_2(t) = -2U_r(t) - U_r(t - 2) + 6U_r(t - 4)$

(c) $g_3(t) = 2U_p(t - 2) - 2U_p(t - 5) - U_p(t - 8) + U_p(t - 14)$

Verify your answer using the graphical approach.

2.2 Sketch the graphs of the following expressions using the graphical approach:

(a) $g_1(t) = U_s(t) + 2U_s(t - 2) + 5U_s(t - 3) + U_r(t - 3) - U_r(t - 5)$

(b) $g_2(t) = U_r(t) - U_r(t - 1) + U_s(t - 2) - (1/2)U_r(t - 2) + (1/2)U_r(t - 6)$

(c) $g_3(t) = 2U_p(t) - 36U_s(t - 4) - 2U_p(t - 6) - 12U_r(t - 6)$

Verify your answer using the mathematical approach.

2.3 Sketch the following function in the interval
0 ≤ t ≤ 20 s.

$$g(t) = 5U_s(t) + U_r(t) + 0.2U_p(t) - 0.2U_p(t - 10)$$

Use the graphical method to obtain the sketch and verify your answer using the mathematical approach.

2.4 Place the function

$$g(t) = 8 \cos 5t + 6 \sin 5t$$

in the form

(a) $g(t) = D \cos(5t - \phi)$

(b) $g(t) = D \sin(5t + \phi)$

2.5 Express the time functions shown in Fig. P2.5 in terms of the appropriate singularity functions.

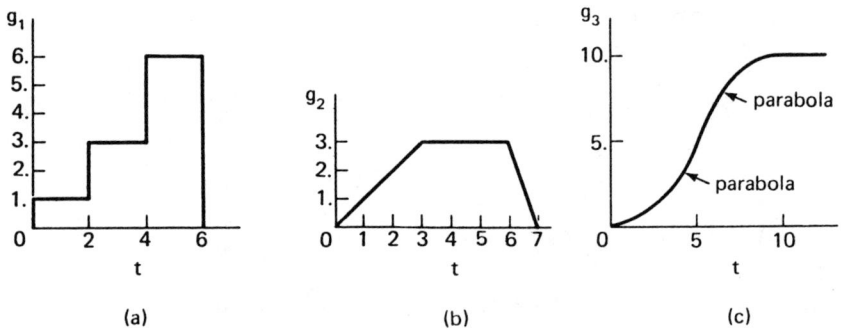

Figure P 2.5

(The slope at the start of each parabola is zero).

Singularity Functions

2.6 (a) Express g(t) of Fig P2.6 in terms of singularity functions.

(b) Find t_1.

(The slope at the start of each parabola is zero).

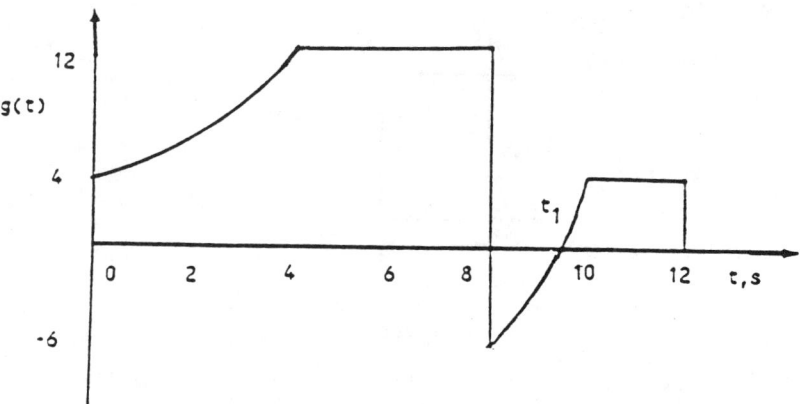

Figure P 2.6

2.7 A function may be expressed in piecewise form as:

$$g(t) = 0 \quad , \quad t \leq 2$$
$$= 1 - \cos \pi t/2 \quad , \quad 2 \leq t \leq 4$$
$$= 2(\cos \pi t/2) - 2, \quad 4 \leq t \leq 6$$
$$= 0 \quad , \quad t \leq 6$$

(a) Sketch the graph of g(t).

(b) Express g(t) in terms of singularity functions.

2.8 A signal variable y(t) has the form shown in Fig. P2.8. If the initial value of the variable at t = 0 [y(0)] equals zero,

(a) Sketch the time integral of y(t).

(b) Repeat part (a), if $T_p = 0.5$.

(c) Repeat part (a), if $T_p \longrightarrow 0$.

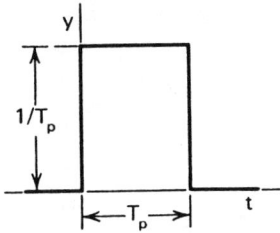

Figure P 2.8

2.9 (a) Express the singularity function

$$g(t) = 10U_p(t - 2)$$

in the functional form (polynomial in time).

(b) For Fig. P2.9, find:

    (i) a graph $[g_i(t)]$ whose integral produces the figure.

    (ii) the expression for $g_i(t)$ in terms of singularity functions.

    (iii) the integral of $g_i(t)$ in terms of singularity functions.

(The slope at the start of each parabola is zero).

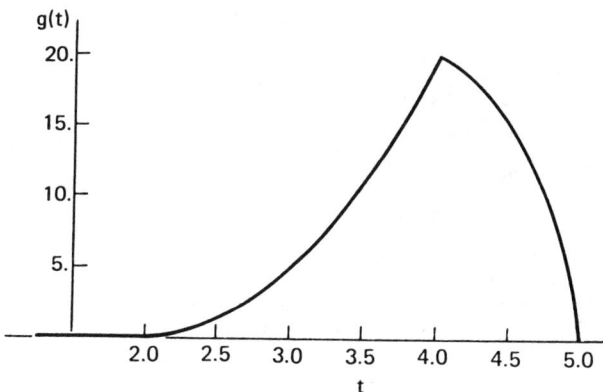

Figure P 2.9

2.10 (a) Express the time functions shown in Fig. P2.10 in terms of the appropriate singularity functions.

(b) Sketch the time integral of each function.

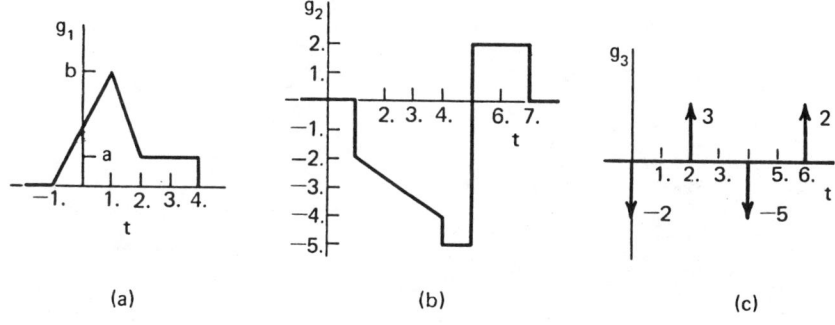

Figure P 2.10

2.11 A four position valve is suddenly moved from position 1 to position 2 where the water flow is 30 cm^3/s. After 10 seconds, the valve is suddenly changed to position 3 (water flow 50 cm^3/s). Then, 20 seconds later, the valve is moved suddenly to position 4 where the flow is 80 cm^3/s. Twenty-five seconds later, the valve is suddenly returned to position 1 where the flow is shut off.

(a) Plot the flow rate against time and express the graph in terms of singularity functions.

(b) Plot the volume of water that has passed through the valve in the time interval from 0 ≤ t ≤ 55 s. Describe this volume by singularity functions.

2.12 (a) Sketch the time integral of the functions $g_1(t)$ and $g_2(t)$ shown in Fig. P2.12.

(b) Compare the results for t greater than 2 s.

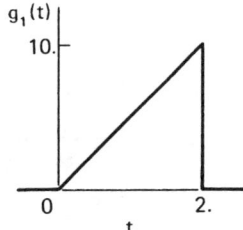
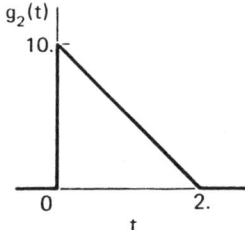

Figure P 2.12

2.13 A dispute arose between the driver of a car and a witness who was looking on. The car had stopped exactly at a wall. The driver claimed:

(a) 135 m from the wall his velocity was 27 m/s.

(b) a constant deceleration to 18 m/s in the first 90 m travelled.

(c) a deceleration of 3.6 m/s^2 thereafter.

The witness claimed:

(a) 81 m from the wall the velocity was 27 m/s.

(b) a deceleration of 3.6 m/s^2 thereafter.

Check each claim separately.

Specifically,

(a) Sketch the acceleration-time relation assuming at t = 0 the car is 135 m from the wall.

(b) Express the acceleration in terms of singularity functions.

(c) Find the velocity of the car in terms of singularity functions.

(d) Sketch the velocity-time relation.

(e) Which account is correct? Why?

2.14 Fig. P2.14 is a function g(t).

(a) Express g(t) in terms of singularity functions.

(b) Sketch the time derivative of g(t).

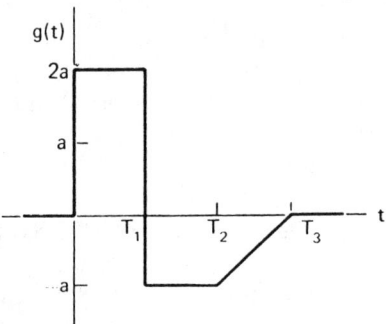

Figure P 2.14

2.15 The velocity of an automobile was measured as shown in Fig P2.15.

(a) Describe v(t) using singularity functions.

(b) Sketch the distance travelled as a function of time in the interval 0 ≤ t ≤ 8 seconds, showing clearly the relevant values on the graph.

(c) Sketch the acceleration-time relation.

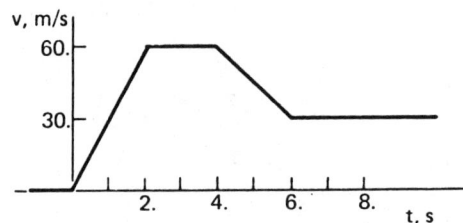

Figure P 2.15

2.16 A function g(t) is the product of two time functions $h_1(t)$ and $h_2(t)$. If

Singularity Functions

$$h_1(t) = u_r(t), \quad \text{and}$$

$$h_2(t) = U_s(t) - 2U_s(t-5) + U_s(t-10),$$

sketch $g(t)$.

2.17 Sketch the following time functions:

(a) $g_1(t) = \varepsilon^{-t} [\text{Im}(\varepsilon^{j2t})] U_s(t)$

(b) $g_2(t) = \varepsilon^{t} [\text{Re}(\varepsilon^{j3t})] U_s(t)$

2.18 (a) Sketch the graph of $g(t) = [\sin t] U_s(t)$.

(b) Sketch the time integral of $g(t)$.

(c) Find the time derivative of $g(t)$ in terms of singularity functions.

2.19 Fig. P2.19 shows a specific function of time, $g(t)$.

(a) Express $g(t)$ in terms of singularity functions.

(b) Sketch the time integral and time derivative of $g(t)$.

(c) An engineer wishes to multiply the function, $g(t)$, by another function, $h(t)$, so that the resulting function is twice as large as $g(t)$ in the interval $1 \leq t \leq 3$ and is zero elsewhere. Express $h(t)$ in terms of singularity functions.

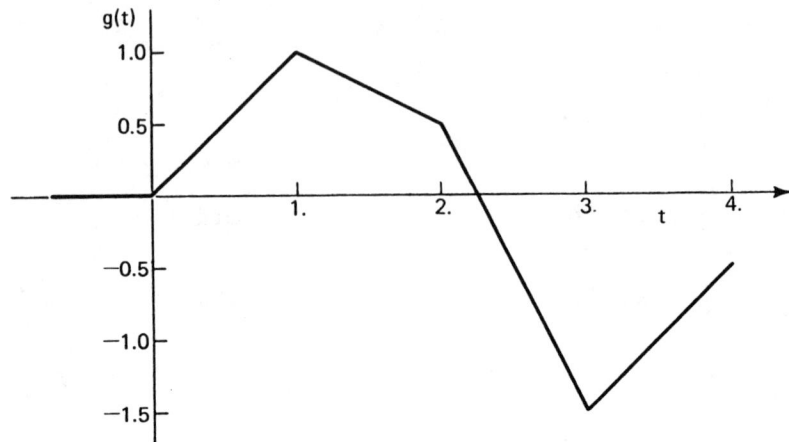

Figure P 2.19

2.20 Make a sketch of the following product functions:

(a) $g_1(t) = (\sin t) [U_s(t - \pi) + U_s(t - 2\pi)$
$+ U_s(t - 3\pi) - 3U_s(t - 4\pi)]$

(b) $g_2(t) = (\cos t) [U_s(t - \pi) + U_s(t - 2\pi)$
$+ U_s(t - 3\pi) - 3U_s(t - 4\pi)]$

2.21 Sketch the time integral and time derivative of the following expressions:

(a) $g_1(t) = [\sin(\pi t/4)] [U_s(t - 4) - U_s(t - 16)]$

(b) $g_2(t) = [\cos t] [U_s(t) - 2U_s(t - \pi)$
$+ U_s(t - 2\pi)]$

Singularity Functions

2.22 Given the proportional plus integral operation:

$$g_o(t) = 2g_i(t) + \int g_i(t)\, dt$$

and the input function

$$g_i(t) = 5U_s(t) + 5U_s(t-5) - 15 U_s(t-10) + 5U_s(t-15)$$

(a) Sketch the proportional part of the output.

(b) Sketch the integral part of the output.

(c) Sketch the complete output.

2.23 The relation between an input function $g_i(t)$ and the output function $g_o(t)$ is given by:

$$g_o(t) = g_i(t) + 2 \int g_i(t)\, dt$$

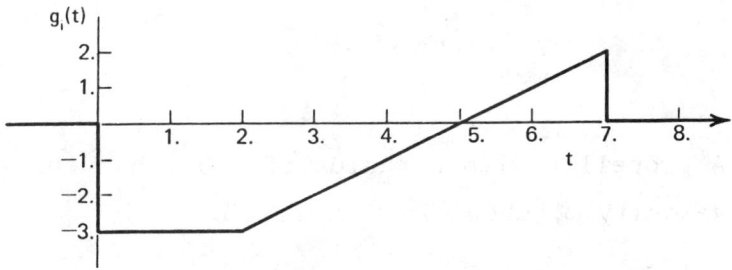

Figure P 2.23

If $g_i(t)$ is as shown in Fig. P2.23,

(a) Sketch $g_o(t)$.

(b) Express $g_o(t)$ in terms of singularity functions.

2.24 A system is designed so that the relation between an input $g_i(t)$ and an output $g_o(t)$ is:

$$g_o(t) = 3g_i(t) + 2\frac{dg_i(t)}{dt}$$

If the input has a form shown in Fig. P2.24, sketch the output as a function of time.

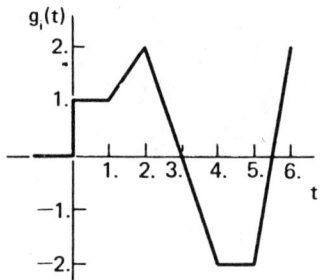

Figure P 2.24

2.25 A propellor with a radius of 0.3 m has the angular velocity ω, shown in Fig. P2.25.

(a) Express $\omega(t)$ in terms of singularity functions.

(b) Find the total distance travelled by a point on the tip of the propellor after $t = 70$ s.

(c) Sketch the angular acceleration $\alpha(t) = [\frac{d\omega}{dt}]$ as a function of time.

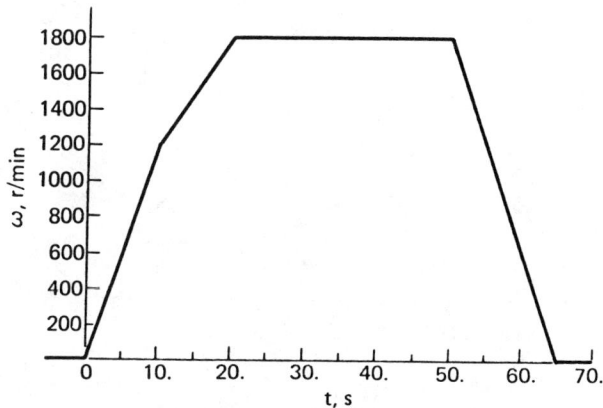

Figure P 2.25

2.26 The relation between the input function $g_i(t)$ [shown in Fig. P2.26] and the output function $g_o(t)$ is given by the relation

$$g_o(t) = g_i(t) + 2 \int g_i(t) + \frac{dg_i(t)}{dt}$$

(a) Express $g_o(t)$ in terms of singularity functions.

(b) Sketch $g_o(t)$

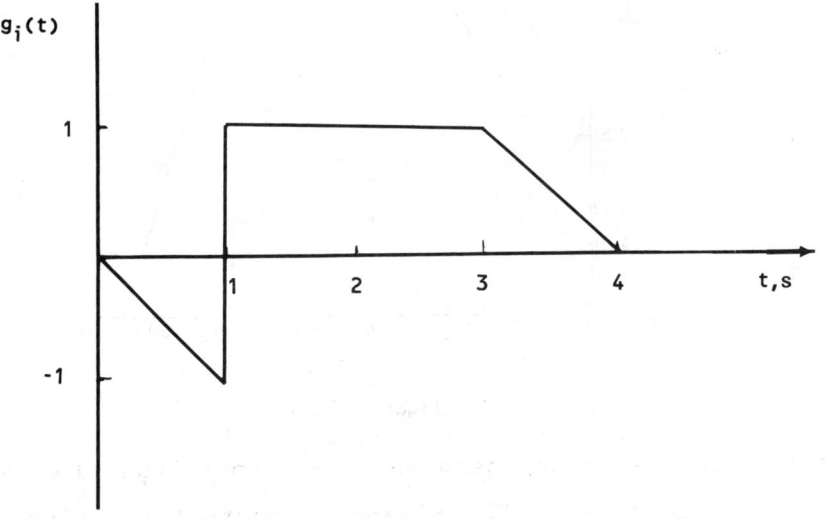

Figure P 2.26

CHAPTER 3

CIRCUIT COMPONENTS

3.1 INTRODUCTION TO CIRCUIT COMPONENTS

In this chapter, we shall consider electric, fluid, mechanical and thermal circuit components. Thus, we deal with four distinct physical situations.

A model of any physical situation requires the use of mathematical variables to represent certain physical quantities associated with its behavior. There are only two basic types of such quantities (Chapter I): (a) those which flow along or through a constrained path, such as water flowing through a pipe, and (b) those which require, whether explicitly or implicitly, a measurement which must be made between two points such as the distance between the points. In connecting an instrument between two points, it must be placed across everything which lies between them and such quantities may reasonably be described by **across** variables. In measuring the flow along a path, it is necessary to cut the path and insert the instrument so that the flow is constrained to pass through it. Such quantities are therefore described by **through** variables. All the media (electric, fluid, mechanical and thermal) will have an across variable and a through variable. The variables for each of the media are different. However, we shall study the cases where the across variables of the different systems are analogous and similarly where the through variables of the different systems are analogous.

The classification of all physical quantities as either across, or through, is the basis of the recognition of analogous situations in different physical media. This grouping is especially useful, if it is done in a manner that ensures that the method of calculating the energy for one medium (e.g., electric) is the same as that used in another medium (e.g.,fluid). This, however, is not a necessary condition. The essential feature in choosing the variables is that the resulting equations for all media shall have the same mathematical form. In this chapter, we shall consider electrical, fluid, mechanical and thermal systems. In the first three of these, the choice of the signal variables is made so that the product of across and through variables represents power. Thermal systems are different, however, in that the through variable itself, the heat flow rate, is power.

In choosing the across and the through variables, it must be kept in mind that the objective in the end is to produce a network representation for a physical system. This simplifies the problem of describing equilibrium conditions. For example, the lines in Fig. 3.1 may represent pipes carrying a liquid. Equilibrium is described in part by noting that the rate at which the liquid leaves the junction through pipes 2 and 3 must equal the rate at which it enters from pipe 1. This is the principle of continuity which in this case derives from the conservation of mass. The same thought process is used to describe the necessary conditions which result from the conservation of charge in electrical systems or the conservation of energy in thermal systems.

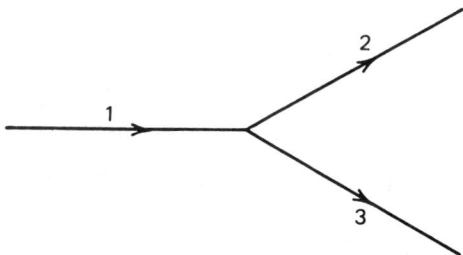

Figure 3.1 Continuity at junction

If the lines represent co-linear forces which are acting in the same or opposite directions at a point or on a rigid body, the equilibrium resulting from the balancing of the forces is described in the same manner, rather more simply than the equilibrium of a set of coplanar forces. Indeed, the junction of connections in a network automatically contains all the information of a free body diagram of mechanics.

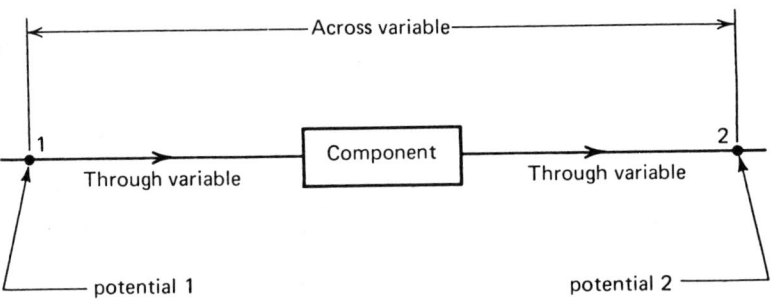

Figure 3.2 Basic two terminal component

However, before considering equilibrium in general, it is necessary to identify certain elemental situations and to derive the elemental equations describing their performance. These correspond to idealized physical situations in which only one physical phenomenon is considered to be present. We model these phenomena with two-terminal elements. Fig. 3.2 is a schematic representation of the basic two-terminal component. We designate the point at which flow enters as terminal 1 and the point at which flow leaves as terminal 2. There is a potential associated with each

Circuit Components

terminal which is a measure of the energy in the circuit at that location. The across variable is the potential difference between terminals 1 and 2. The through variable, representing a transmitted quantity, passes through the component without change. There are only three basic elemental relations and they are expressed mathematically as:

$$av(t) = K_1 \, tv(t) \qquad ..(3.1a)$$

$$av(t) = K_2 \frac{d}{dt}[tv(t)] \qquad ..(3.1b)$$

$$av(t) = K_3 \int [tv(t)] \, dt \qquad ..(3.1c)$$

where $\qquad av(t) =$ Across Variable $\qquad ..(3.1d)$

and $\qquad tv(t) =$ Through Variable $\qquad ..(3.1e)$

In the above equations, K_1, K_2 and K_3 are proportionality constants that depend on the geometric configuration and the material properties. It should be emphasized that (3.1) is merely a general form for circuit components and the variables are functions of time. The basic physical laws of electricity, fluid mechanics, mechanics and heat transfer can be placed in this format. It is these laws which determine the appropriate signal variables and the specific values of K_1, K_2 and K_3. In the following sections, we will introduce the basic elements in electrical, fluid, mechanical and thermal systems.

3.2 ELECTRIC CIRCUIT COMPONENTS

3.2.1 Electric Signal Variables

Although electric charge (Q) is perhaps the most basic electrical quantity from a theoretical viewpoint, the most useful variables for electric circuits are current and voltage. Current 'i' is the rate of flow of charge and is clearly identified as a through variable. The unit of charge is the **coulomb** (abbreviation C) and an electric current is the rate of flow of this charge in coulombs per second or **amperes** (abbreviated A). Thus, in mathematical terms,

$$i = \frac{dQ}{dt} \qquad \qquad ..(3.2)$$

The voltage difference, $e_1 - e_2$, (sometimes called potential difference) is the across variable. Potential difference is the work done per unit charge in moving the charge between two points at potentials e_2 and e_1. The fact that two points are required in the determination of a voltage difference (or voltage drop) clearly identifies it as an across variable. The unit for electric potential is the **volt** (abbreviated V). This is the potential when the work involved in moving one coulomb is one joule. Thus the work, W, done in moving a charge, Q, between two points from potentials e_2 to e_1 is $W = Q(e_1 - e_2)$. The rate at which this work

Circuit Components

is expended is the power, P. The instantaneous power is:

$$P = \frac{dW}{dt} = (e_1 - e_2)i \qquad ..(3.3)$$

Note that the power is the product of the across variable and the through variable.

Since the rate of change of work is defined as power, this is the well-known general expression for power in an electric circuit. The unit of power is the **watt** (abbreviated W) which is one joule per second. It also follows that determination of energy or work is simply a matter of integrating (3.3):

$$W = \int_{t_1}^{t_2} (e_1 - e_2) \, i \, dt \qquad ..(3.4)$$

There are three elemental situations which are found in circuits (see Eqs. 3.1) which do not involve a source of energy to be put into the system. These are commonly known as the ideal resistor, inductor, and capacitor.

Although real components only approach the ideal, they often can be modeled as a combination of two or more of these ideal components. All the components must originate from physically realizable phenomena and obey physical laws.

3.2.2 Electrical Resistance

From experience on many metallic circuit parts, we observe that the voltage difference established in such a circuit is directly proportional to the current. This observation may be expressed in mathematical form as:

$$e_1 - e_2 = R\, i \qquad \ldots (3.5)$$

This relation is known as Ohms's law. Note that (3.5) is identical in form to the general elemental relation given in (3.1a) with $R = K_1$. The quantity symbolized by R is called the resistance and is a property of the material and geometry of the circuit part.

The ideal resistor has the voltage/current characteristic shown in Fig. 3.3. In this case, the voltage difference is directly proportional to current as indicated by the line of constant slope. The slope of the line is the resistance, R.

The unit of resistance is the **ohm** (abbreviated Ω). There are occasions however, when the reciprocal, G, is more convenient. This is the conductance for which the unit is the **siemens** (abbreviated S).

Note that the ideal resistance element is bidirectional or bilateral. If potential 1 is greater than potential 2, then the across variable, $e_1 - e_2$, is positive and this defines the positive direction of current. When potential 2 is greater than potential 1,

Circuit Components

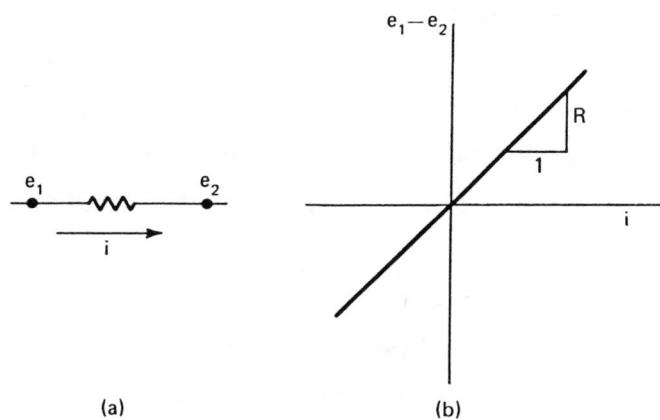

Figure 3.3 Characteristics of ideal resistance

the across variable is negative and the current reverses. In a bilateral resistive element, a given positive or negative voltage difference causes a current of the same magnitude, but in opposite directions.

All objects have electrical resistance. However, in electrical circuits that are used to produce some desired effect on the signal variables, we are dealing with an electric component that is especially made for the purpose. In general, resistance is obtained from the resistivity and geometry of the wire material. The resistance of a wire is expressed as:

$$R = \frac{\rho l}{A}, \; \Omega \qquad \qquad ..(3.6)$$

where ρ is the resistivity ($\Omega.m$ - ohm-meter) of the wire material, l is the wire length in meter (m), and A is the wire cross-sectional area in meters^2 (m^2).

There are many other forms of resistors available in the market. The exact value of the resistance of any resistor is dependent on the material and the geometry. In addition, there are devices which exhibit unilateral resistances, in that the slope of the v-i characteristic will have one value when the potential difference is positive and will have another value when the potential difference is negative. Also, these devices can be temperature-sensitive, in that the actual value of the resistor will be dependent on the ambient temperature. Such resistors can also be modeled using the concept of ideal resistors discussed above. In such cases, the range of validity of the model has to be clearly obtained. This will not be discussed further, as this is beyond the scope of the present text.

The energy dissipated in a resistance can be determined by rearranging (3.3) in the form

$$W = \int P \, dt \qquad \qquad ..(3.7)$$

When the voltage and current are not changing with time, the energy is:

$$W = R \, i^2 \, t = (e_1 - e_2)^2 \, t/R \qquad ..(3.8)$$

Note that the energy increases with time. The

Circuit Components

resistance is, therefore, an energy dissipative element.

3.2.3 Electrical Inductance

The most common form of inductor is a coil of wire wrapped around a magnetic core. The core contains a high level of magnetic flux due to current in the conducting wire coil. When the wire coil contains more than one loop, each loop contributes to the flux linkage, λ_{12}, between the two terminals of the coil. When the physical arrangement is such that the flux-linkage is a single-valued function of current, the element is called an inductor. This is expressed as:

$$\lambda_{12} = L\, i \qquad \qquad ..(3.9)$$

where λ_{12} is the flux-linkage in webers and L is the inductance in henrys (abbreviated H). Thus, the ideal inductor has the flux-linkage/current characteristic shown in Fig. 3.4. Faraday's law states that the voltage difference induced in coils by a changing magnetic field is:

$$e_1 - e_2 = \frac{d\lambda_{12}}{dt} \qquad \qquad ..(3.10)$$

The combination of (3.9) and (3.10) yields a relation between the signal variables, that is:

$$e_1 - e_2 = L\, \frac{di}{dt} \qquad \qquad ..(3.11)$$

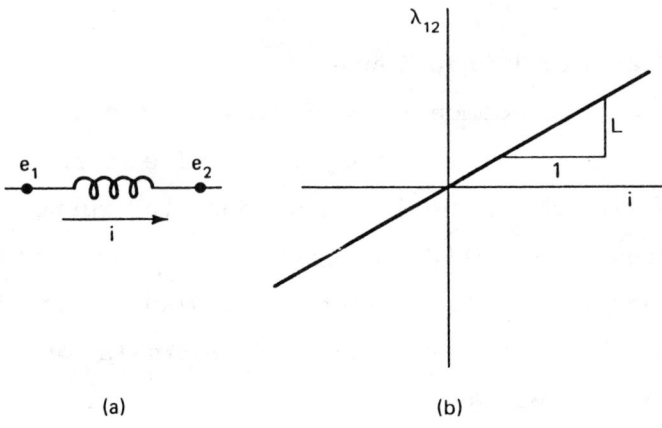

Figure 3.4 Characteristics of ideal inductance

Eq. (3.11) has the same form as given in the general elemental relation in (3.1b). In this case, K_2 = L. From the ideal characteristics shown in Fig. 3.4, we may observe that the ideal inductor is also a bilateral component.

The magnitude of inductance for a wire wound over a cylindrical core is:

$$L = \frac{4\pi (10^{-7}) \mu_r A N^2}{l}, \text{ H} \quad ..(3.12)$$

where μ_r is the relative permeability of the core material, A is the area of the core (m^2), l is the length of the core (m), and N is the number of coils. For an air core, the permeability $\mu = 1$. When an iron rod or ferrite core is used, the relative permeability may be as high as 10,000, the actual value depending on

Circuit Components

the flux density.

In actual electrical elements, it is not possible to achieve the ideal characteristics given for the resistance and inductance. In both cases, the element consists of a length of wire. Thus the actual inductance always has some resistance and the actual resistance has some inductance. However, we will assume throughout the text that the components are ideal and represent only one phenomenon.

The energy into an inductor is from (3.3) and (3.11):

$$W = L \int_{i_o}^{i} i' \, di' = \frac{L}{2} (i^2 - i_o^2) \qquad ..(3.13)$$

where i_o is the initial current. For the inductor, the energy is not a function of time directly. The reason for this is that the inductor stores energy rather than dissipates it. The energy stored by the inductor is due to the flow of current and the expression given in (3.13) is valid whether or not the current is constant. The energy stored is not directly dependent on the voltage difference. Indeed (3.13) applies even when the voltage difference is zero. The inductor is in a class of energy storage elements known as T-type (for through variable) storage.

3.2.4 Electrical Capacitance

Two conductors separated from each other by an insulating material (called the dielectric) form a capacitor. In the capacitor, an electric field is established between the conductors. When the physical arrangement is such that the charge on the conductors is a single-valued function of the voltage difference, the element is called a pure capacitor. This is expressed as:

$$Q = C(e_1 - e_2) \quad \quad ..(3.14)$$

where Q is the charge in coulombs and C is the capacitance in farads (abbreviated F). Thus, the pure or ideal capacitor has the charge/voltage difference characteristic shown in Fig. 3.5. In terms of the signal variables, we may replace the charge Q in (3.14) by the integral of the current so that:

$$e_1 - e_2 = \frac{1}{C} \int i \, dt \quad \quad ..(3.15)$$

For this type of component, the governing equation (3.15) has the same mathematical form as the general elemental relation of (3.1c) with $K_1 = \frac{1}{C}$. The ideal capacitor is a bilateral component in the same way as the ideal resistor and ideal inductor.

For a parallel plate capacitor, the magnitude of the capacitance is:

$$C = 8.84 \times 10^{-12} \frac{\varepsilon_r A}{d}, \, F \quad \quad ..(3.16)$$

Circuit Components

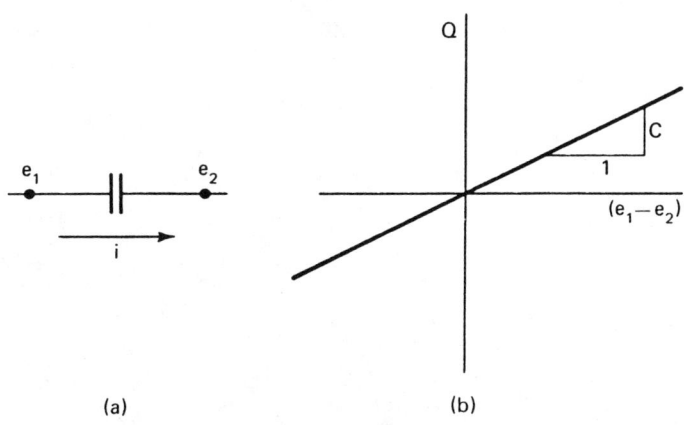

Figure 3.5 Characteristics of ideal capacitance

where ε_r is the permittivity of the dielectric in farads per meter, A is the cross-sectional area (m^2) of the plates, and d is the thickness (m) of the dielectric.

The energy associated with a capacitance may be determined from (3.3) and (3.5) as:

$$W = C \int_{\Delta e_o}^{\Delta e} e' \, de' = \frac{C}{2} (\Delta e^2 - \Delta e_o^2) \qquad ..(3.17)$$

where Δe is the voltage difference, $e_1 - e_2$ and Δe_o is the initial voltage difference. Since a capacitor stores energy, it is also an energy storage component. In this case, energy is stored due to the presence of a voltage difference. The expression for energy given in (3.17) is valid whether or not the voltage difference is constant. The energy stored is not directly dependent on the current. Indeed (3.17) applies even when the current is zero. The capacitor is in the class of energy storage devices known as A-type (for across variable) storage.

The energy stored by an inductance (Eq. 3.13) and by a capacitance (Eq. 3.17) may be formulated in a similar way for the analogous storage elements in fluid and mechanical systems.

The normal symbols for the three circuit elements are shown in Figs. 3.3, 3.4 and 3.5. The arrow associated with the current shows the positive direction of conventional flow through the element. The terminal at which the positive current enters is positive for each of the elements and polarity indicators must be set up as in the figures, if the positive signs implied in the elemental equations are to apply.

At this stage, it can be noted that each of the three elements is significantly different. One of them, the resistor, dissipates energy which cannot be recovered by the circuit. The others store energy which can be recovered, when the circuit conditions are suitable. However, in one case, the capacitor (A-type storage), the energy is stored by virtue of there being a voltage between the two terminals irrespective of whether or not there is current flow. In the other case, the inductor (T-type storage), the opposite statement holds: the energy is stored by virtue of there being a current flow through the element

Circuit Components   117

irrespective of whether or not there is a voltage across it.

In this chapter, the circuit models of the storage elements do not include initial storage. That is, i_o, for the inductor and Δe_o, for the capacitor have been assumed to be zero. The analogous fluid, mechanical and thermal storage elements will also be introduced without initial storage. We will consider the circuit models for storage elements with initial storage in Chapter IV.

It is also indicated in the above study that across the two terminals, there is only one property present, namely a resistance or an inductance or a capacitance. Such a study is known as **lumped circuit modeling**. However, there exist circuits where it is not possible to separate these properties so that the same can be modeled using the above concept of lumped circuit modeling. Such a circuit is known as **distributed circuit** and its modeling will not be discussed here. However, it must be noted that distributed circuits are also modeled using lumped network elements and this is beyond the scope of the present text.

3.3 FLUID CIRCUIT COMPONENTS

To develop models for components with fluid flow, we must apply the equations of fluid mechanics to these components and try to adapt them to the forms given for the basic elements in (3.1). In some cases, the components analyzed by the fluid mechanics equations do not match the basic forms exactly. When that happens, restrictions are placed on the resulting fluid components. Let us be aware, therefore, that the analogy between the fluid components and the basic electric components is not perfect.

3.3.1 Fluid Signal Variables

The quantity that is conserved in the flow of fluid at a junction is the mass flow, M, in kg/s. Strictly speaking, this ought to be the through variable. However, it is more convenient for measurement purposes to use the volume flow, q, in m^3/s as the through variable. The relation between mass flow and volume flow is

$$M = \rho . q \qquad \qquad ..(3.18)$$

where ρ is the fluid density in kg/m^3.

Thus the volume flow will be proportional to the mass flow, when the fluid density remains constant throughout the circuit. Under this condition, the volume flow becomes an exact through variable. The constant density condition is approximated very closely, when we have a gas medium with low velocity or a liquid medium. We, therefore, restrict the fluid components to liquid flow or incompressible gas flow, when we select volume flow as the through variable.

The choice of a suitable across variable is

Circuit Components   119

suggested by the condition that the product of through and across variables should represent power. From this, we determine that the units of an appropriate across variable are Newtons/meter^2 (N/m^2) or Pascals (abbreviated Pa). Thus, the pressure difference, $p_1 - p_2$, is clearly indicated as the across variable in fluid systems.

One interesting result is that it is possible to define pressure in a manner which is directly analogous to the definition of voltage. That is, the difference in pressure, Δp, is the work done, ΔW, in moving unit volume, ΔV, between two points. This is illustrated by considering the movement of an incremental volume of fluid between two points along a pipe of uniform cross-section A, as shown in Fig. 3.6. The work done according to this alternate definition is given by:

$$\Delta p = \frac{\Delta W}{\Delta V} \qquad \qquad ..(3.19)$$

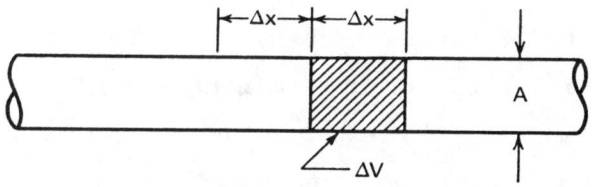

Figure 3.6 Elemental volume within a pipe

However, the volume may be expressed as $\Delta V = A \cdot \Delta x$, and the work done is the net force on the incremental volume multiplied by the distance moved. That is, $\Delta W = (\Delta F)(\Delta x)$. The substitution of the work and volume into (3.19) yields:

$$\Delta p = \frac{(\Delta F)(\Delta x)}{A(\Delta x)} = \frac{\Delta F}{A} \qquad \ldots (3.20)$$

In the limit, this is the force per unit area which is the normal definition of pressure.

Just as there are three elemental physical situations with electric circuits, there are three directly analogous situations in problems of the flow of fluids. However, depending on the fluid involved, the linear relationships between the through and the across variables do not exist over as wide range of values and the resulting models are therefore not quite as accurate. Nevertheless, a significant amount of information can be derived from such models leading to a considerable insight into how fluid systems behave. Since the amount of effort involved in obtaining solutions for linear systems is significantly less than that required for nonlinear systems, the linear model is an attractive tool for determining the general nature of the performance of fluid systems. It is, of course, necessary to have some appreciation of the conditions under which the linear model ceases to be accurate, and part of the following discussion is directed towards making this assessment.

3.3.2 Fluid Resistance

The fluid resistance models frictional effects which impede the flow of fluid along a pipe. These exist at the interface between the fluid and the pipe, and also between layers of fluid which are not moving at the same velocity. In certain circumstances, the flow of all the particles of fluid is in the direction of the tube such as the one indicated in Fig. 3.7, this form of flow being termed **laminar**. The profiles of fluid velocity are shown at three locations in the pipe.

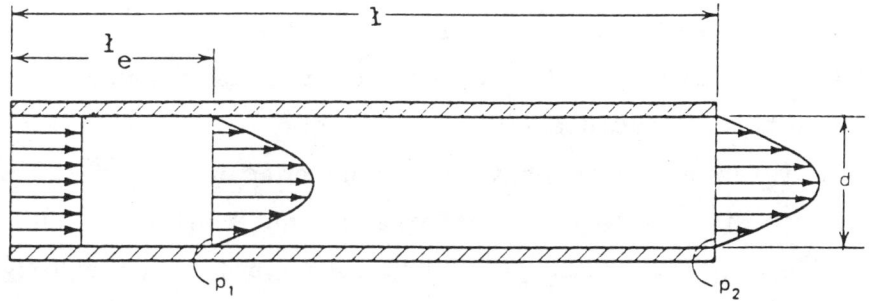

Figure 3.7 Pipe flow with viscous friction

At the pipe entrance, the velocity profile is uniform. The viscous properties of the fluid cause shear forces which act to retard the velocity at the boundaries and increase it along the center line. At some distance down the pipe, a stabilized parabolic shape profile is reached. This distance is known as the entrance length, l_e, and usually varies from about 80 to 120 diameters. Thereafter, the velocity profile remains unchanged. In the region where the profile is not changing, the steady state momentum equation of fluid mechanics yields:

$$p_1 - p_2 = Rq \qquad ..(3.21a)$$

$$R = \frac{128\,\mu(l - l_e)}{\pi d^4} \qquad ..(3.21b)$$

where μ is the dynamic viscosity of the fluid in Pa.s
and   d is the diameter of the pipe in m.

Thus, for laminar flow only, the pressure drop along a pipe is proportional to the flow rate and the fluid resistance is defined in the same manner as the electrical resistance. The linear portion of the pipe resistance occurs only after the entrance length. Since there is no way to eliminate the nonlinear entrance effects, the usual procedure is to select pipe lengths, l, that are at least four times the entrance length. In this way, entrance length effects are minimized. Actually, the fluid resistance is usually approximated by replacing the distance, $l - l_e$, in (3.21b) with the length, l, alone. Thus, as a good approximation when l is much larger than l_e, we have

$$p_1 - p_2 = Rq \qquad ..(3.22a)$$

$$R = \frac{128\,\mu l}{\pi d^4} \qquad ..(3.22b)$$

where p_1 is now the pressure at the entrance to the pipe. We should emphasize that (3.22) holds only for laminar flow.

When the flow rate increases beyond some critical value, the flow breaks up and contains circulating currents called **vortices**. This type of flow is called **turbulent** and results in an increased rate at which energy is dissipated within the fluid. The critical point between laminar and turbulent flow is not precisely known, but may be estimated by means of the Reynolds number (RE) which is a nondimensional expression involving several properties of the fluid and the geometry of the pipe:

$$Re = \frac{4 \rho q}{\pi \mu d} \quad \quad ..(3.23)$$

Normally, if the Reynolds number is less than 2000, the flow is laminar, but becomes fully turbulent when the Reynolds number is 5000.

No special name has been given to the unit of fluid resistance and probably the most convenient unit is N.s/m. The laminar flow fluid resistance given in (3.22b) is applicable for both liquid and gas flows. We will limit our consideration to two common fluid media, water and air. At 20° C, the properties of water and air are:

	Water	Air
Viscosity	$\mu = 10.1(10^{-4})$ Pa·s	$\mu = 0.18(10^{-4})$ Pa·s
Density	$\rho = 1000$ kg/m^3	$\rho = 1.2$ kg/m^3

A fluid resistance is bilateral if a flow in either direction develops the same magnitude of pressure difference between the end terminals. In fluid systems, we may assume at once that the resistances are bilateral if only one fluid medium is present. However, in cases where liquid pipes empty into air, the resistance is not bilateral. Consider, for example, Fig. 3.8.

When the resistance is connected between the pump and the tank (Fig. 3.8a), it is bilateral. There is only one medium, the water. Now, on the other hand, when the resistance is connected between water pump and air (Fig. 3.8b), the resistance is not bilateral. A reversal of pump will draw in air. We may still apply our circuit model, however, if we know in advance that the flow always remains in the proper direction from pump to atmosphere.

The power dissipated by a fluid resistance is equal to the product of the pressure drop and flow rate associated with the component. This may also be expressed as Rq^2 or $(p_1 - p_2)^2/R$.

Circuit Components

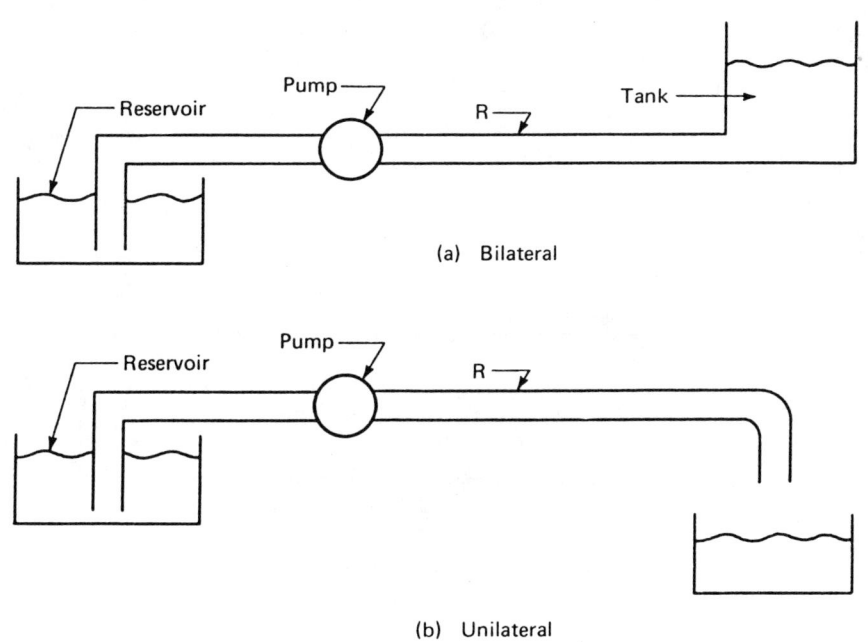

Figure 3.8 Fluid resistances

3.3.3 Fluid Inertance

Another physical situation in fluid systems relates to the inertia effect in the fluid. This is of particular importance with liquids, although it is not necessarily negligible in pneumatic systems. Consider a tube as indicated in Fig. 3.9, in which the fluid volume rate is q m^3/s.. If we restrict our attention to a section of length l and uniform cross-sectional area A, the mass of the fluid is $\rho A l$. The velocity of the fluid is q/A and hence the momentum is $\rho l q$. From the momentum equation of fluid mechanics for frictionless flow, the rate of change of momentum is equal to that of the net force acting on the mass of fluid. This net force is equal to the product of pressure difference and area. Thus

$$p_1 - p_2 = \frac{\rho l}{A} \frac{dq}{dt} \qquad ..(3.24)$$

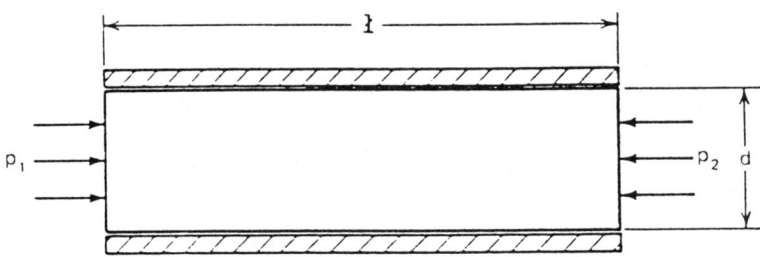

Figure 3.9 Fluid inertance model

Eq. (3.24) has the same form as the general relation given in (3.1b). Pressure difference, $p_1 - p_2$, as the across variable is directly analogous to voltage difference and q is directly analogous to current. We therefore note that this equation has identically the same form as the elemental equation for inductance [Eq. (3.11)]. The effect can therefore be called fluid inductance, although usually the term inertance is used, since it is the inertia effect which is being modeled.

The fluid inertance of a pipe is calculated from

$$L = \frac{\rho l}{A} \qquad ..(3.25)$$

The unit for fluid inertance is $N.s^2/m^5$.

The energy stored by a fluid inertance is dependent upon the flow rate (through variable) and inertance in the same form as given for the electrical inductance in (3.13).

Fluid inertance has been derived under the assumption of the flow of a frictionless fluid in a pipe. Previously fluid resistance was also derived in a pipe for a fluid with friction but without inertia. Both components share the same physical geometry: the pipe. For analysis purposes, it is accepted to separate the friction and inertia effects in the pipe. In any practical case, however, both effects exist simultaneously. Thus, in a real pipe, there is both fluid resistance and fluid inertance. There is only one flow, q, passing through the pipe. The complete model for a pipe is therefore:

$$p_1 - p_2 = Rq + \frac{dq}{dt} \qquad ..(3.26)$$

The pressure difference, $p_1 - p_2$, is the sum of the pressure differences caused by fluid resistance and fluid inertance. This model is shown in Fig. 3.10. We use the same symbols for fluid components that we previously used for the electrical components. The model of a fluid line or fluid pipe consists of a resistance and inertance in series. The order of the components does not matter for the fluid line, because there is no definable pressure terminal between the

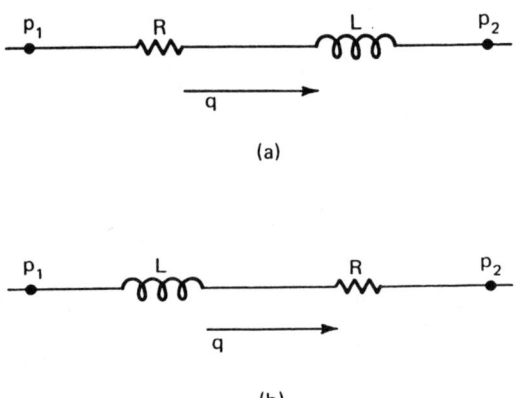

Figure 3.10 Fluid line circuit models

resistance and the inertance. This occurs because the effects are not separable in a physical sense, although we did separate them analytically. Sometimes one of the effects is much smaller than the other and can be neglected. We will indicate in each problem the appropriate model for each line component. However, in general, both effects must be considered.

3.3.4 Fluid Capacitance

The momentum equation of fluid mechanics was used to define resistance and inertance. Fluid capacitance is due to an effect we may observe in the continuity equation of fluid mechanics. A simple way to demonstrate the formulation of the signal variables in the form of (3.1c) is to consider the closed chamber [Fig. 3.11(a)].

Circuit Components

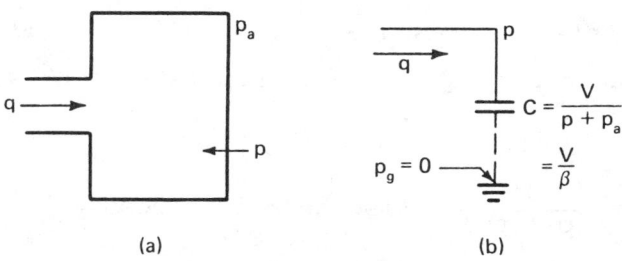

Figure 3.11 Closed tank capacitance

The mass of fluid, M, in the chamber is equal to the product of density and volume ($M = \rho V$). If we take the time derivative of the mass, we obtain an expression for the mass flow. In terms of the volume flow, the result is:

$$q = \frac{dV}{dt} + \frac{V}{\rho}\frac{d\rho}{dt} \qquad ..(3.27)$$

Eq. (3.27) is a form of the continuity equation of fluid mechanics. When the volume of the chamber is fixed, the first term on the right-hand side is equal to zero. The resulting equation relates volume flow to fluid density. However, our signal variables are volume flow and pressure difference. To find a relation between the signal variables, we must specify the connection between pressure and density. This connection depends on the fluid. For a fluid, we may apply Hooke's law to obtain:

$$\frac{d\rho}{\rho} = \frac{dp}{\beta} \qquad ..(3.28)$$

where β is the bulk modulus of the liquid in Pa. For water, $\beta = 2.14\ (10^9)$ Pa. When the fluid is a gas, the relation between pressure and density depends on the thermodynamic process. The appropriate results are:

$$\frac{d\rho}{\rho} = \frac{dp}{(p + p_a)} \quad \text{(isothermal process)} \quad \quad ..(3.29a)$$

$$\frac{d\rho}{\rho} = \frac{dp}{\gamma(p + p_a)} \quad \text{(adiabatic process)} \quad \quad ..(3.29b)$$

where p_a is the atmospheric pressure and γ is the ratio of specific heats.

We emphasize the atmospheric pressure here, because the absolute pressure is required in (3.29) and p is usually the gage pressure. If (3.27) and (3.28) are substituted into (3.29) and an integration is performed, we get

$$p = \frac{\beta}{V} \int q\ dt + p(0) \quad \text{(liquid)} \quad \quad ..(3.30a)$$

$$p = \frac{(p + p_a)}{V} \int q\ dt + p(0) \quad \text{(gas, isothermal)} \quad ..(3.30b)$$

$$p = \frac{\gamma(p + p_a)}{V} \int q\ dt + p(0) \quad \text{(gas, adiabatic)} \quad ..(3.30c)$$

where p(0) is the value of the pressure in the chamber at t = 0.

Circuit Components

If we assume that $p(0) = 0$, (3.30) has almost the desired form given in (3.1c). The difference between (3.30) and (3.1c) is that there is only one pressure in (3.30) and a pressure difference is required to fit the standard form. This is a consequence of the fact that the closed tank has only one terminal which is the internal pressure, p. In effect, then, we are trying to make a two-terminal component model out of a physical component that has only one terminal. We accomplish this by assuming a fictitious terminal, p_g, at zero potential and expressing (3.30) as:

$$p - p_g = \frac{1}{C} \int q \, dt \qquad \ldots (3.31)$$

where
$$C = \frac{V}{\beta} \text{ for a liquid} \qquad \ldots (3.32a)$$

and
$$C = \frac{V}{(p_a + p)} \text{ or } \frac{V}{\gamma(p_a + p)} \qquad \ldots (3.32b)$$

for a gas. The unit of fluid capacitance is m^5/N.

Thus capacitance for a gas is a function of pressure, p. When the system operates with small pressure changes, the quantity $(p + p_a)$ does not change appreciably and the capacitance is essentially constant. When there are large pressure transients, the capacitance value is variable. In general, a component that varies as a function of one of the signal variables is called a nonlinear component. Such components are beyond the scope of this text. For our purposes, all problems relating to gas operation will be restricted to small pressure transients. Then we can treat the capacitance as a constant.

The energy stored by a fluid capacitance is dependent on the pressure difference (across variable) and has the same form as that given for the electrical capacitance in (3.17).

The fictitious terminal adopted to complete the two-terminal form is at zero potential. Thus, a closed tank always represents a capacitance to ground. However, since the ground terminal is fictitious, it cannot be located physically. We denote a fictitious ground terminal by dashing the circuit line going to the ground. Fig. 3.11(b) shows the circuit representation for a closed tank capacitor. Notice that the liquid filled closed tank is not bilateral. If liquid is removed from the tank, cavitation occurs and we are no longer dealing with a homogeneous medium. A gas filled tank is bilateral up to a point. It ceases to be bilateral, when all the gas has been removed.

Fig. 3.12(a) shows a fixed volume tank with inlet and outlet flow. The previous reasoning is still applicable. There is only one pressure terminal in the tank and our model [Eq. 3.1] requires two pressure terminals for each component. Analysis of the tank with inlet and outlet flows q_1 and q_2 leads to:

$$p - p_g = \frac{1}{C} \int (q_1 - q_2)\, dt \qquad \ldots (3.33)$$

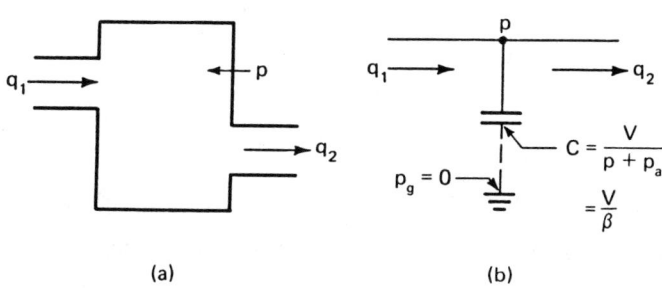

Figure 3.12 Closed tank capacitance with inlet and outlet

The circuit equivalent for this tank is shown in Fig. 3.12(b). The tank again represents a capacitance to ground.

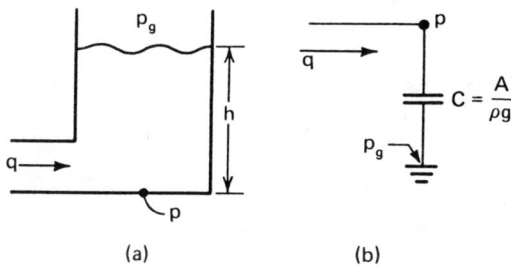

Figure 3.13 Liquid open tank capacitance

Consider now the open tank in which a liquid is stored, as indicated in Fig. 3.13. In this case, the pressure p is indicated at the bottom of the tank. The pressure is gravitational force per unit area of the column of liquid above it. If the horizontal area of the tank is A, then the weight of the column of height h is:

$$w = (p - p_g)A = \rho g A h \qquad ..(3.34a)$$

from which
$$p - p_g = \rho g h \qquad ..(3.34b)$$

where g is the gravitational constant (= 9.81 m/s^2) and p_g is the pressure above the liquid column.

Before proceeding to identify the capacitive effect, it may be noted that (3.34) provides the justification for using the height of the column of liquid as an alternate measure of pressure, commonly called **head**. However, from the discussion at the beginning of this section, it should be evident that the product of head and flow rate cannot possibly result in power as expressed in any acceptable unit. The resulting confusion with expressions involving fluid resistance, capacitance and inductance is such that head is not used as an across variable in the fluid circuits analyzed in this text.

Now, from (3.27), when density is not changing, but volume is changing, q = dV/dt. In addition, since V = Ah, we may express (3.34) in terms of pressure difference and flow rate as:

$$p - p_g = \frac{\rho g}{A} \int q \, dt \qquad ..(3.35)$$

Comparison with the standard form in (3.1c) shows that the capacitance of the open tank is

$$C = \frac{A}{\rho g} \qquad ..(3.36)$$

This component is essentially a two-terminal component without the assumption of a fictitious terminal. One terminal exists at the bottom of the tank. The other is the pressure over the liquid column. Usually, this pressure will be zero gage pressure and the open tank like the closed tank will represent a capacitance to ground [Fig. 3.13(b)]. However, this is not necessarily the case.

Consider Fig. 3.14(a), in which a closed tank is partially filled with liquid. Another way of viewing this is that we put a cover on the open tank. In the partially filled tank, the addition of flow, q, compresses the gas trapped above the liquid column. As a result, the pressure above the liquid column is no longer constant. The equivalent circuit for this circumstance is shown in Fig. 3.14(b). From the fluid circuit point of view, there are two capacitances in series. The resulting capacitance is reduced. Thus, sealing an open tank will reduce the capacitance.

The liquid open tank capacitance is bilateral only if there is liquid in the tank. When there are flow reversals, the liquid level in the tank must be checked to ensure that the tank has not emptied.

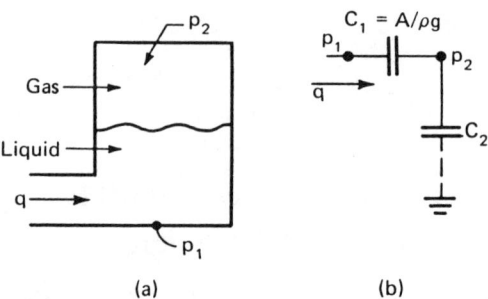

Figure 3.14 Liquid tank capacitance closed at top

There is still another type of fluid capacitance. This type features a moving part between two chambers. Fig. 3.15(a) shows a spring loaded piston separating two chambers at p_1 and p_2.

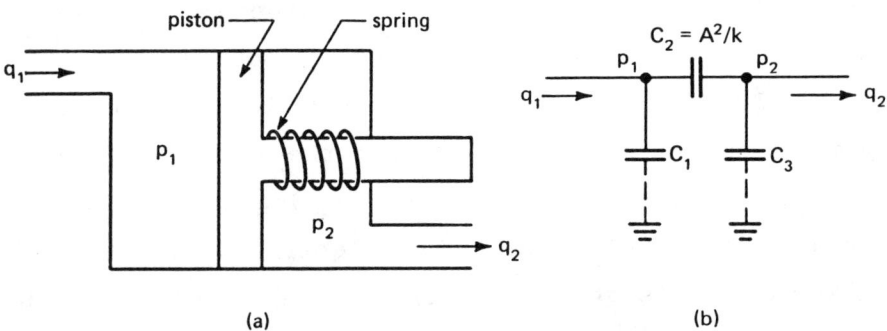

Figure 3.15 Spring-loaded piston type capacitance

The change in volume depends on the pressure difference and spring constant k. If the area of the chamber is A, the volume change may be expressed as:

$$\Delta V = \frac{A^2}{k} (p_1 - p_2) \qquad \ldots (3.37)$$

and from (3.26) (for no density change), we may relate pressure difference and volume flow for this component as:

$$p_1 - p_2 = \frac{k}{A^2} \int q \, dt \qquad \ldots (3.38)$$

Now, once again, comparison with the standard form (3.1c) shows that this is a two-terminal capacitance and has a capacitance value of

$$C = \frac{A^2}{K} \qquad \ldots (3.39)$$

Fig. 3.15(b) shows the equivalent circuit. The capacitance calculated above is designated as C_2. There is a compressibility capacitance associated with each chamber. These are C_1 and C_3. Most often, the capacitance C_1 and C_3 will have small values. In that case, the spring-loaded piston represents a two-terminal fluid capacitance.

3.4 MECHANICAL CIRCUIT COMPONENTS

The procedure for determining circuit models for mechanical components is similar to that used for fluid components. In this case, we apply Newton's Laws to mechanical systems and then rearrange the resulting equations into the standard two-terminal forms. Here again, the equations do not always fit the standard form exactly, and we must use fictitious terminals.

3.4.1 Mechanical Signal Variables

The key to the network representation of mechanical systems is the recognition of force as the through variable for mechanics of translation and torque for the mechanics of rotation. Here, we note that it is quite common to speak of forces being transmitted through certain parts or of torques being transmitted along shafts. This transmission does not require any motion. In the case of rotation, the torque is transmitted in the axial direction and is not directly associated with the rotary motion. Also, there is no quantity of fluid or charge which is being transmitted and in this sense, the mechanical system is rather different from the electrical and fluid systems considered so far. Nevertheless, the concept of a **flow** of force or torque is a great help in providing a systematic view of mechanical systems.

Consider the blocks A, B, C, (Fig. 3.16(a)) in equilibrium on a smooth table. A horizontal force, f, is applied to block A, as shown. To demonstrate the transmissibility of force, we draw free body diagrams of blocks, A, B, and C in Fig. 3.16(b) (vertical forces have been neglected for clarity). Since the blocks are in equilibrium, there must be no unbalanced force acting on them. The applied force f, acting to the right on

Circuit Components

block A, must therefore be balanced by a force of the same magnitude acting to the left on block A. This latter force is applied to block A by block B. From Newton's third law, every action has an equal and opposite reaction. Since block B pushes against block A, there is a reaction of block A against block B. Thus, on the free body diagram of block B, the force from block A acts to the left. To maintain block B in equilibrium requires a force from block C. The process of action and reaction shows that the force is transmitted through the blocks to the wall.

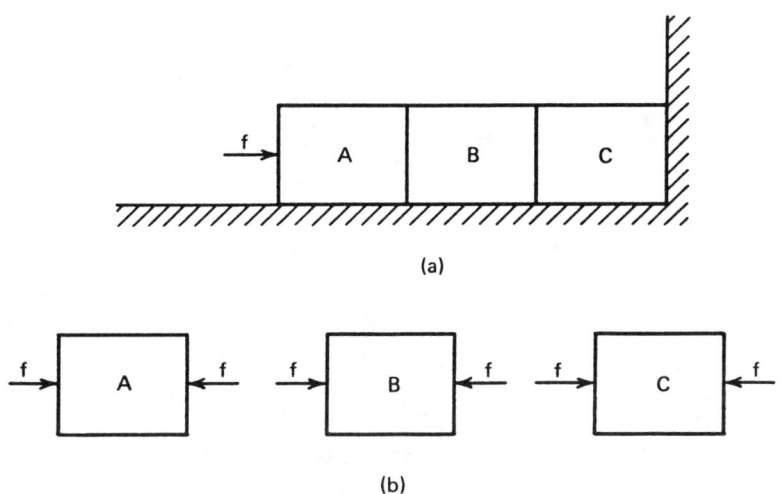

Figure 3.16 Transmission of force

In a similar way, we may demonstrate the through variable nature of torque. Fig. 3.17(a) shows three disks, A, B and C connected together by dowel pins. A torque, T, is applied to disk A. The free body diagrams of each disk [Fig. 3.17(b)] show that the torque T = fd is transmitted through the disks to the wall. The balancing torques on each disk come from consecutive application of the action and reaction principle.

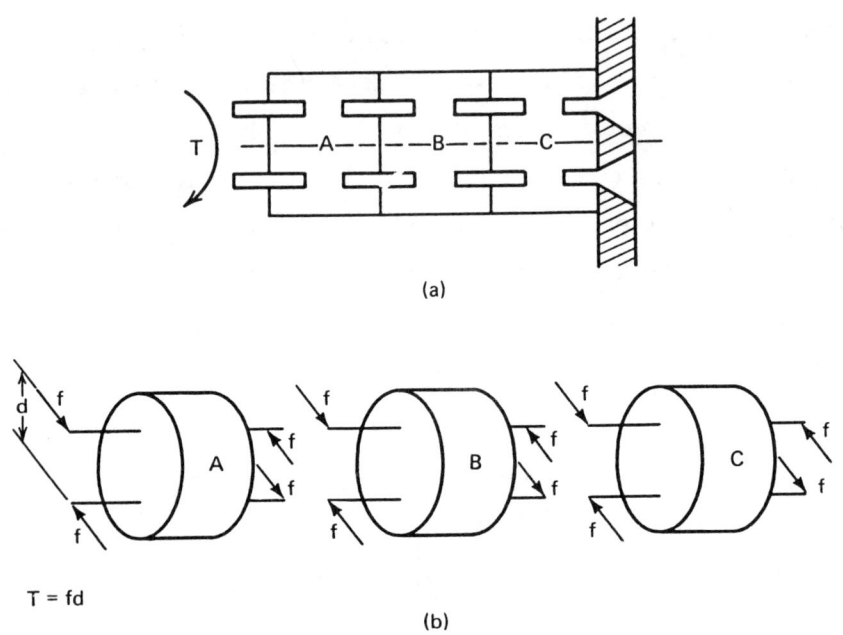

Figure 3.17 Transmission of Torque

The across variable in mechanical systems can be distance (or displacement) or velocity or acceleration, the choice being made to ensure that the three elemental physical situations are modeled by elemental equations of the same mathematical form as those in the electric and fluid circuits, and that the product of the through and across variables is power.

The appropriate across variable is therefore velocity difference and the dimensional analysis of the

Circuit Components

product is shown for confirmation:

$$[\Delta V][f] = \frac{m}{s} \cdot N = \frac{J}{s} = W \qquad \ldots (3.40)$$

Thus, the product of velocity in m/s and force in N gives the power directly in watts. For rotation, the product of angular velocity difference ($\Delta \omega$) in radians per second and torque in N.m is also in watts.

3.4.2 Damping (Mechanical Resistance)

The elemental mechanical situation in which energy is dissipated is that involving friction. Unfortunately, friction effects are varied, although generally, three basic forms are recognized:

(a) Static

(b) Coulomb

(c) Viscous

Static friction is the effect whereby the force required to move a stationary object is greater than that required to move the same object, when it is moving very slowly. Clearly this does not conform to the linear relationship which is analogous to the electrical resistance. Similarly, Coulomb friction, (that is force independent of velocity) does not conform to the linear situation in which force is proportional to the velocity and will be considered in Section 3.6. However, there are frictional effects and devices in which the force is proportional to velocity, this being termed viscous friction. These devices are usually called dampers, although the most common example is probably the shock absorber on an automobile.

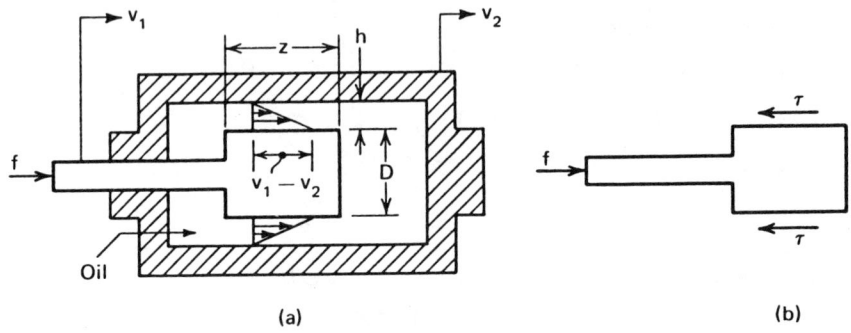

Figure 3.18 Schematic of translational damper

A schematic diagram of a translational damper is shown in Fig. 3.18(a). The damper has a piston-cylinder arrangement. However, the damping effect is due to a viscous liquid (oil) in the cylinder. The piston and the cylinder may move at different velocities. Thus, the damper is a two-terminal device, one terminal being at piston velocity and the other terminal being at cylinder velocity. The shear force developed in the damper is proportional to the difference in velocities of piston and cylinder. From the free body diagram of Fig. 3.18(b), we find that

$$f = A\tau = \frac{(\pi Dz)\mu}{h}(v_1 - v_2) \qquad ..(3.41)$$

where D is the piston diameter, z is the piston length, h is the piston clearance, τ is the shear stress, and μ is the viscosity of the oil.

The usual way of defining the damping, b, is to express this proportional response as

$$f = b(v_1 - v_2) \qquad \qquad ..(3.42)$$

Consideration of the analogous electric circuit variables shows that the damping, b, is therefore analogous to the electrical conductance, 1/R. From (3.1a) and (3.41), we may describe damping as:

$$b = \frac{1}{R} = \frac{\pi D z \mu}{h} \qquad \qquad ..(3.43)$$

As with the fluid resistance, there is no special unit for damping which is most conveniently expressed in N.s/m in translational systems.

In rotational systems, the rotational damping, B, is defined by

$$B = \frac{1}{R} = \frac{\pi D^3 z \mu}{4h} \qquad \qquad ..(3.45)$$

The unit of rotational damping is N.m.s.

Fig. 3.19 shows the symbols used for damping in translational and rotational systems. The symbols on

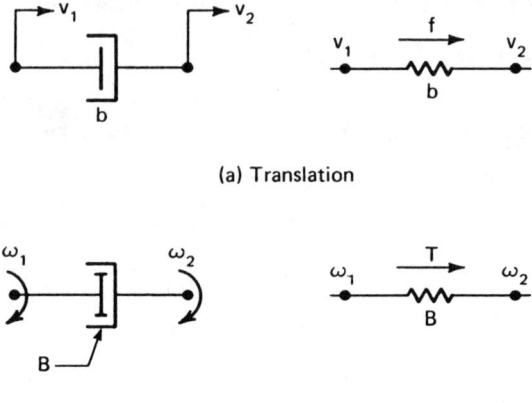

(a) Translation

(b) Rotation

Figure 3.19 Symbols for damping

the left are descriptive indications of the process. As a circuit component, however, both dampers are represented by the symbol for resistance. Remember, however, that B is analogous to conductance (1/R).

In general, frictional effects do not conform exactly to any one of the three frictional models. However, other than the further consideration of Coulomb friction (Section 3.6), at this stage, it is quite reasonable to approximate the actual friction characteristic with the nearest equivalent (viscous) damping. When this is done, the analytical techniques used to predict system behavior can then be the same as those used in electric circuits.

The power dissipated by a damper is equal to the product of the across and through variables associated

3.4.3 Spring (Mechanical Inductance)

Another mechanical situation is that of a spring (Fig. 3.20). The ideal spring (negligible mass) is a direct application of Hooke's law; the extension or compression is directly proportional to the force producing it. Normally, this is expressed in the form

$$f = k(x_1 + \Delta x_1 - x_2 - \Delta x_2) \qquad ..(3.46)$$

where k is the spring stiffness of the translational spring and x_1 and x_2 are the positions of the ends of the spring relative to some fixed reference, when the force is zero. Δx_1 and Δx_2 are the displacements of the ends, when the force is applied.

If we differentiate (3.46) with respect to time and since x_1 and x_2 are constants, we obtain

$$\frac{df}{dt} = k \frac{d}{dt}(\Delta x_1 - \Delta x_2) = k(v_1 - v_2) \qquad ..(3.47)$$

This is the same form as (3.1b). Thus, a spring is the analogous situation to an inductor, although it must be noted that the stiffness, k, is analogous to the reciprocal of inductance, 1/L. The basic unit of stiffness is N/m, which is rather small for most useful springs and therefore k is more likely to be expressed in kN/m or MN/m. For a helical spring, the spring constant may be calculated from the formula

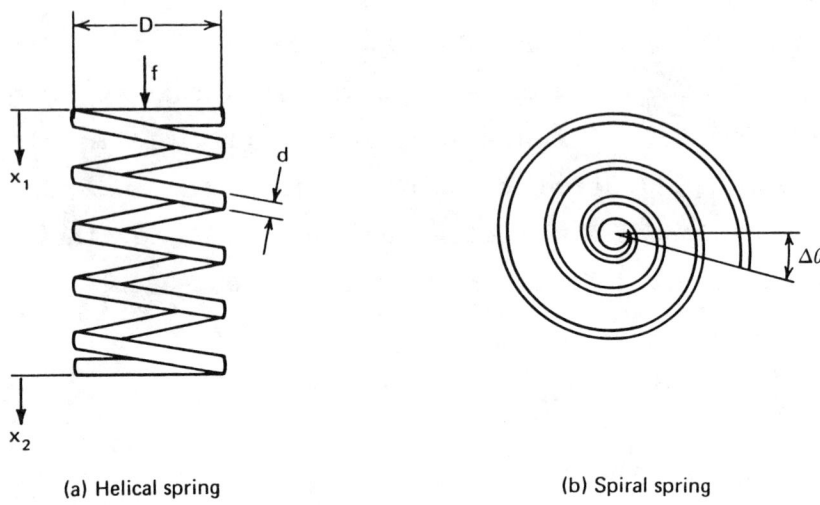

(a) Helical spring (b) Spiral spring

Figure 3.20 Typical spring configurations

$$k = \frac{1}{L} = \frac{Gd^4}{8ND^3} \quad \quad ..(3.48)$$

where G is the shear modulus of elasticity, d is the wire diameter, D is the coil diameter, and N is the number of coils.

For a torsional spring (Fig. 3.20b), the torque is related to the change in angular displacement $\Delta\theta$, as

$$T = K(\Delta\theta) \quad \quad ..(3.49)$$

Thus the spring constant for the torsional spring has the same units as torque (N.m). Now, the derivative of (3.49) with respect to time yields

Circuit Components

$$\frac{dT}{dt} = K(\omega_1 - \omega_2) \qquad ..(3.50)$$

where ω_1 and ω_2 are the angular velocities of the inner and the outer terminals of the spiral spring. Note, that the helical spring may also be used in torsion. The spring constant for a torsional spring with round wire is:

$$K = \frac{1}{L} = \frac{E\pi d^4}{64 l} \qquad ..(3.51)$$

Where E is the modulus of elasticity, d is the wire diameter and l is the total length of the spring.

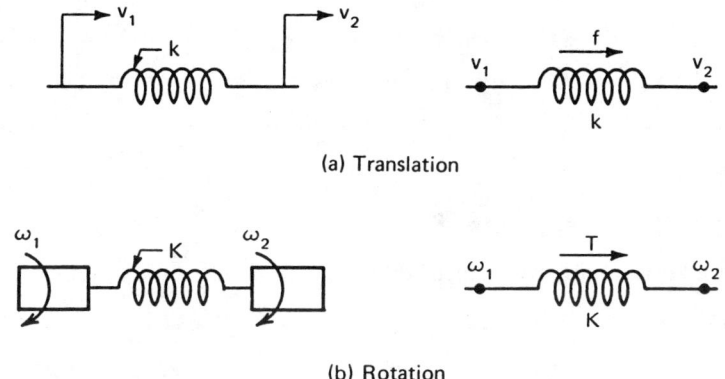

(a) Translation

(b) Rotation

Figure 3.21 Symbols for springs

Fig. 3.21 shows the mechanical and circuit symbols for springs. The circuit representation for both types is the same as the electric inductance and the fluid inertance. Note, however, that K is analogous to 1/L.

The energy stored by a spring depends on the through variable and the spring constant. It has the same form as given for the energy stored in the electrical inductance [Eq. 3.13].

The spring produces a force proportional to displacement. For some special geometries, a force proportional to displacement can be obtained without a spring. Consider the harmonic motion shown in Fig. 3.22. When a mass is connected to a spring [Fig. 3.22(a)], the system will oscillate, if the mass is initially displaced and then released. The spring creates a restoring force that is always proportional to the displacement. A similar restoring force may occur due to position without a spring present. Fig. 3.22(b) shows a mass suspended from a string in the form of a pendulum. Now, if the mass is displaced and released, the mass will oscillate. The position of the mass is providing the restoring force in this case. This force can be calculated by considering a free body diagram of the mass [Fig. 3.23(a)]. From a summation of forces [Fig. 3.23(b)], the relation between restoring force and displacement is:

$$f_R = (mg/l)x \qquad \qquad ..(3.52)$$

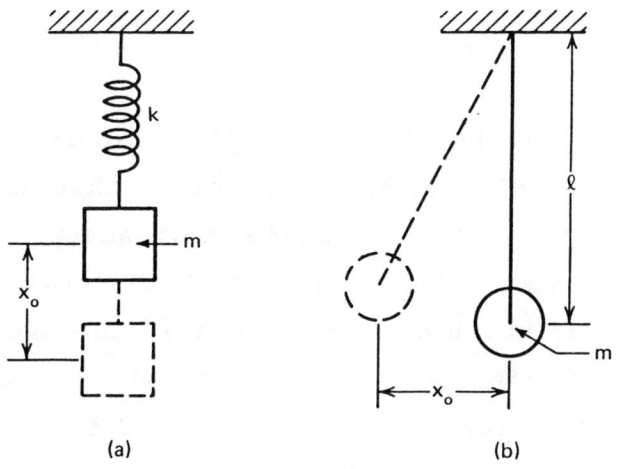

Figure 3.22 Harmonic motion

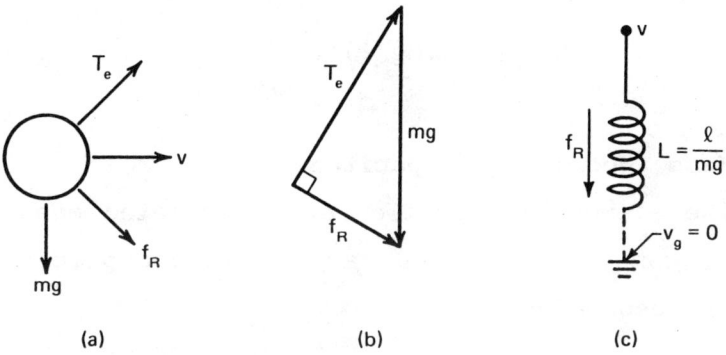

Figure 3.23 Pendulum restoring force

where ℓ is the length of the string in meters, and m is the mass in kilograms.

If (3.52) is differentiated with respect to time,

the result is:

$$\frac{df_R}{dt} = (mg/\ell)v \qquad ..(3.53)$$

The restoring force and the velocity have the relationship indicated by (3.1b) except that here again, we have only one terminal. We must add a fictitious terminal to make this fit into the two-terminal model. In this case, we choose the ground at zero velocity as the other terminal. Fig. 3.23(c) shows the circuit representation for restoring force due to pendulum position. The component is shown dashed, because of the assumption of a fictitious terminal. The equivalent spring constant for the pendulum is:

$$k = \frac{1}{L} = (Mg/\ell) \qquad ..(3.54)$$

3.4.4 Mass (Mechanical Capacitance)

The elemental situation of an isolated mass [Fig. 3.24(a)] can be considered, as a simple application of Newton's second law:

$$f = m\frac{dv}{dt} \qquad ..(3.55)$$

or integrating with respect to time

$$v = \frac{1}{m}\int f\, dt + v(0) \qquad ..(3.56)$$

Circuit Components

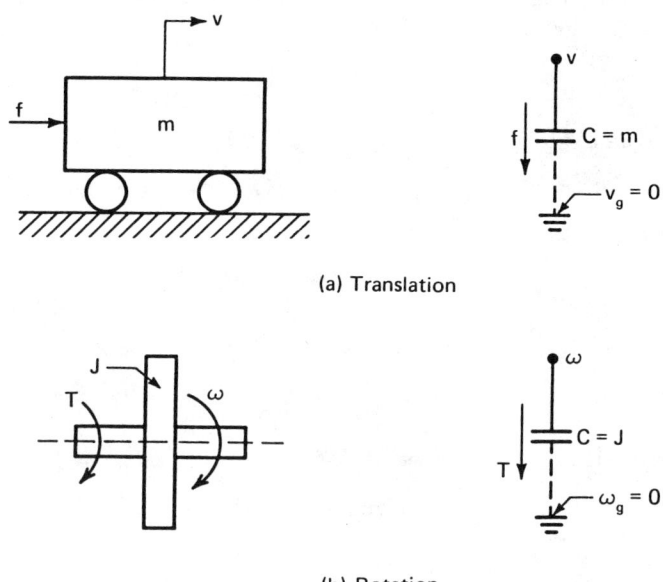

(a) Translation

(b) Rotation

Figure 3.24 Symbols for mass

When the initial velocity $v(0)$ is zero, (3.56) almost fits the model described in (3.1c). The difference is that the mass is not a two-terminal device. There is only one velocity associated with the mass. To place this in a component form, we must add the ground terminal at zero velocity to (3.56). Then, (3.56) becomes

$$v - v_g = \frac{1}{m} \int f \, dt \qquad ..(3.57)$$

For a translational system, the mass is directly analogous to capacitance. From a circuit viewpoint, the mass always represents capacitance to a fictitious ground terminal. Thus, it is shown dashed in Fig. 3.24(a).

A rotating mass, Fig. 3.24(b), is also related to its signal variables by a specialization of Newtons's second law. Thus:

$$T = J \frac{d\omega}{dt} \qquad ..(3.58)$$

where J is the polar moment of inertia (kg.m^2). With a fictitious nonrotating ground terminal, (3.58) becomes:

$$\omega - \omega_g = \frac{1}{J} \int T \, dt \qquad ..(3.59)$$

and the rotating mass also represents capacitance to a fictitious ground terminal. For the rotational system, the capacitance equals the polar moment of inertia. Fig. 3.24 shows the circuit representation.

The energy stored by a mass in a translational system depends on velocity and mass. For a rotational system, the energy stored is a function of angular velocity and moment of inertia. The formulation for the energy is the same as that given for the electrical capacitance in (3.17).

3.5 THERMAL CIRCUIT COMPONENTS

Since the dissipation of energy in all of the

Circuit Components 153

systems considered so far appears as heat, it would be logical to extend this modeling process to thermal circuits. The transfer of heat is a complex process which may occur through a combination of conduction, convection and radiation. Conduction is heat transfer by direct molecular communication. In convection, heat is transported by the movement of fluid masses. Radiation heat transfer, such as the transfer of heat from the sun to the earth, operates by a wave motion. We restrict our consideration here to conduction and convection, since these processes fall more naturally into the linear circuit models given in (3.1). In this case, we apply Fourier's law of heat conduction and Newton's law of cooling to obtain the appropriate network models.

3.5.1 Thermal Signal Variables

We may recognize immediately that the flow of heat, ϕ, (for example through the walls of a building) is a through variable. However, it is significantly different than the other through variables that we have encountered in the electrical, fluid and mechanical circuits. The difference is that the heat flow being the rate at which heat energy is transferred has the units of power by itself. As a result, the product of the through and across variables is of no particular significance and the across variable is chosen without any regard to it. The most convenient across variable is temperature difference, $\Delta\theta$, expressed in degrees Celsius.

Despite the differences noted above, the analogy between the thermal and electric circuit is very real and useful. The quantity of heat is directly analogous to the quantity of charge. Thus, the rate at which heat flows corresponds to the rate at which charge flows, namely the electric current. An additional similarity is that temperature is proportional to the energy per unit of heat whereas voltage is the energy per unit of charge.

3.5.2 Thermal Resistance

Thermal resistance models the relationship between temperature difference and heat flow rate for a material of a specific geometry, when there is no storage of heat taking place. The absence of storage is merely an approximation which depends on the properties of the material. We will discuss this point more thoroughly at the end of the section. From Fourier's law of heat conduction in the steady state (no storage), we find that

$$\phi = \frac{kA}{l}(\theta_1 - \theta_2) \qquad \ldots (3.60)$$

where ϕ is the heat flow in watts, θ is the temperature in $^\circ C$, k is the thermal conductivity of the material in $W/m.^\circ C$, A is the area through which heat flows in m^2, and l is the length of the material, m.

Circuit Components

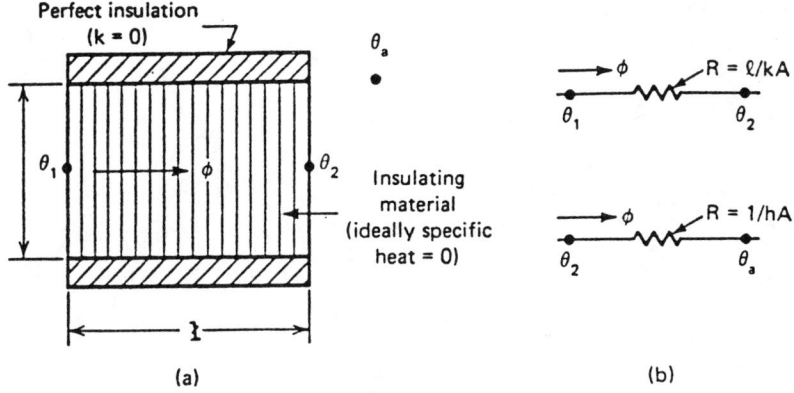

Figure 3.25 Thermal resistance

Let us clarify thermal resistance by referring to the specific geometry shown in Fig. 3.25. Here, a round piece of insulating material has a diameter, d, and length, ℓ. The piece is surrounded by a collar of perfect insulation that is shown cross hatched. The purpose of the collar is to keep the heat flow in the axial direction. Though perfect insulation is not physically realizable, it can be closely approximated. In this case, it is a valuable concept, because it helps to define clearly the area through which heat flows. The left face of the piece is at temperature, θ_1, the right face at temperature, θ_2, and the ambient temperature is at θ_a. For this configuration, (3.60) can be placed in the form of (3.1a) as:

$$\theta_1 - \theta_2 = \left[\frac{4\ell}{\pi k d^2} \right] \phi \qquad \ldots (3.61)$$

Thus the thermal resistance of the cylindrical piece with axial conduction heat transfer is:

$$R = \frac{4l}{\pi k d^2}, \quad (^\circ C/W) \qquad \qquad ..(3.62)$$

There is heat transferred from the right face of the piece to the ambient air at temperature θ_a. This heat transfer takes place by convection. From Newton's law of cooling, we may express this heat transfer by

$$\phi = hA(\theta_2 - \theta_a) \qquad \qquad ..(3.63)$$

where h is the surface coefficient (or film conductance) in $W/m^2 \, ^\circ C$. For the convection heat transfer, the thermal resistance has the form

$$R = \frac{1}{hA} = \frac{4}{\pi h d^2}, \quad (^\circ C/W) \qquad \qquad ..(3.64)$$

The surface coefficient, h, depends on the motion of the fluid medium and is given in heat transfer texts for various geometric configurations and surface conditions. Fig. 3.25(b) shows the resistance circuit for the solid piece and for the film. Together they may be modeled as two resistances in series.

3.5.3 Thermal Inertance

From the signal variables for thermal circuits and (3.1b), we require a relation of the type

$$\theta_1 - \theta_2 = K_2 \frac{d\phi}{dt} \qquad \ldots (3.65)$$

This equation would define thermal inertance. However, there is no physical situation known to produce such a relation between the temperature difference, $\theta_1 - \theta_2$, and the heat flow rate, ϕ. As a result, there is no thermal component analogous to an inductor.

3.5.4 Thermal Capacitance (Heat Storage)

The storage of heat within a mass has long been recognized as a physical process in which the heat energy is regarded almost as a fluid, or as charge as noted previously. As a result, the terms **thermal mass** and **thermal capacity** appear frequently. The basic equation of heat storage in a mass is:

$$\phi = \rho c V \frac{d\theta}{dt} \qquad \ldots (3.66)$$

where ρ is the mass density in kg/m^3, c is the specific heat of the mass in $W.s/kg.^{\circ}C$, and V is the volume of the mass in m^3.

The temperature, θ, in (3.66) refers to the temperature of the mass. For pure heat storage components, the temperature is uniform throughout the mass. Thus there is only one temperature associated with a mass storage element. To put the mass into the form required for two terminal capacitive components, we must assume a fictitious ground terminal at constant temperature, θ_g. Then, (3.66) is written as:

$$\theta - \theta_g = \frac{1}{\rho c V} \int \phi \, dt \qquad \qquad ..(3.67)$$

and by comparison with (3.1c), we find that the thermal capacitance, C, is equal to

$$C = \rho c V, \quad (W.s/^\circ C) \qquad \qquad ..(3.68)$$

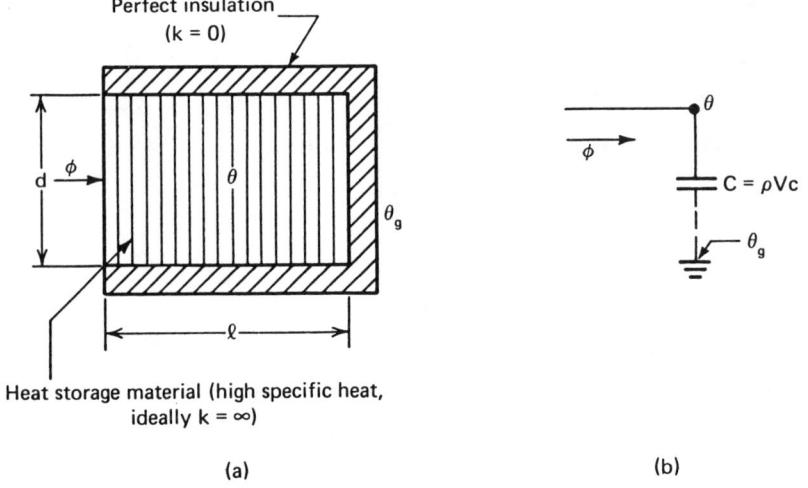

Figure 3.26 Thermal capacitance completely insulated

Consider the cylindrical mass of material enclosed by perfect insulation in Fig. 3.26(a). For pure heat storage, the temperature of the mass is uniform throughout. This means that materials used for heat storage will be very good heat conductors. The thermal conductivity of such materials will be high and they

Circuit Components

will have very low thermal resistance. If heat is applied to the cylindrical mass, the temperature of the mass increases in accordance with (3.67). Fig. 3.62(b) shows the analogous circuit component. Once again, the assumption of a fictitious potential terminal leads to a capacitance to ground and is indicated by a dashed line. All thermal capacitance has this fictitious terminal and therefore, always represents capacitance to ground in the circuit sense. Note that because the terminal is fictitious, there is no heat transferred to ground.

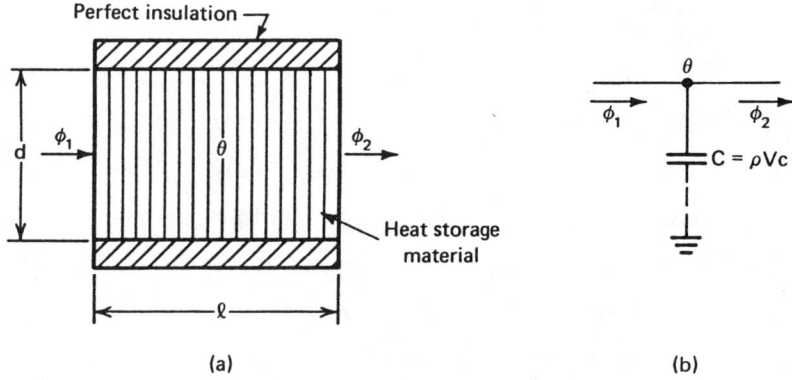

Figure 3.27 Thermal capacitance partially insulated

If the insulation does not completely surround the heat storage component as shown in Fig. 3.27(a), the capacitance effect is still to ground. The heat flow that acts to raise the temperature, θ, is now $\phi_1 - \phi_2$ and the component may be represented as shown in Fig. 3.27(b).

3.5.5 General Thermal Components

Actually, all thermal components have both resistance and capacitance. The magnitude of the resistance is inversely proportional to the thermal conductivity of the material. The magnitude of the capacitance depends on the density-specific heat product. The physical size of the component is a factor in both resistance and capacitance. Some combinations of material and configuration are predominantly resistive or predominantly capacitive. These are the models previously described in the sections on Thermal Resistance and Thermal Capacitance.

Fig 3.28, however, shows the general circuit model for all thermal components. The temperatures θ_1 and θ_2 are the temperatures of the left and the right boundaries and θ is the mid-plane temperature. The general circuit model is a T-junction with half the resistance on each side of the mid-plane and a capacitance to ground from the mid-plane temperature. When the thermal material has a large thermal conductivity, the component reduces to the circuit shown in Fig. 3.27(b). In this case, the resistance is negligibly small and the component is a capacitance to ground. On the other hand, a material with lower thermal conductivity has resistance. If the specific heat-density product is low, the capacitance will be small and the component is purely resistive as shown in Fig. 3.25(b).

Circuit Components

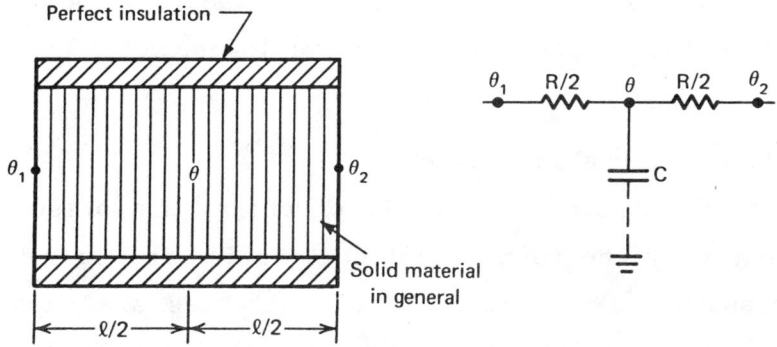

Figure 3.28 General thermal circuit model

In the general case, however, both the thermal conductivity and the density-specific heat product must be considered to model a thermal component.

3.6 SOURCES AND LOADS

The elemental situations and devices discussed so far have implied a source of energy which is quite independent of the element. These elements have been either dissipative or storage devices which can only act on the energy while it is within a system and they are therefore called **passive elements**. The passive elements (resistance, inductance and capacitance) are often also called **loads**. In contrast, **active** elements are devices which imply situations where energy is supplied to or taken from the system. Often active elements involve some energy conversion process such as takes place in a battery, pump or furnace. That is, the source of energy for any physical system is modeled by means of an active element. Thus, in general, a source supplies energy and a load receives energy. In this section, we will discuss various types of sources and loads.

3.6.1 Across Variable Sources

The details of the origin of energy are often not required; only the form in which it is fed into a system is necessary. For example, most batteries are capable of supplying energy to an electric circuit with the voltage difference between its terminals changing only slightly if the current changes. The ideal situation would presumably be one in which the voltage difference would be constant, irrespective of the current flowing from the battery. In electric circuit analysis, this leads to the idea of the **ideal** voltage source usually referred to simply as a **voltage source**. Pursuing our across and through variable analogies, we would expect to find ideal pressure sources, velocity sources, and temperature sources, which for convenience can be called **across variable** or **potential** sources. Fig. 3.29(a) shows the static characteristics (across variable - through variable relation) of an ideal across variable source. Such a source can supply flow without changing potential. The circuit representation of an ideal potential source [Fig. 3.29(b)] is a circle with polarity indicated by a plus sign. When the polarity is reversed, the device is sometimes termed a **sink**. However, we may also look on this situation as a negative source.

Circuit Components

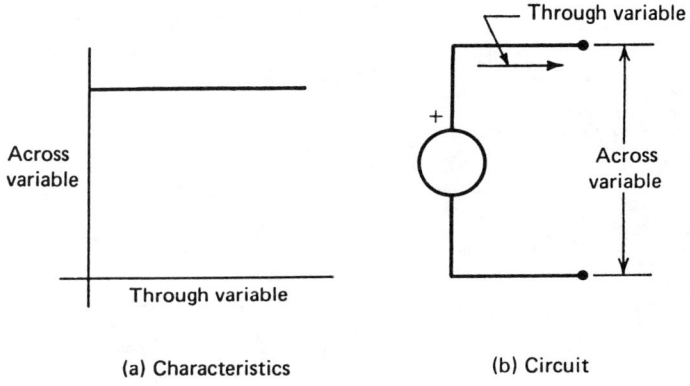

(a) Characteristics (b) Circuit

Figure 3.29 Ideal across variable (potential) source

Of course, real sources would deviate from the ideal in that a normal pressure pump, for example, does not quite maintain constant pressure, irrespective of the flow rate. However, just as a **real** battery is regarded as a constant voltage with the **internal resistance** connected in series (modeled by a voltage source and series connected resistance), the analogous energy sources can be modeled by an across variable source and series dissipative element. Fig. 3.30(a) shows the static characteristics of a real across-variable source. The equivalent circuit representation of the real source is shown in Fig. 3.30(b). The circuit representation has an ideal source and an internal source, R_i, in series.

It should be noted that for dynamic studies the model of an energy source may require at least one storage element in addition to the across variable

164   Modeling and Analysis of Linear Physical Systems

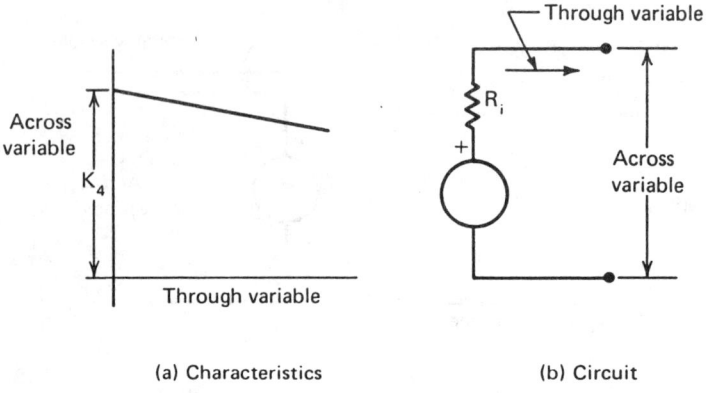

Figure 3.30 Real across variable (potential) source

source and dissipative element. This should become evident, when Thevenin's and Norton's network theorems are considered in Chapter VII.

Situations involving an ideal velocity source are not often encountered in elementary mechanics. The main point is to look for whether the velocity at a point in the system is constrained to some specific value or function of time. The situation where a body is struck and carried by a much larger mass which is already moving at a constant velocity can be modeled by a velocity source, since the velocity of the larger mass is virtually unchanged after making contact with the smaller mass. The resulting forces may be sufficiently large to cause damage, and therefore great care is required when subjecting any part of a mechanical system to a suddenly applied constraint of constant velocity.

The ideal temperature source is a reasonable model of a thermostatically controlled oven, provided the item to be heated can be inserted without any change in oven temperature. This is inevitably very difficult to achieve exactly. Nevertheless, the times involved are such that in many situations, the models of the temperature-controlled oven as an ideal temperature source gives reasonably accurate results.

In general, an ideal across variable or potential source is the correct model for any situation in which the across variable is constrained to some specified value or function of time. This is true whether the constraint is produced naturally or whether it is the result of some automatic control scheme. If the end result is a value or functional expression for the across variable which is independent of the through variable, it is appropriate to use an ideal potential source as model.

The characteristic equation of a potential source may be written as:

$$(\text{Across Variable}) = K_4 - R_i (\text{Through Variable}) \quad ..(3.69)$$

where K_4 is the value of the across variable when there is no through variable. Note that for the ideal potential source, $R_i = 0$, and the across variable is constant and independent of the through variable.

3.6.2 Through Variable Sources

As might be expected, situations arise in which the value of a through variable is constant or is some specified function, independent of across variable. Most elementary mechanics are based on the premise that it is possible (at least theoretically) to apply a force which is independent of the velocity of the point at which it is applied. Such situations can be modeled as an ideal source usually called simply a **force source.** In electrical engineering, there are very few situations where the current from a source is naturally constant, but it is quite common for the current from a DC power supply to be controlled electronically such that these supplies may be represented by current sources. The ideal current source is one where the current is entirely independent of the voltage across it. Fig. 3.31(a) shows the static characteristics of an ideal through variable or flow source. The through variable remains constant for all values of the across variable.

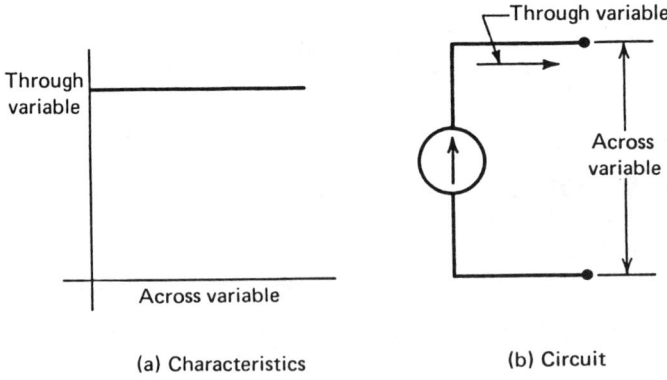

Figure 3.31 Ideal through variable (flow) source

The circuit symbol to represent the ideal flow source is a round circle with an arrow indicating the flow direction along the diameter.

At this stage, it may be wise to emphasize the fact that a **source of current** and a **current source** have significantly different meanings. A **source of current** has absolutely no implication as to whether the voltage or current or neither is constant. However, a **current source** unequivocally implies an ideal flow source, namely one in which the current is constant or some function independent of voltage. Similarly, a **source of force** must not be confused with a **force source**.

For convenience, sources in which the through variable is constrained are called **through variable sources** or **flow sources**. This latter term is particularly applicable to hydraulic pumps in which clearly identifiable quantities of liquid are passed through the pump. Provided such pumps are driven at constant speed, the rate at which the liquid is delivered is constant. In the case of thermal systems, an ordinary heater without any thermostatic control produces heat at a rate which is reasonably independent of the temperature. An ideal flow source is therefore used to model this situation.

In general, a flow source is used to model any situation in which the through variable is constrained to some specified value or function of time. This is true whether the constraint is produced naturally or whether it is the result of some automatic control scheme. If the end result is a value or functional expression for the through variable which is independent of the across variable, it is appropriate to use an ideal flow source as a model for this situation.

Actually, there are no ideal flow sources. Any real flow source is affected by the across variable. Thus for example, a constant displacement pump would slow down if the pressure difference got too large. Fig. 3.32(a) shows the static characteristics of a real flow source. As a circuit component, the real flow source may be considered an ideal flow source in parallel with an internal resistance, R_i. This is shown in Fig. 3.32(b). The equation of the real flow source is:

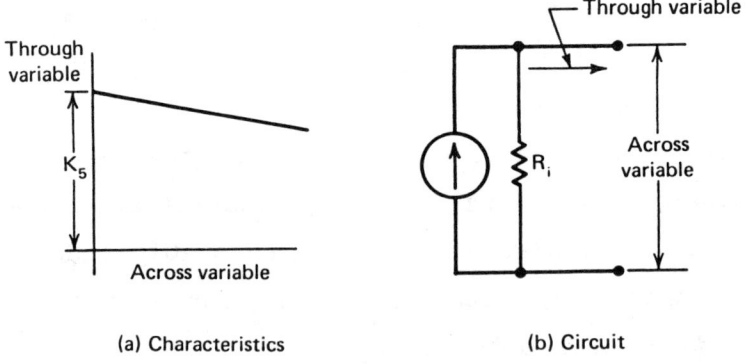

(a) Characteristics   (b) Circuit

Figure 3.32 Real through variable (flow) source

Circuit Components

$$\text{Through Variable} = K_5 - \frac{1}{R} [\text{Across Variable}] \quad ..(3.70)$$

where K_5 is the value of the through variable when the across variable is zero. In this case, an ideal flow source has an infinite value of internal resistance. Note that under this condition the through variable does not depend on the across variable.

The characteristics of real flow sources and real potential sources are very similar. [Compare Figures 3.30(a) and 3.32(a)]. For a unity value of the internal resistance, (3.69) and (3.70) are in the same form and can be identical when $K_4 = K_5$. In this case, we would not designate the component as a **flow** source or a **potential** source. We would merely designate as a **source**.

In general, any active element can be used to model either a source or a sink of energy. If the device is acting as a source, positive flow will emerge from the positive terminal for both the potential and flow sources. If the physical situation is one which absorbs energy, positive flow will enter at the positive terminal. Although not entirely consistent, it is nevertheless convenient to call the active element a source, whether it actually is acting as a source of energy or whether it is acting as a sink.

3.6.3 Loads

Resistance, inductance and capacitance are all load elements. However, resistance is the only load that has a static characteristic. Inductance and capacitance generally have time dependent signal variables. Fig. 3.33(a) shows the characteristics of a resistive load. For a linear resistance, the characteristic is a straight line passing through the origin. The slope of the line is the value of resistance. Fig. 3.33(b) indicates the analogous circuit for the load resistance, R_L. The relation between the variables for a resistive load can be obtained from [3.1(a)] for this case as:

$$\text{Across Variable} = R_L (\text{Through Variable}) \quad \ldots (3.71)$$

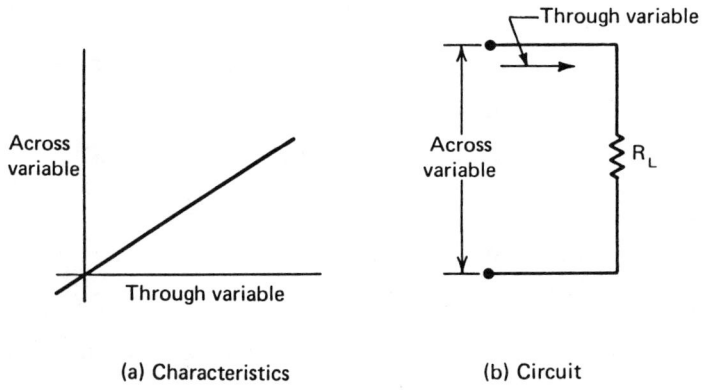

(a) Characteristics   (b) Circuit

Figure 3.33 Resistive load

The interaction of a source and a load involves some principles of circuit theory. Although circuit theory is not considered till Chapter IV, we may gain some insight now by superimposing source and load characteristics. For example, suppose an ideal potential source and resistive load characteristics are superimposed [Fig. 3.34(a)]. This is equivalent to the circuit connecting an ideal source to a resistive load [Fig. 3.34(b)]. Let us recall that we may interpret a source as representing the locus of all possible operating states of the energy element. Similarly, the load represents the locus of all possible values of the signal variables for a resistance. The circuit configuration is such that both loci must be applicable

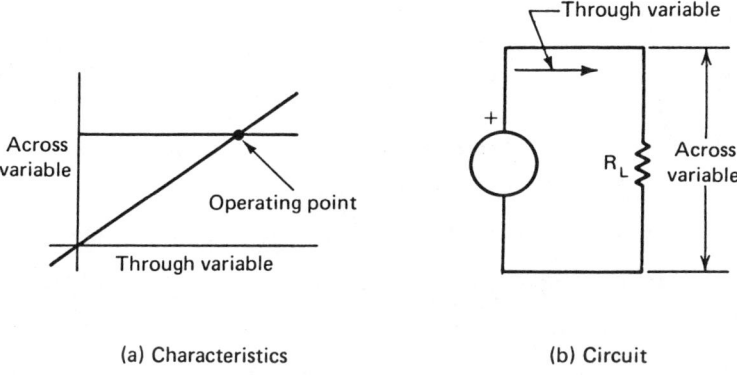

(a) Characteristics   (b) Circuit

Figure 3.34 Source-load interaction

simultaneously. The only point that permits both loci to coexist is the intersection point of the characteristics. This point is called the operating point, because it shows the value of the signal variables in the circuit. This is a graphical interpretation of the solution of two simultaneous equations (3.69) and (3.71).

The source-load method is especially valuable in nonlinear circuits where the results are more difficult to obtain analytically.

3.6.4 Modeling of Coulomb Friction

When considering friction, it has been noted that the characteristic of Coulomb friction is that the

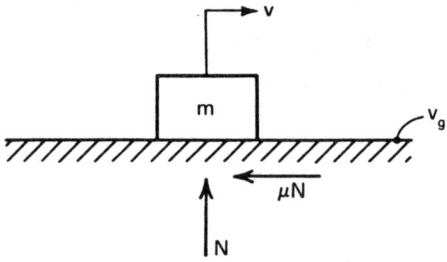

Figure 3.35 Sliding friction

frictional force is independent of velocity. Accordingly, the appropriate model for this situation is a force source (torque source, if rotating) connected such that the flow of force enters at its positive terminal. This model is exact provided motion or rotation is restricted to one direction only. If the motion reverses, the polarity of the force source must be reversed since the frictional force reverses.

Let us examine, for example, the case of a mass, m, moving along a rough horizontal surface with a velocity, v (Fig. 3.35). The normal force between the

Circuit Components

mass and surface, N, is equal to the weight, mg. The frictional force retarding motion, f, is μN (or μmg) where μ is the coefficient of sliding or kinetic friction. Fig. 3.36 shows a plot of the velocity

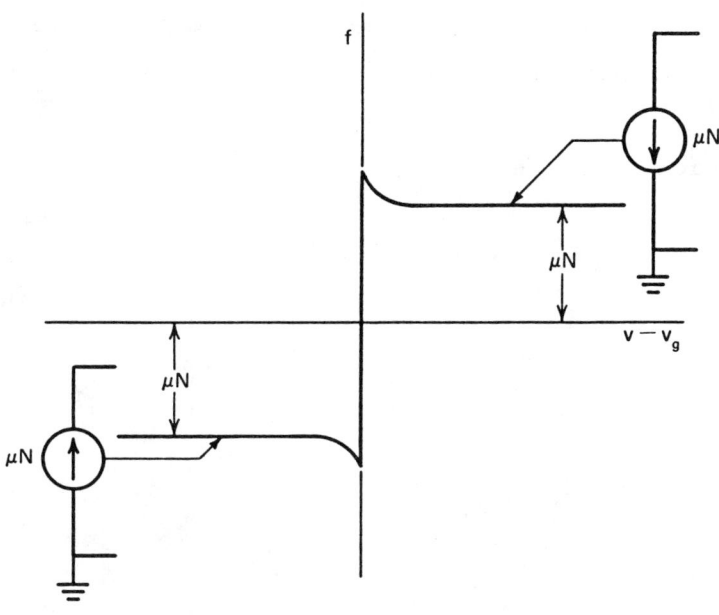

Figure 3.36 Modelling of Coulomb friction by force (flow) source

difference-force characteristics of the friction. When the velocity difference is zero (no motion), the frictional force is somewhat larger than when the mass is moving. This occurs because the coefficient of static friction is larger than the coefficient of kinetic friction. Actually, the frictional force decreases slightly with increased velocity difference. Notice, however, that the characteristic shown in Fig. 3.36 is quite similar to the ideal flow source characteristic shown in Fig. 3.31(a). As an approximation, therefore, we use a force source circuit model to represent Coulomb friction. If the motion reverses in direction, the force source is reversed. This is a nonlinear effect which is outside the scope of this text, the analysis of such a situation usually being introduced in texts on automatic control systems.

3.6.5 Modeling of Weight

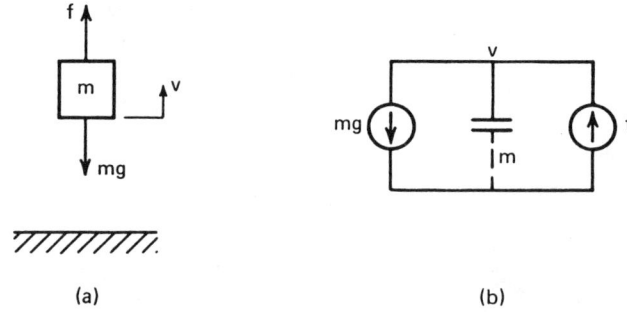

Figure 3.37 Modelling of weight

Another situation involving a constant force which is not necessarily a source of energy is the weight of a mass subject to vertical motion. In Fig. 3.37(a), an applied force, f, is required to balance the weight, if the mass is to remain stationary, this being the means by which we are aware of the gravitational force on an object. However, the weight of the object is the gravitational force, mg, which should be expressed in newtons. Provided distances from the center of mass of the earth do not vary significantly, the weight is constant and can therefore be modeled by means of a force source. The polarity of the source will depend on the direction chosen for positive velocity. If velocity is positive when the motion is upwards, the polarity is that shown in Fig. 3.37(b).

3.7 ELEMENTS AS MATHEMATICAL OPERATORS

From the elemental equations of all the passive elements, it can now be noted that the through variable is either a multiple, or time-integral or time-derivative of the across variable. That is, the voltage drop across a resistor, for example, is simply the current multiplied by a constant (i.e., the resistance). This is true for ideal resistance no matter how the current is varying and therefore the appearance of a graph of voltage drop against time is the same as that of a graph of current against time, other than the scale factor which the value of resistance introduces. The voltage difference across an inductor is the time derivative of the current flowing through it with the scale factor of the value of the inductance included. If the current is constant, the voltage difference across an ideal inductance is zero. If the current is some function of time, the voltage difference is then the derivative of that function multiplied by the value of the inductance as a scale factor. The voltage difference across a capacitor, on the other hand, is the time integral of the current flowing through it with the scale factor of the capacitance included. If the current is constant, the voltage drop across the capacitor will increase at a uniform rate. If the current is some function of time, the voltage drop is then the integral of that function divided by the capacitance.

The process of integration generally requires evaluation of the constant of integration. However, in this context, the value of the integral is usually known at the instant of time at which the process is considered to start, i.e., t = 0 in most cases. By using the format of a particular integral, the value of the constant of integration usually is obtained automatically. For example, if we consider a capacitor which is initially charged when a current of $I.U_s(t)$ passed through it, the resulting voltage drop across the capacitor may be developed from the basic equation (3.1c):

$$\Delta e(t) = \frac{1}{C} \int_{-\infty}^{t} i \, dt$$

$$= \frac{1}{C} \int_{-\infty}^{0} i \, dt + \frac{1}{C} \int_{0}^{t} i \, dt \qquad \ldots (3.72)$$

The first term of the right-hand side represents the voltage drop (Δe_0) resulting from the initial charge on the capacitor. Thus:

$$\Delta e(t) = \Delta e_0 + \frac{1}{C} \int_{0}^{t} i \, dt$$

$$= \Delta e_0 + It/C \qquad \ldots (3.73)$$

Circuit Components

Circuit models for initial conditions will be presented in Chapter IV. However, we should be aware that time integration always adds the integral in an interval to the value of the variable at the beginning of the interval.

Example 3.1

If the effective driving force on an automobile has the form shown in Fig. 3.38(a), determine the resulting velocity, if the road is horizontal and wind resistance is neglected, and the mass of the automobile is 1500 kg. The automobile is initially at rest.

Solution

The graph of force against time has three basic segments where it has a value other than zero. The first segment, from $t = 0$ to $t = 5$ has a constant slope 240 N/s, the second from $t = 5$ to $t = 15$ has a constant value of 1200 N, and the third from $t = 15$ to $t = 20$ has a constant slope of -240 N/s.

The elemental equation describing the situation is:

$$f = m \frac{dv}{dt} \quad \text{or} \quad v = \frac{1}{m} \int f \, dt \qquad \ldots (3.74)$$

Thus the velocity is the integral of the function shown in Fig. 3.38(a) divided by the scale factor 1500. This is a case where the form of the velocity is

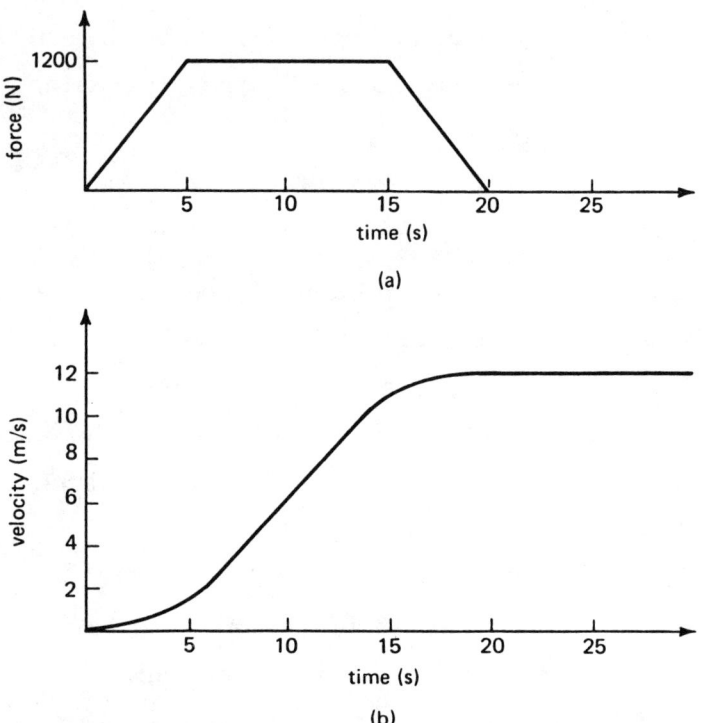

Figure 3.38 Force and velocity diagrams (example 3.1)

obtained quite simply by noting that the integral of a straight line of nonzero slope is a parabola whose change in ordinate over the period under consideration equals the area under the existing function. Thus, the velocity at t = 5 s is:

$$v(5) = 0 + \frac{1}{1500}(1/2)(5)(1200) = 2 \text{ m/s} \qquad ..(3.75)$$

since the vehicle was initially at rest.

Circuit Components   179

The velocity during the second segment is a straight line of slope 1200/1500 or 0.8 m/s^2, the value at the end of the period being obtained from the increase in the area underneath the graph of the force.

$$v(15) = 2 + \frac{1}{1500} (1200)(10) = 10 \text{ m/s} \qquad ..(3.76)$$

Similarly, the velocity during the third segment is a parabola, concave down, the final value being

$$v(20) = 10 + \frac{1}{1500} (1/2)(5)(1200) = 12 \text{ m/s}$$
$$..(3.77)$$

After t = 20 s, there being no further change in the area under the graph of force, there is no further change in velocity, which, with the assumption of no wind resistance, would remain constant. Hence, the sketch of velocity is that shown in Fig. 3.38(b).

Example 3.2

A frictionless pipe of 1.0 m diameter carries water from a reservoir to a turbine over a distance of 10 km. The flow rate is 20 m^3/s and the flow is cut off in a period of 10 seconds, as shown in Fig. 3.39(a). Determine the pressure difference due to the inertia of the water.

Solution

The fluid inertance is given by

$$L = \frac{\rho l}{A} = \frac{(1000)(10)(10^3)}{(\pi)(1^2)/4} = 1.27\,(10)^7 \text{ N.s}^2/\text{m}^5$$

..(3.78)

While the flow is being cutoff, the rate of change is:

$$\frac{dq}{dt} = -\frac{20}{10} = -2 \text{ m}^3/\text{s}^2$$

..(3.79)

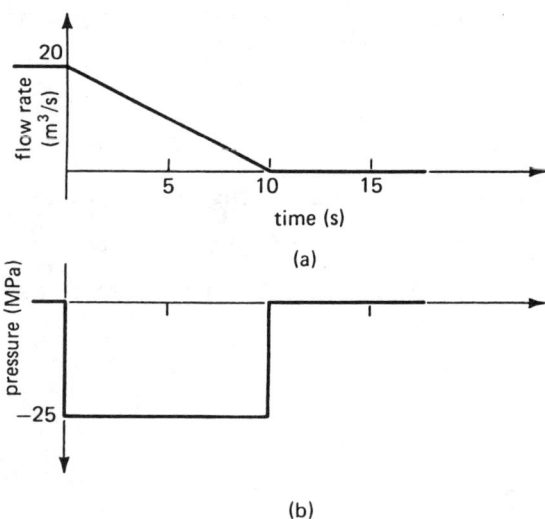

Figure 3.39 Flow rate and pressure drop across frictionless fluid line (example 3.2)

The pressure difference due to the inertia effect is, therefore:

$$\Delta p = L\frac{dq}{dt} = (1.27)(10^7)(-2) = -25.4\,(10^6) \text{ Pa}$$

..(3.80)

The pressure drop is shown in Fig. 3.39(b).

It should be noted that in effect this is the pressure acting on the gate or valve as it is being closed. Just before the final closure, it is therefore acting over almost the entire cross-sectional area of the pipe, thus applying the force of

$$f = (\Delta p)A = \frac{(25.4)(10^6)(\pi)(1^2)}{4} = 20~(10^6)~N$$

..(3.81)

which is the same as the gravitational force on a mass of 2.04 (10^6) kg. This extremely high force indicates that damage to valves may result from rapid closure. Indeed, unless special precautions are taken, the closing of a valve in a pipe which has a moving liquid generally results in forces which may be large enough to displace the pipe with the possibility of resulting damage. This phenomenon is usually called **water hammer**.

3.8 SUMMARY AND DISCUSSIONS

The process of determining analogous physical situations can now be seen to be in two main parts. The first has been to identify the analogous through and across variables. Although at first, selected on the basis that the product would be power, there is no reason not to use power as the through variable, if the elemental equations turn out to have the same form, as in the thermal system. The final choice is shown in Table 3.1.

TABLE 3.1 THE ANALOGOUS THROUGH AND ACROSS VARIABLES

	Through	Across	Product
Electric Circuit	Current (A)	Voltage (V)	Power (W)
Fluid	Volume Flow Rate (m^3/s)	Pressure (Pa)	Power (W)
Mechanical (Rotation)	Torque (N.m)	Angular Velocity (rad/s)	Power (W)
Mechanical (Translation)	Force (N)	Velocity (m/s)	Power (W)
Thermal	Heat Flow Rate (W)	Temperature Difference ($^\circ C$)	---

The elemental physical situations are modeled by means of the circuit components. With the systems having power as the product of the through and across variables (electric, fluid and mechanical), there are only three components. One dissipates energy, usually as heat, and the other two store energy. One of the storage elements stores energy when the across variable exists, whether or not the through variable has some value. For convenience, these components can be categorized as A-type storage elements. The other element stores energy by virtue of the through variable whether or not the across variable has some value. For convenience, these components can be categorized as T-type storage elements. For thermal systems, there are only two basic elements: (a) dissipative and (b) A-type storage. The basic form of elemental equations are given in (3.1). The appropriate interpretation of K_1, K_2 and K_3 is summarized in Table 3.2.

TABLE 3.2 THE ANALOGOUS ELEMENTS

	Dissipative K_1	T-Type K_2	A-Type K_3
Electric	R	L	1/C
Fluid	R	L	1/C
Mechanical (Rotation)	1/B	1/K	1/J
Mechanical (Translation)	1/b	1/k	1/m
Thermal	R	---	1/C

A word of caution is in order. That is, the limitations of these models shall be clearly noted. While modeling the range of validity plays a highly important part.

3.9 PROBLEMS

3.1 The graphs shown in Fig. P3.1 represent the through

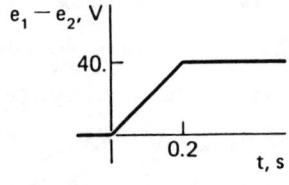
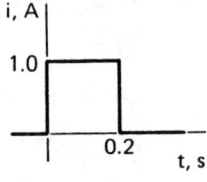

(a)

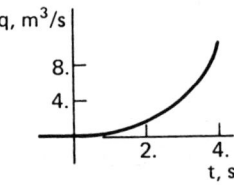

(b)

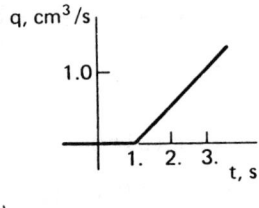

(c)

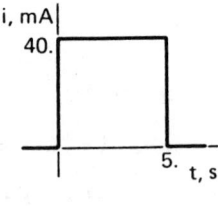

(d)

Figure P 3.1

and across variables for particular components. Identify the components and determine their magnitude.

3.2 Some data is presented below from four elemental physical situations. In each case, identify the element and state whether the analogous electric circuit element is resistance, inductance or capacitance. Give the magnitude of the component.

(a) Element A

Torque, N.cm	Angular Velocity, rad/s
60	1
120	2
180	3

(b) Element B

Force, N	Deflection, cm
10	1.25
20	2.50
50	6.25

(c) Element C

Pressure Difference, N/cm^2	Volume, cm^3
4	160
12	480
16	640

(d) Element D

Force, N	Time, s	Velocity, m/s
9	0	0
9	1	4
9	2	8
9	3	12

3.3 The voltage drop, $e_1 - e_2$, across a 2 henry inductor is shown in Fig. P3.3. Find the value of the current when t = 3.36 s. What is the energy stored at this time?

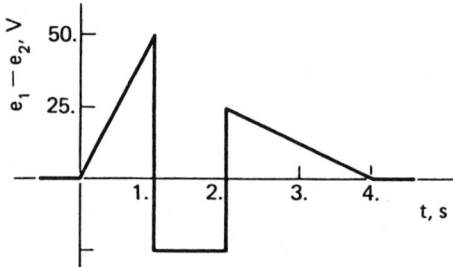

Figure P 3.3

3.4 The current passing through an inductance is sin ωt, A. The value of the inductance is 0.1 H. Plot a graph of voltage amplitude across the inductance against the angular frequency, ω, which varies from 0 to 20 rad/s.

3.5 The voltage drop across a 1 μF capacitor is shown in Fig. P3.5. Plot the current in the capacitor against time and find the energy stored when t = 5 s.

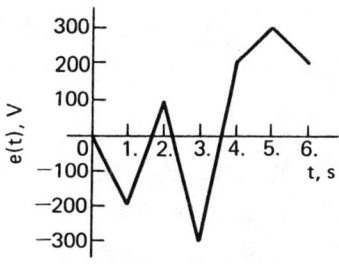

Figure P 3.5

3.6 Water flows into an open tank at $q_1(t)$ and out of the tank at $q_2(t)$ as shown in Fig. P3.6. The initial height of the water in the tank is 0.25 m and the cross-sectional area of the tank is 49 cm^2. Plot a graph representing the pressure, $p(t)$, at the bottom of the tank.

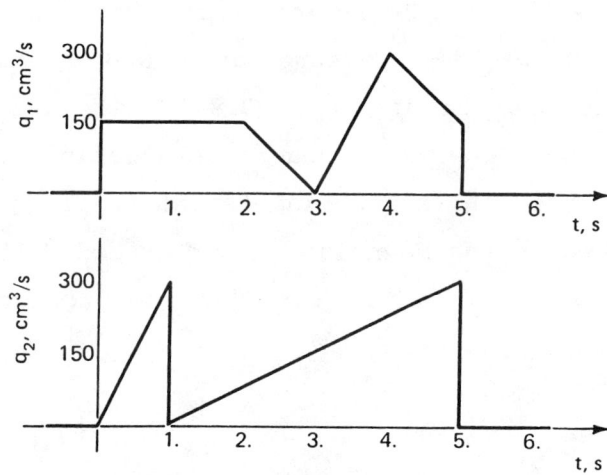

Figure P 3.6

3.7 An open water tank has a square cross-section 7 m on

each side. The tank is initially empty when the flow input shown in Fig. P3.7 is applied. The initial value of the flow input is q_o. If the tank is half-full at t = 14.7 s and completely full at t = 27 s, find the value of q_o and the height of the tank. Make a sketch of the pressure at the bottom of the tank in the time interval from 0 to 30 s.

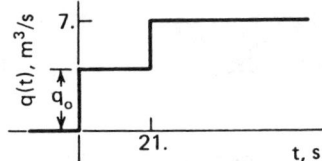

Figure P 3.7

3.8 Measurements on a circular fluid line with water flowing show the pressure drop and volume flow functions plotted in Fig. P3.8. The line may be modeled as an inertance and resistance in series.

(a) Find the length and diameter of the line.

(b) When does the Reynolds number exceed 2,000?

(c) How much energy is stored in the inertance when laminar flow ceases?

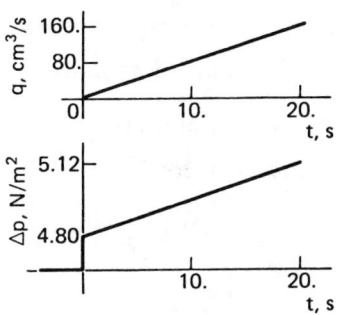

Figure P 3.8

3.9 The graphs shown in Fig. P3.9 represent the through variable in a mechanical component. Suppose that the following mechanical components are available.
(1) spring, k = 80 N/m
(2) mass, m = 150 kg
(3) damper, b = 40 N.s/m
Sketch the resulting across variable for each component subject to (a), (b) and (c). Find the value of the across variable at t= 8 s.

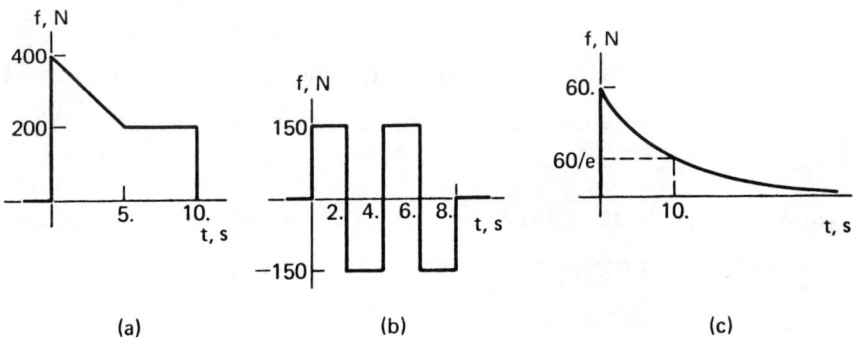

Figure P 3.9

3.10 An angular velocity input is applied to the free end of a rotational spring. The other end is firmly attached to a wall. The spring constant k = 8 N.m and the input function is shown in Fig. P3.10. Find the magnitude of the torque developed in the spring at t = 12 s.

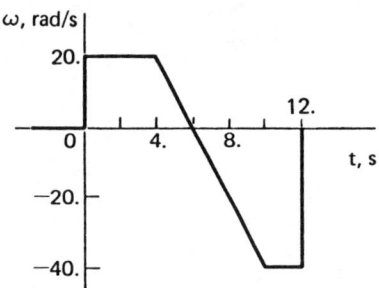

Figure P 3.10

3.11 A mass, m, is falling freely. When it is at a distance, x, above ground, its velocity is v_o. Prove that

$$x = v_o t_1 + \frac{1}{2} g t_1^2$$

where t_1 is the time for the mass to reach the ground. Assume that air resistance is negligible.

3.12 A 90 kg mass rests 165 m from a wall. There is no friction between the mass and the ground. A series of impulses in force, shown in Fig. P3.12, are applied to move the mass towards the wall. Find the time for the mass to reach the wall and its velocity at this time.

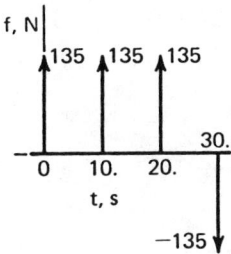

Figure P 3.12

3.13 A flywheel has a polar moment of inertia of 70 kg.m^2. A torque is applied to the flywheel as shown in Fig. P3.13. If the diameter of the flywheel is 1 m, find the final distance traveled by a point on the outer edge in the time interval from 0 to 60 s. How much energy is stored in the flywheel at t = 36 s?

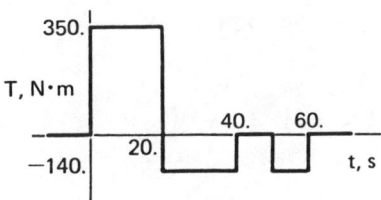

Figure P 3.13

3.14 A flywheel with moment of inertia, J, turns through an angle θ(t), where

$$\theta(t) = 8t^2 U_s(t) - (12t^2 - 120t + 300) U_s(t-5)$$
$$+ (7t^2 - 140t + 700) U_s(t-10), \text{ rad.}$$

When the angular velocity of the flywheel is 200 rad/s, the torque required is 72 N.m. Find the value of J.

3.15 A 20 kg mass is moving at a constant velocity of 40 m/s. If the energy stored in the mass is all transferred to a spring with k = 20 N/m, what force would be developed in the spring?

3.16 A mechanical system consists of a mass and a spring as shown in Fig. P3.16. The velocity at the left-end of the spring $v = 2.5\, U_s(t)$, m/s. The difference in velocities at each end of the spring is

$$v_1 - v_2 = (2.5 \cos 4t)\, U_s(t), \text{ m/s}$$

The spring constant k = 80 N/m. Find:
(a) the force in the spring;
(b) the value of the mass, m.

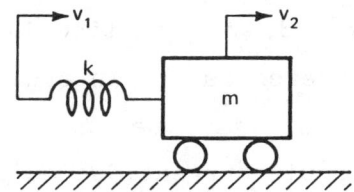

Figure P 3.16

3.17 A perfectly insulated tank is filled with 0.05 m^3 of water at 20°C. An immersion heater in the water is suddenly turned on and the heat input to the water is shown in Fig. P3.17. If the specific heat of water is 4180 joules/(°C.kg), find the temperature in the tank as a function of time. To prevent the water from exceeding a temperature of 50°C, what is the longest time the heater may remain on?

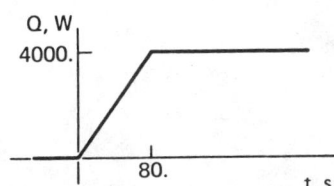

Figure P 3.17

3.18 A car battery with clean terminals can be considered as an ideal voltage source delivering 12 volts. The battery is connected to the ignition circuit of the car through the ignition switch. Assume that the ignition circuit is a pure resistance, $R = 3 \, \Omega$. If the battery terminals become corroded, an additional resistance of $1 \, \Omega$ may be considered as the internal resistance of the source. What is the percentage decrease in ignition circuit current due to corrosion?

CHAPTER 4

BASIC NETWORK MODELS

There are two basic ways to represent an electric circuit schematically. In one, the details of the connections are shown. Diagrams of this type are difficult to follow. In the other, the details of the connections are ignored. Instead, the components are arranged for maximum clarity to show how currents and voltages combine or separate. Schematic diagrams of this type are the ones normally used for analyzing the performance of an electric circuit. There are standard techniques for writing equilibrium equations by direct reference to these diagrams. This enables the formulation of the differential equations to represent the electrical signal variables.

The interactions of the elemental physical components in the mechanical, fluid and thermal media may also be represented by network diagrams having exactly the same form as the electrical network diagrams. This serves to make the same analytical techniques available to these media. The result is that equilibrium equations for systems of mechanical, fluid and thermal components can be written with equal facility and reliability.

4.1 FORMULATION OF CIRCUITS

The objective of circuit formulation is to place mechanical, fluid and thermal component arrangements into circuit diagrams. The essence of the problem is to identify those points in the system that may have different values of the across variable. In addition to these points, each circuit must have a datum or reference value for the across variable. The datum is usually called the **ground** of the circuit. The other network points are called **nodes**. For mechanical systems, the number of nodes (other than the datum node) is equal to the number of **degrees of freedom.**

The process of circuit formulation requires the identification of the basic elemental effects. Specifically, we must first find all the across variable nodes and then determine the appropriate element or elements that connect them together. The nodes therefore become the junctions of the circuit branches. Each branch indicates one element and corresponds to one elemental physical situation. The process is illustrated in the following examples.

Example 4.1

Consider the mass-spring-damper system shown in Fig. 4.1(a) in which someone applies a force, f(t), to one end of the damper.

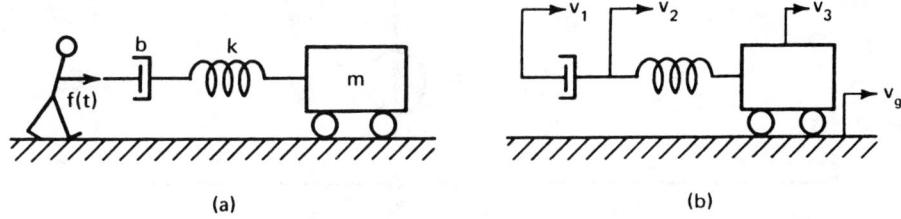

Figure 4.1 Mechanical arrangement (example 4.1)

The first step is to identify all parts of the system which are constrained to move at the same velocity and to indicate the direction in which these velocities will be considered as positive. This step is shown in Fig. 4.1(b). All the velocities are taken to be positive in the same direction. This facilitates the formulation of equilibrium equations and makes the results significantly easier to interpret. For the circuit under consideration, there are three velocities that are not necessarily the same. These are the velocity at the force end of the damper (v_1), the velocity at the point between the damper and the spring (v_2) and the velocity of the mass (v_3). Therefore, the equivalent circuit diagram will have three nodes or junctions in addition to the one representing the datum (v_g). These junctions, shown in Fig. 4.2(a), are the framework on which the elements are inserted. The junctions correspond to the variables v_1, v_2, v_3 and v_g. For convenience and clarity, the ground terminal v_g may be extended along a line at the same potential.

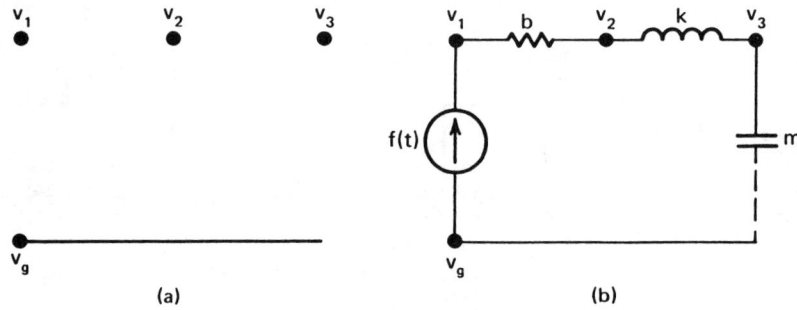

Figure 4.2 Mechanical circuit (example 4.1)

The passive components are placed in the circuit diagram by noting the velocity difference (across variable) that is required for each elemental equation. For example, the damping, b, models an effect which is dependent on the difference between v_1 and v_2.

Thus the damping element is connected between junctions 1 and 2. Damping, as discussed in Chapter III, is mechanical resistance and is a dissipative component. We represent it schematically by a resistance symbol. The spring stiffness, k, also models an effect that depends on velocity difference. In this case, the velocities at the ends of the spring

Basic Network Models   199

are v_2 and v_3. The spring element is therefore connected between these two nodes. Recall from Chapter III that the spring is analogous to the electrical inductance and has the same circuit symbol. Now, the physical equation that describes the motion of the mass shows that it is dependent on only one velocity. In this case, the velocity is v_3. Since all circuit components require two terminals, we add a terminal that does not change the physical relations. This terminal is the ground junction at zero velocity. When a fictitious terminal is used, the circuit diagram contains a dashed line to show this. The mechanical mass is analogous to an electrical capacitance with one terminal always connected to ground. Fig. 4.2(b) shows the circuit diagram for the mass-spring-damper system.

The driving force exerted by the person is modeled by a force source connected between junctions v_1 and v_g. Since the force is applied in the direction of positive velocity, the flow must be shown as entering junction v_1. Note that the force representation is complete, because it also indicates the necessary reaction force on the datum in the opposite direction (away from junction v_g).

The electrical analog of this particular mechanical circuit is a series RLC circuit with a current source input. However, it must be emphasized that other mass-spring-damper systems in which the elements appear in tandem do not necessarily result in a series circuit.   This point is illustrated in Example 4.2.

Example 4.2

Fig. 4.3(a) shows a mass-spring-damper arrangement in which the force is applied to the mass.

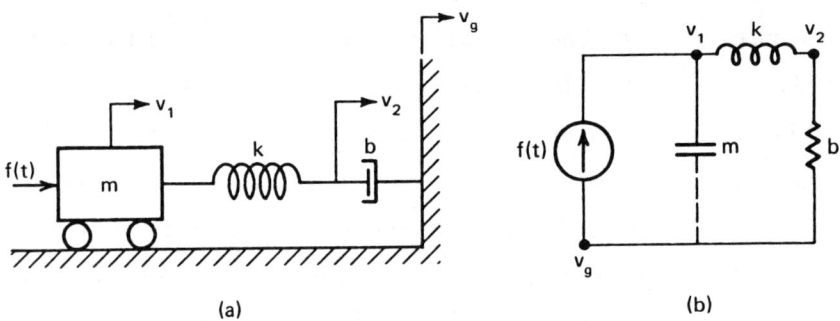

Figure 4.3 Mechanical circuit (example 4.2)

Here again, the mechanical arrangement has the components in tandem. However, in this case, the order is different. With this arrangement, there are now only two velocities that can be identified as independent. These are the velocity of the mass (designated v_1) and the velocity of the point between the spring and the damper (designated as v_2). Once again, the reference terminal is the velocity of the ground (v_g). The mass represents a capacitance between v_1 and v_g. The spring is equivalent to an inductance between v_1 and v_2, and the damper is a resistance that goes from v_2 to v_g. The resulting circuit is shown in Fig. 4.3(b).

Basic Network Models   201

In this case, the passive elements do not form a series circuit. Thus, it is important not to jump to conclusions about series or parallel equivalences solely on the basis of mass-spring-damper sketches. The analogous circuit representation can be obtained only by determining the terminals and connecting them with the appropriate components.

Before proceeding to more complicated examples with more than one source, it will be useful to discuss further the polarity of sources. Polarity concepts are especially pertinent, when the sources are not all acting in the same direction. Consider the isolated mass shown in Fig. 4.4, in which several possible situations are indicated.

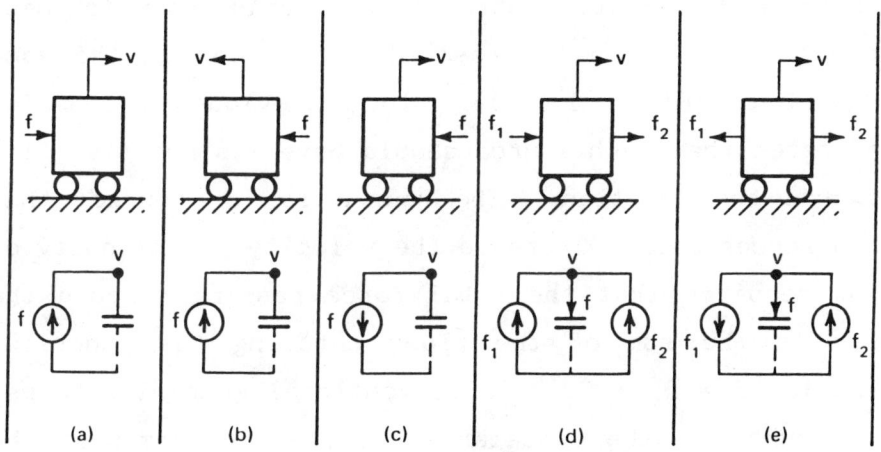

Figure 4.4 Polarity of sources

In all the cases, the direction selected for velocity (v) is conceptually positive. Then the application of the elemental equation [f = m (dv/dt)] requires that the force, f is positive from the terminal v through the element to the ground terminal. Thus, the situations presented in Fig. 4.4(a) and (b) are identical in the sense that the force applied to the mass acts in the same direction as the assumed positive direction for velocity. In other words, the force tends to increase the velocity. On the other hand, the force in Fig. 4.4(c) is directed to the left while the assumed positive velocity direction is to the right. As a result, the corresponding circuit representation shows a force source that is directed away from the terminal.

When two forces act on a mass, the circuit has two sources and therefore contains three branches which meet at a junction. In the case of two forces in the same direction [Fig. 4.4(d)], the previous discussion indicates that each source should have its positive flow towards the junction. The reason is, of course, that both forces act to increase the velocity. Continuity of flow requires that the total force passing through the mass is the sum of the flows entering the junction. That is, $f = f_1 + f_2$. In conventional mechanics terms, the sum of the unbalanced forces determines the acceleration of the mass. The circuit approach therefore is merely another way of recognizing and developing basic physical relations.

Suppose now that the forces are acting in opposite directions, as in Fig. 4.4(e). The force source in the direction of positive velocity (f_2) is modeled by a flow source directed towards the terminal v. This is the force that tends to increase the velocity. The opposite force, f_1, is modeled by a flow source directed away from the terminal and has a flow at the terminal that yields $f = f_2 - f_1$.

Both situations [Figs. 4.4(d) and (e)] are in exact agreement with physical reality. The junction of the circuit representation corresponds to the free body diagram of mechanics. However, for complicated mechanical arrangements, several free body diagrams may be required. An advantage of the network representation is that all the junctions are interconnected and the interactions between different parts of a mechanical arrangement are completely described by only one diagram.

For situations which are modeled by velocity sources, the procedure is simply to observe whether the source is in the positive or negative direction. If positive, the positive terminal of the velocity source is connected to the junction. When a velocity source is directed opposite to the assumed positive velocity direction, the negative terminal of the source is connected to the junction. These points are illustrated in the following examples.

Example 4.3

Fig. 4.5(a) shows an automobile approaching a bump in the road.

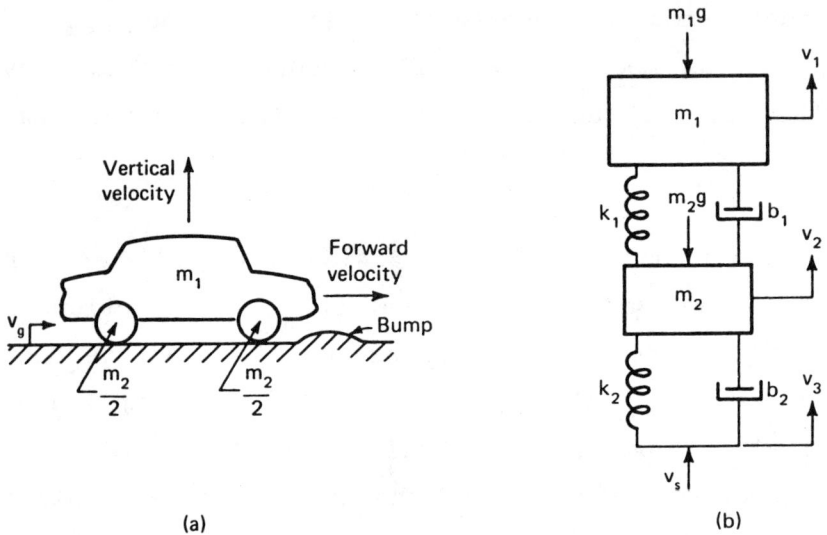

Figure 4.5 Mechanical arrangement (example 4.3)

The suspension system of the automobile supports the car body by means of a set of springs (k_1) and shock absorbers (b_1). The tires exhibit a spring effect (k_2) which is close to the ideal spring, but the hysteresis normally associated with rubber springs is modeled by an equivalent damper (b_2). The constraint at the point of contact between the tire and the road is that the position of this point is defined. If the car is driven at constant forward velocity along the road, the vertical velocity of the point of contact is defined. This vertical velocity constraint is, therefore,

Basic Network Models

appropriately modeled by a velocity source (v_s). In addition, there are force sources to model the weight of the car and the weight of the wheels. These considerations lead to the mass-spring-damper arrangement shown in Fig. 4.5(b).

To obtain the network diagram, we must first determine the number of independent velocities associated with the mechanical sketch. Each mass may move independently and this leads to terminals representing v_1 and v_2. The point of contact with the ground is v_1. The velocity of the ground is v_g. Thus, there are three terminals plus a ground on the circuit diagram (Fig. 4.6). The connection of components between the terminals is obtained by referring to the sketch in Fig. 4.5(b). The velocity source, v_s, acts in the direction of positive velocity. As a result, the positive terminal of this source is connected to v_1. This means that $v_s = v_1 - v_g$. Had the reference velocity direction been selected as down, the negative terminal of v_s would have been attached to v_1. In that case, $v_s = v_g - v_1$. The force sources due to the weight of the car ($m_1 g$) and the weight of the wheels ($m_2 g$) are directed away from their respective terminals (v_1 and v_2). Obviously, if the reference velocity was down, these sources would be pointed towards the terminals.

Example 4.4

Winds apply large forces to building structures. These forces must be included in the structural design.

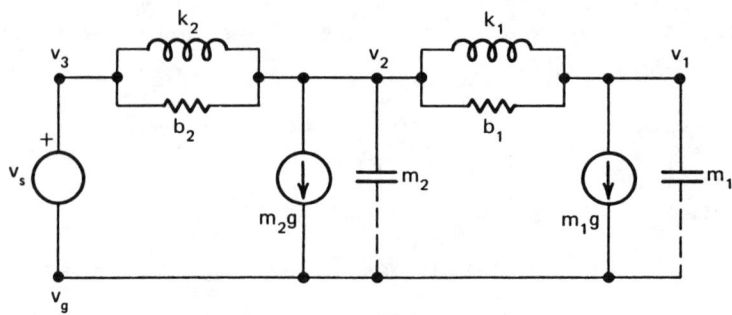

Figure 4.6 Mechanical circuit (example 4.3)

If the wind is to be treated as a variable force source, the dynamic problem posed is rather difficult. However, a lumped model of the structure can simplify the problem appreciably. Fig. 4.7(a) shows a schematic drawing of a three floor building subjected to a wind force. We wish to obtain a circuit diagram for this situation.

In this example, the first step is to develop a mechanical sketch. To do this, we must have some idea about the distribution of mass in the structure. We will assume here that most of the mass of the building is concentrated in the slabs between floors and ceilings. Thus m_3 lumps the mass of the roof with a portion of the mass of the supporting columns. Similarly, m_2 lumps the slab, between the ceiling of the second floor and the floor of the third floor, with a portion of the column. The mass, m_1, is also lumped in a similar manner. Now, the slab on which the building is constructed may be modeled as m_g. However, this mass is fully constrained by the ground. Thus, we may model

Basic Network Models

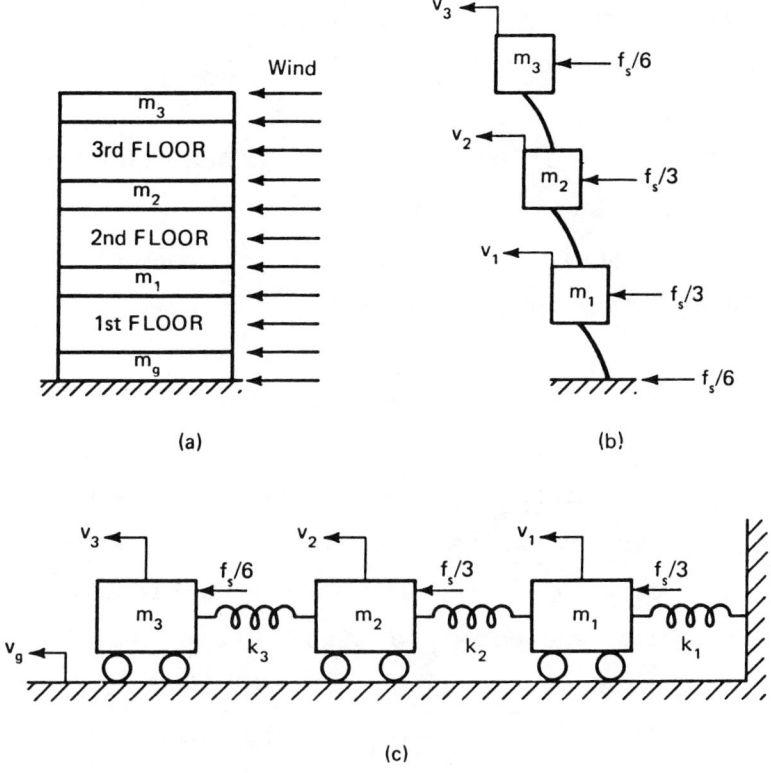

Figure 4.7 Building model for wind gusts (example 4.4)

the structure by three lumped masses m_1, m_2 and m_3. The columns connecting these masses are considered as cantilever beams which have an equivalent (spring) stiffness.

Fig. 4.7(b) is an exaggerated sketch of the building subjected to a total wind force of f_s. The wind force is assumed to be uniformly distributed across the entire outside wall. In relation to the concentration of mass, however, we assume that m_1 and m_2 are each acted upon by $f_s/3$. The remaining $f_s/3$ is assumed to be equally distributed between m_3 and m_g. Thus, m_3 is acted upon by $f_s/6$.

Fig. 4.7(c) emphasizes the fact that the motion under consideration is entirely transverse. The cantilever springs are modeled by ordinary extension springs of equal stiffness.

Once the sketch shown in Fig. 4.7(c) is made, we may proceed as before to identify the independent velocities. There is always a velocity associated with each mass and so we have three terminals plus ground to form our circuit. Fig. 4.8 shows the circuit model for the building subjected to wind. Note that the masses are always connected to ground. The wind forces act in the assumed positive velocity direction and therefore the arrows in the flow sources point towards the terminals.

4.2 BASIC CIRCUIT LAWS

The objective of circuit analysis is to derive mathematical relations between the output and the input variables. The circuit representation of a physical system permits us to accomplish this analysis through the application of two basic circuit laws. One of these laws, **Kirchhoff's Current Law (KCL)**, refers to conditions at any junction and is expressed as:

Basic Network Models

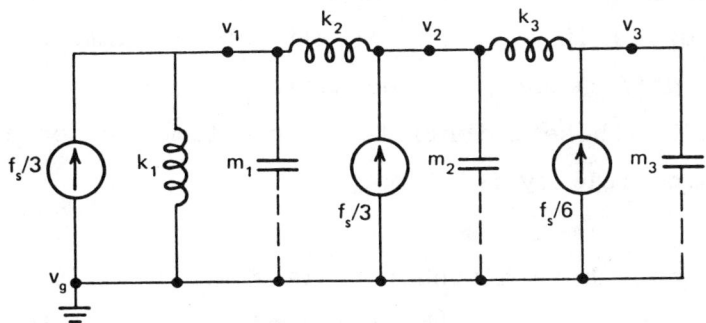

Figure 4.8 Mechanical circuit (example 4.4)

$$\sum (\text{Through Variables}) = 0 \qquad ..(4.1)$$

Thus, the summation of currents at a junction of an electric circuit is analogous to the summation of forces in a mechanical circuit or to the summation of fluid flow rates in a fluid circuit. The fact that the sum (algebriac or vectorial, as the case may be) of each of these variables at a junction is zero is simply an application of the basic ideas of continuity and equilibrium. Since the through variable in mechanical systems is force, KCL can be recognized as being analogous to D'Alembert's principle.

The other circuit law, **Kirchhoff's Voltage Law (KVL)**, refers to conditions around a closed path in the circuit and is:

$$\sum (\text{Across Variables}) = 0 \qquad ..(4.2)$$

210   Modeling and Analysis of Linear Physical Systems

Thus, the summation of voltage drops in an electrical circuit is analogous to the summation of pressure drops in a fluid circuit or the summation of velocity differences in mechanical circuits. The summation of these potentials is sometimes referred to as the compatibility relation.

Although both Kirchhoff's laws are usually stated (Eqs. 4.1 and 4.2) as the algebraic or vectorial sum being zero, it is often simpler to equate one of the variables to the sum of the others. For example, the flow into a junction may be set equal to the flow out of the junction. When the basic circuit laws are used in this way, it is usually possible to set up the equilibrium equations in a form which is suitable for direct solution. The generalization of this process for complex circuits is considered in Chapter VII. However, for simple series, parallel and series-parallel circuits, we may apply the basic laws in an easy fashion.

Example 4.5

Fig. 4.9 shows a series RL electrical circuit with a voltage source, e_s.

(a) Find the differential equation for the current in terms of the components and voltage source.

(b) Find a differential equation to represent the voltage across the inductor.

(c) Find an analogous mechanical circuit. Express the force and the velocity in differential equation

Basic Network Models

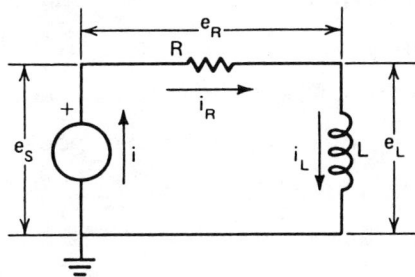

Figure 4.9 Electrical circuit (example 4.5)

form.

(a) Since the circuit contains one mesh only, it is easier to write the KVL equation. Thus the application of KVL around the closed electric circuit path yields

$$-e_s + e_R + e_L = 0 \qquad ..(4.3)$$

The next step is to apply the component relations, giving

$$-e_s + R\, i_R + L\frac{di_L}{dt} = 0 \qquad ..(4.4)$$

In a series circuit, there is only one current. This can be confirmed by applying KCL at any junction. For example, at the junction between the resistor and the inductor, KCL yields

$$-i_R + i_L = 0 \quad \text{or} \quad i_R = i_L \qquad ..(4.5)$$

where we have used the convention that flows entering a junction are considered negative and flows leaving a junction are considered positive. This is an arbitrary selection. The opposite convention would result in the same relations. However, in this book, we shall adopt the convention used in the formulation of (4.5). In a similar way, we could find that $i = i_R$. Since all the currents are the same in the series circuit, the differential equation above may be expressed in the form

$$\frac{L}{R}\frac{di}{dt} + i = \frac{1}{R} e_s \qquad ..(4.6)$$

This format is used throughout this book.

The dependent variable (the output) is in this case the current, i. It will always appear on the left-hand-side of the differential equation. The independent variable (the input) is in this case the voltage source and will always be placed on the right-hand side of the equation.

(b) To determine the voltage, e_L, begin with the current relation at the junction between resistance and inductance, $i_R = i_L$. Then, employ the component relations so that:

$$\frac{e_R}{R} = \frac{1}{L} \int e_L \, dt \qquad ..(4.7)$$

Basic Network Models

Now, use the voltage relation (4.3) to eliminate e_R. The result is:

$$\frac{e_s - e_L}{R} = \frac{1}{L} \int e_L \, dt \qquad \qquad ..(4.8)$$

Now differentiate and rearrange (4.8) into the standard format

$$\frac{L}{R} \frac{de_L}{dt} + e_L = \frac{L}{R} \frac{de_s}{dt} \qquad \qquad ..(4.9)$$

(c) Fig. 4.10(a) shows the mechanical arrangement that has the same analogous circuit. It consists of a damper and spring in tandem. A velocity source, v_s, is applied to the free end of the damper. The analogous circuit diagram is shown in Fig. 4.10(b).

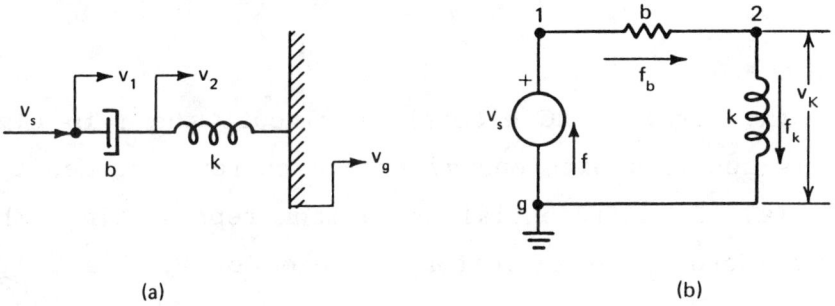

Figure 4.10 Analogous mechanical circuit (example 4.5)

Thus:

$$f_b - f_k = 0 \qquad \text{(KCLA)}$$

$$-v_1 + (v_1 - v_2) + v_2 = 0 \qquad \ldots (4.10)$$

$$-v_s + v_b + v_k = 0 \qquad \text{(KVLA)}$$

where the designations KCLA and KVLA refer to the analogs of the basic circuit laws. The substitution of the elemental equations yields:

$$\frac{b}{k}\frac{df}{dt} + f = b\,v_s \qquad \ldots (4.11a)$$

and

$$\frac{b}{k}\frac{dv_k}{dt} + v_k = \frac{b}{k}\frac{dv_s}{dt} \qquad \ldots (4.11b)$$

Example 4.6

The parallel RC electrical circuit shown in Fig. 4.11 is supplied with energy from a current source, i_s. Find (a) A differential equation representing the voltage across the capacitor in terms of R, C and i_s, and (b) An analogous fluid circuit. Express the pressure as a differential equation.

Basic Network Models

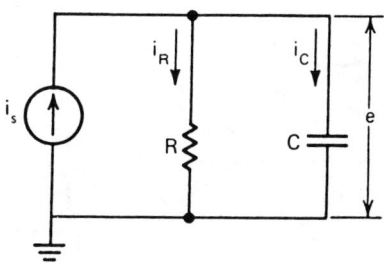

Figure 4.11 Electrical circuit (example 4.6)

(a) Since we are interested in a differential equation for voltage, we begin with the basic current law

$$i_c + i_R = i_s \qquad \text{(KCL)} \qquad ..(4.12)$$

Substituting the elemental equations gives

$$C \frac{de}{dt} + \frac{1}{R} e = i_s \qquad ..(4.13a)$$

or in standard form

$$RC \frac{de}{dt} + e = R\, i_s \qquad ..(4.13b)$$

(b) Fig. 4.12(a) shows the fluid component arrangement that produces the same analogous circuit. This circuit is illustrated in Fig. 4.12(b). The pressure, p, is the one that exists at the bottom of the tank. The line from the tank is modeled in this case, by a pure resistance element. The procedure for determining the pressure, p, is exactly the same as the

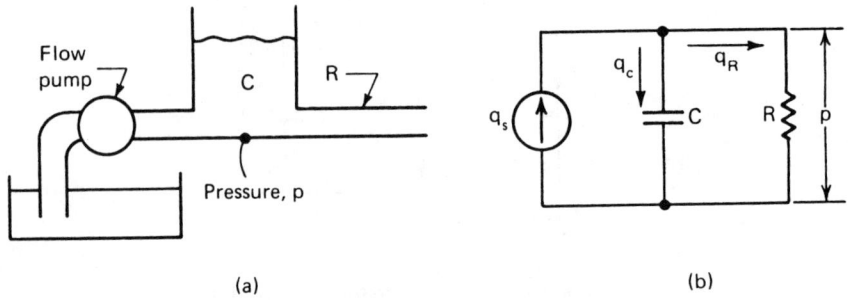

Figure 4.12 Analogous fluid circuit (example 4.6)

procedure for finding, e, in the electrical circuit.

Thus, we may write at once that:

$$q_c + q_R = q_s \quad (KCLA) \quad \quad ..(4.14)$$

and with the use of the component equations

$$RC \frac{dp}{dt} + p = R\, q_s \quad \quad ..(4.15)$$

4.3 PROPERTIES OF STORAGE ELEMENTS

Any physical system with energy storage elements requires a differential equation to describe its equilibrium. To predict the response of the system, we must find a solution of the differential equation. As will be shown in Chapters V, VI and IX, this necessitates a foreknowledge of the initial conditions. These initial conditions are not arbitrary mathematical constraints, but are rather conditions which result from

Basic Network Models

the interaction between each energy storage element and the rest of the circuit at the time of any sudden change in input. Many possible input functions were derived from the basic step function $U_s(t)$. Accordingly, we now proceed to investigate the properties of the storage elements when they are subjected to a sudden change. The properties to be considered now are the initial response $y(0^+)$ (that is, the response immediately after the change has been applied) and the final response $y(\infty)$ (after all transient conditions have disappeared). The complete response, $y(t)$, includes the transient period. This will be determined in Chapters V, VI and IX.

4.3.1 Initial Conditions of T-type and A-type Elements

(a) T-type Elements

The elemental relationships for T-type storage elements are given by Eqs. (3.11), (3.24) and (3.47) for the electric, fluid and mechanical systems respectively. There is no corresponding element in thermal systems. These elemental equations give the relation between the across variable (av) and the through variable (tv) as:

$$av(t) = L \frac{d}{dt}\left[tv(t)\right] \qquad ..(4.16)$$

The initial response of these elements is more easily explained, however, by considering the inverse relation:

$$tv(t) = \frac{1}{L} \int_{-\infty}^{t} av(t') \, dt' \qquad \qquad ..(4.17)$$

Eq. (4.17) is valid, of course, for all values of t. However, the initial response depends only on the value of the integral over the infinitesimally small interval from just before the driving function is applied ($t = 0^-$) to just after it is applied ($t = 0^+$). As a result, the change in the value of the through variable at $t = 0$ is given by

$$\Delta tv(0) = \frac{1}{L} \int_{0^-}^{0^+} av(t) \, dt \qquad \qquad ..(4.18)$$

If the conditions in the system or the circuit are such that the across variable is finite (which is normally the case), the value of this integral must always be zero. Therefore, the basic property of the T-type element may be expressed as

$$tv(0^-) = tv(0^+) \qquad \qquad ..(4.19)$$

If there is no energy stored in the element just before $t = 0$ [that is $tv(0^-) = 0$], the element provides the constraint that the through variable remains zero as the driving function is being applied. Thus, when there is no energy initially stored in the T-type element, we may write

Basic Network Models   219

$$tv(0^+) = 0 \qquad \qquad ..(4.20)$$

In the case of an inductor, this means that it has the effect of keeping the current zero over the interval from $t = 0^-$ to $t = 0^+$. As a result, the inductor may be considered to be acting at $t = 0$ as an open circuit. The process of finding the initial response of any circuit in which an inductor (without any initial energy storage) is present can therefore be simplified by redrawing the circuit diagram with the inductor replaced by an open circuit.

The analogous property of inertance in a fluid circuit constrains the flow rate to remain unchanged in the interval from $t = 0^-$ to $t = 0^+$. Thus inertance can also be replaced at $t = 0$ by an open circuit. Similarly, the force or torque in translational or rotational springs cannot change suddenly at $t = 0$ and they, too, behave initially as open circuits.

Example 4.7

A series RL circuit (Fig. 4.13 (a)) is subjected to a step input $e = 10\ U_s(t)$ volts. Find $i(0^+)$, $e(0^+)$ and $\frac{di}{dt}(0^+)$.

Initially the circuit behaves as if the inductor is an open circuit. This condition is shown by the circuit of Fig. 4.13 (b). We may refer to this circuit as the initial condition circuit.

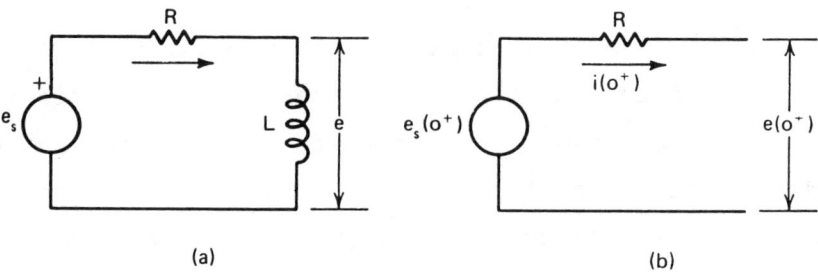

Figure 4.13 Initial conditions (example 4.7)

The current through the initial condition circuit is obviously zero. Thus $i(0^+) = 0$. The voltage $e(0^+) = e_s(0^+) = 10$ v. To find $\frac{di}{dt}(0^+)$, we use the elemental inductor equation $= L\frac{di}{dt}$. We may adopt the equation for $t = 0^+$ as:

$$e(0^+) = L\frac{di}{dt}(0^+) \quad\quad ..(4.21a)$$

from which

$$\frac{di}{dt}(0^+) = \frac{e(0^+)}{L} \quad\quad ..(4.21b)$$

(b) A-type Elements

The elemental relationships for A-type storage elements are given in Eqs. (3.15), (3.28), (3.57) and (3.58) for electrical, fluid, mechanical and thermal systems respectively. All these relationships have the form

$$av(t) = \frac{1}{C} \int_{-\infty}^{t} tv(t') \, dt' \qquad ..(4.22)$$

The initial response of the A-type elements depends only on the value of this integral over the infinitesimally small interval from $t = 0^-$ to $t = 0^+$.

Thus, the change in the value of the across variable at $t = 0$ is given by:

$$\Delta av(0) = \frac{1}{C} \int_{0^-}^{0^+} tv(t) \, dt \qquad ..(4.23)$$

If the conditions in the system or circuit are such that the through variable is finite, (which is normally the case), this integral must be zero. The basic initial property of the A-type element may therefore be expressed as

$$av(0^+) = av(0^-) \qquad ..(4.24)$$

When there is no initial energy storage in the element [$av(0^-) = 0$], the across variable remains zero as the driving function is being applied. Thus, we may write:

$$av(0^+) = 0 \qquad ..(4.25)$$

For the case of an electrical capacitor (without

initial charge), the voltage drop across it remains equal to zero in the interval from $t = 0^-$ to $t = 0^+$. Therefore, the electrical capacitor may be considered to be acting at $t = 0^+$ as a short circuit. To find the initial response of a circuit with an electrical capacitor, the circuit diagram can be simplified by redrawing it with the capacitor replaced by a short circuit.

The analogous property of a fluid capacitor keeps the pressure constant in the interval from $t = 0^-$ to $t = 0^+$. This is represented on the analogous circuit diagram by replacing the fluid capacitor with a short circuit at $t = 0^+$. Similarly, the mass and the moment of inertia in translational and rotational mechanical systems do not permit any sudden change in velocity or angular velocity. In addition, thermal mass inhibits an instantaneous change in temperature. All of these components behave like short circuits at $t = 0^+$.

Example 4.8

A mass is connected to a damper as shown in Fig. 4.14(a). The mass is initially at rest ($v(0^-) = 0$). A step input of force, $f_s = 5U_s(t)$ N is applied to the mass. Find $v(0^+)$ and $\frac{dv}{dt}(0^+)$.

The analogous circuit for the mechanical arrangement is shown in Fig. 4.14(b). The circuit has an across type storage device and a dissipative element

Basic Network Models

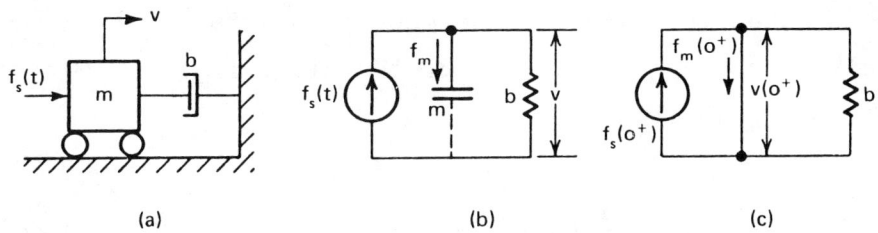

Figure 4.14 Initial conditions (example 4.8)

in parallel. The input is represented by an ideal force source. We may convert the analogous circuit to an initial condition circuit by **shorting out** the A-type storage device. Fig. 4.14(c) shows the resulting initial condition circuit for this case. To determine the required values $v(0^+)$ and $\frac{dv}{dt}(0^+)$, we analyze the initial condition circuit. Evidently, $v(0^+) = 0$, since the analogous circuit behaves as if the mass (capacitor) is a short circuit to ground. The relationship

$$f_m = m \frac{dv}{dt} \qquad ..(4.26)$$

is the equation for the mass element and holds at every instant of time. For the specific case of $t = 0^+$, we find that

$$\frac{dv}{dt}(0^+) = \frac{f_m(0^+)}{m} \qquad ..(4.27)$$

Thus, the problem of determining $\frac{dv}{dt}(0^+)$ is reduced to the simpler problem of finding $f_m(0^+)$. Initial condition circuit [Fig. 4.14(c)] indicates that the force taken initially by the mass $f_m(0^+)$ is equal to the entire force applied at t = 0, that is $f_m(0^+) = f_s(0^+) = 5$. As a result:

$$\frac{dv}{dt}(0^+) = \frac{5}{m} \qquad \qquad ..(4.28)$$

Remember that although the mass is behaving as a short-circuit, the mass element is still located between the velocity terminal and ground. Thus, all the force from the ideal source **flows** through the mass at $t = 0^+$.

4.3.2 Final Conditions with Step Inputs

We have observed that the initial values of the response to a sudden change such as a step input can be obtained directly from a circuit representation that is valid only at $t = 0^+$. In a similar way, when the input is a step, we may obtain the final values from a circuit equivalent that is valid only at $t = \infty$. As we shall see in Chapters V, VI and IX, the final values provide the particular solutions for the circuit differential equations, when a step input is applied.

One of the main characteristics of any linear system is that the particular solution is the part of the complete solution that is due to the forcing or driving function (that is, the input). The particular solution must have the same shape as the forcing function. The only difference between forced response (output) and the driving function (input) shape are a

Basic Network Models

scale factor and, depending on the driving function, a timing factor. For a step input, the forced response is a constant after the step is applied. Therefore, there is no timing factor. On the other hand, the forced response to a sine wave input has both a scale factor (magnitude) and a timing factor (phase difference).

The scale factor for a step input is obtained by noting that the shape of the input is not changing after $t = 0^+$. The forced response of a linear system is, therefore, also characterized by an unchanging value.

One way to derive the final value of the circuit variables is to determine the final condition circuit. As in the case of the initial condition circuit, this requires the replacement of the storage elements with equivalent open and short circuits.

(a) T-type Elements

The final state of a T-type element under the conditions imposed by step excitation is more easily derived by returning to the general elemental relationship given in (4.16). In this case, we note that the final condition of a step is characterized by zero rate of change. For a linear system, this zero rate of change must also be a characteristic of the final state of all the elements within the system. Thus (4.16) may be rewritten at $t = \infty$ as:

$$av(\infty) = 0 \qquad \qquad ..(4.29)$$

In the case of an inductor, this means that the final value of the voltage across it is zero, and it may therefore be represented by a short-circuit. This, of course, applies only to the ideal inductor which it should be recalled has no resistance. Similarly, the final pressure difference associated with inertance is zero, and the final difference in velocity between the ends of a spring is zero. The process of finding the final state of a system or circuit in which T-type elements are present is therefore quite simple, if the circuit diagram is redrawn with T-type elements replaced by short-circuits.

(b) A-type Elements

The final state of an A-type element, when it is part of a system subjected to a step excitation is determined by considering the inverse form of (4.22):

$$tv(t) = C \frac{d}{dt} [av(t)] \qquad ..(4.30)$$

Again, we note that with a step function the final rate of change is zero, and therefore the final value of the through variable associated with an A-type element is:

$$tv(\infty) = 0 \qquad ..(4.31)$$

In the case of an electrical capacitor, this means that there is no flow of current through it, when the voltage across it is constant. This situation is simply represented on a circuit diagram by redrawing it with

Basic Network Models 227

the capacitor replaced by an open circuit. Similarly, the final flow rate into a fluid capacitor is zero, the final force or torque acting on a mass or inertia is zero, and the final flow of heat into thermal capacitance is zero. For the final state with step excitation, all of these situations are easily considered by redrawing the analogous circuit diagram with the A-type elements replaced by open-circuits.

Example 4.9

Fig. 4.15(a) shows a mechanical model that consists of two masses, m_1 and m_2, two dampers, b_1 and b_2, and one spring, k. A step function $f_s(t) = 2U_s(t)$ N is applied to mass, m_1. The velocity of mass m_1 is v_1 and the velocity of mass m_2 is v_2. Find

(a) $v_1(0^+)$, $v_2(0^+)$, $\dfrac{dv_1}{dt}(0^+)$, $\dfrac{dv_2}{dt}(0^+)$

(b) $v_1(\infty)$, $v_2(\infty)$.

Fig. 4.15(b) shows the analogous circuit representation of the mechanical model. There are two terminals v_1, and v_2, plus ground. Notice that the mass components are always between a terminal and ground. Once the terminals have been numbered and the masses have been positioned, it is a simple matter to connect the other components. Thus, the spring connects the two mass terminals together and the dampers go from a mass to ground. To determine the initial and the final

228 Modeling and Analysis of Linear Physical Systems

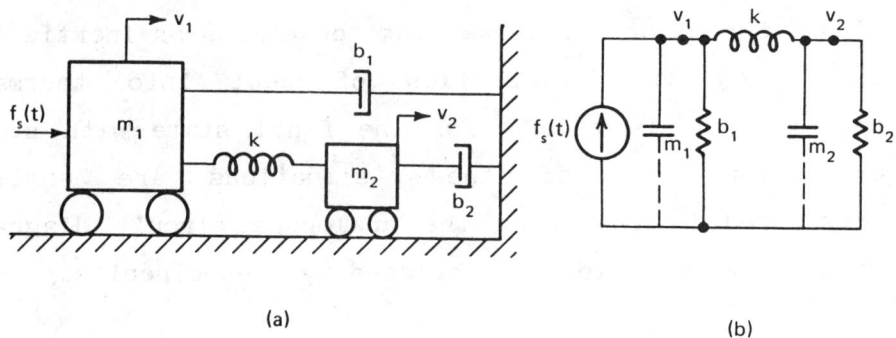

Figure 4.15 Mechanical circuit (example 4.9)

conditions sought, it is convenient to draw initial and final condition circuits. These circuits contain only dissipative elements and sources.

(a) Fig. 4.16(a) shows the initial condition circuit. This circuit is obtained by replacing the A-type storage devices m_1 and m_2 by short-circuits and replacing the T-type storage device, k, by an open-circuit. The dissipative elements, b_1 and b_2 remain in place. Now we analyze the initial condition circuit to determine the initial values of the variables. Since the masses are represented by short-circuits to ground, $v_1(0^+)$ and $v_2(0^+)$ have the same velocity as ground. Thus

$$v_1(0^+) = v_2(0^+) = 0 \qquad \qquad ..(4.32)$$

To determine the initial values of the

Basic Network Models

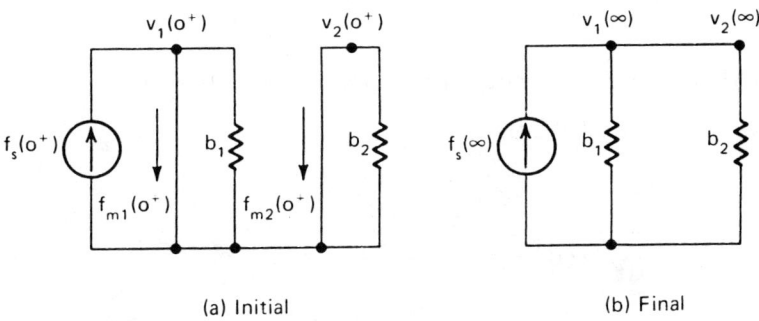

Figure 4.16 Initial and final condition circuits (example 4.9)

derivatives, we must find a component in the original circuit [Fig. 4.15(b)] that has an elemental equation containing the derivative. Thus from m_1, we know that:

$$f_{m_1} = m_1 \frac{dv_1}{dt} \qquad ..(4.33a)$$

and from m_2 that:

$$f_{m_2} = m_2 \frac{dv_2}{dt} \qquad ..(4.33b)$$

These equations are valid for all times. Thus, for the special case of interest here, we may write that:

$$\frac{dv_1}{dt}(0^+) = \frac{f_{m_1}(0^+)}{m_1} = \frac{2}{m_1} \qquad ..(4.34)$$

where
$$f_{m_1}(0^+) = f_s(0^+) = 2 \qquad ..(4.35)$$

is obtained by noting the initial condition circuit behavior. In addition

$$\frac{dv_2}{dt}(0^+) = \frac{f_{m_2}(0^+)}{m_2} = 0 \qquad ..(4.36)$$

since there is no source in the loop of the initial condition circuit that encompasses m_2.

(b) The final conditions can be obtained by drawing the circuit equivalent at $t = \infty$ as shown in Fig. 4.16(b). As in the case of the initial condition circuit, the elements in the final condition circuit are all dissipative and sources. However, in this circuit, the masses behave like open circuits and the spring behaves like a short circuit. It is evident from the resulting final condition that the two masses must have the same velocity and that the dampers are effectively in parallel. Thus

$$f_s(\infty) = (b_1 + b_2) v_1(\infty) \qquad ..(4.37a)$$

or
$$v_1(\infty) = v_2(\infty) = \frac{2}{b_1 + b_2} \qquad ..(4.37b)$$

A word of caution is required regarding the use of the designation **parallel**. Its use in this example to

Basic Network Models

describe the constraints on the dampers b_1 and b_2 is limited to the final state where the difference in velocity across the spring is zero. At all other times, the two masses have different velocities. Thus, these two dampers must not be considered as being in parallel as far as the general problem of describing the system behaviour is concerned. It must be remembered that before any two elements in any system may be considered to be in parallel, the circuit representation must show a definite constraint of equal across variables.

Example 4.10

The electrical circuit shown in Fig. 4.17 has $R_1 = 4$ ohms, $R_2 = 3$ ohms, $C = 0.2$ F and $L = 1.0$ H. If a current source $i(t) = 2\, U_s(t)$ A is impressed, find:

(a) $e_1(0^+)$, $e_2(0^+)$, $e_3(0^+)$, $\dfrac{de_3}{dt}(0^+)$, $\dfrac{di_2}{dt}(0^+)$,

(b) $e_1(\infty)$, $e_2(\infty)$, $e_3(\infty)$, $i_1(\infty)$, $i_2(\infty)$.

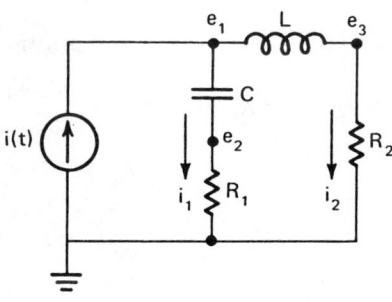

Figure 4.17 Electrical circuit (example 4.10)

(a) To find the initial conditions ($t = 0^+$), we make an initial condition circuit. This is shown in Fig. 4.18(a). From the initial condition circuit, we observe that $i_1(0^+) = 2$ A and $i_2(0^+) = 0$. Thus

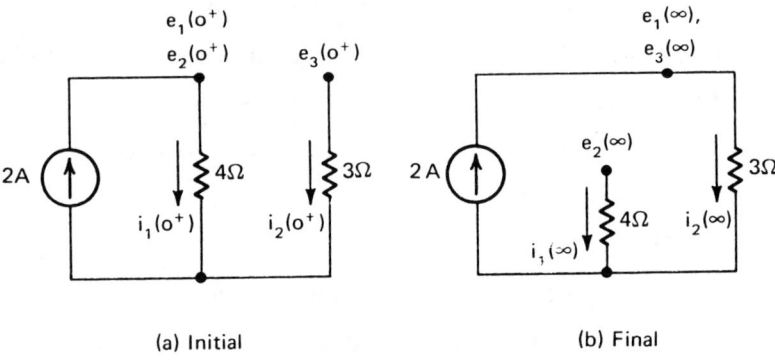

(a) Initial   (b) Final

Figure 4.18 Initial and final condition circuits (example 4.10)

$$e_1(0^+) = e_2(0^+) = R_1 \, i_1(0^+) = 8 \text{ V}$$
$$e_3(0^+) = R_2 \, i_2(0^+) = 0 \text{ V}$$
.. (4.37)

To evaluate $\frac{di_2}{dt}(0^+)$. we seek a component that has an elemental equation containing this term. Note from Fig. 4.17 that:

$$e_1 - e_3 = L \frac{di_2}{dt}$$
.. (4.38)

This equation holds good for all time. For the special case of $t = 0^+$, we have

$$\frac{di_2}{dt}(0^+) = \frac{1}{L}\left[e_1(0^+) - e_3(0^+)\right] = 8.0 \text{ A/s} \qquad ..(4.39)$$

There remains now to determine $\frac{de_3}{dt}(0^+)$. In this case, there is no component that contains this term. Thus we must formulate this term from another component by differentiation. From Fig. 4.17:

$$e_3 = R_2 i_2 \qquad ..(4.40)$$

and by differentiation:

$$\frac{de_3}{dt} = R_2 \frac{di_2}{dt} \qquad ..(4.41)$$

This relationship must hold for all times. At $t = 0^+$, we have

$$\frac{de_3}{dt}(0^+) = R_2 \frac{di_2}{dt}(0^+) = 24.0 \text{ V/s} \qquad ..(4.42)$$

(b) Fig. 4.18(b) shows the final condition circuit representation for this problem. From the final condition circuit, we see that $i_1(\infty) = 0$ and $i_3(\infty) = 2.0$ A. Thus:

$$e_1(\infty) = e_3(\infty) = R_2 i_2(\infty) = 6.0 \text{ V}$$
$$e_2(\infty) = R_1 i_1(\infty) = 0 \text{ V} \qquad ..(4.43)$$

It must be emphasized that the foregoing method for determining the final conditions is applicable only when the input is a step function.

4.4 MATHEMATICAL APPROACH TO INITIAL CONDITIONS

There is a mathematical method to determine the initial conditions in a circuit, when the input function is a singularity function and the energy storage elements had no stored energy prior to the application of the input. We will discuss this method here only briefly. However, this should be sufficient to permit us to check the initial condition results obtained from the initial condition circuit.

To use the mathematical approach, we must first find the differential equation in terms of the dependent variable, whose initial value we wish to know. Suppose we designate the variable as y(t) and the governing equation as

$$\frac{d^2y}{dt^2} + b_1 \frac{dy}{dt} + b_0 y = a_0 g(t) + a_1 \frac{dg}{dt}(t) \quad ..(4.44)$$

where $g(t)$ is the input function and b_0, b_1, a_0 and a_1 are constants. We want to find $y(0^+)$ and $\frac{dy}{dt}(0^+)$. The procedure is to integrate (4.44), until the left-hand side has only one non-integral term. After one integration, (4.44) becomes:

$$\frac{dy}{dt} + b_1 y + b_0 \int_{-\infty}^{t} y(t') \, dt' = a_0 \int_{-\infty}^{t} g(t) + a_1 g(t)$$

..(4.45)

where t' is the variable of integration.

There are two non-integral terms on the left-hand side of (4.45). Thus, we must perform another integration to reach:

$$y(t) + b_1 \int_{-\infty}^{t} y(t') \, dt' + b_0 \int_{-\infty}^{t} \left[\int_{-\infty}^{t} y(t'') \, dt'' \right] dt'$$

$$= a_0 \int_{-\infty}^{t} \left[\int_{-\infty}^{t'} g(t'') \, dt'' \right] dt' + a_1 \int_{-\infty}^{t} g(t') \, dt'$$

..(4.46)

where t" is also a variable of integration. Now remember that (4.46) is valid for all times. Thus, we may evaluate (4.46) at $t = 0^+$ directly, since g(t) is made out of the singularity functions [$U_i(t)$, $U_s(t)$, $U_r(t)$, $U_p(t)$].

Since the input is applied at t = 0, we have

$$\int_{-\infty}^{0^-} y(t') \, dt' = 0 \qquad (4.47a)$$

Further. y(t) is finite and hence

$$\int_{-\infty}^{0} y(t') \, dt' = \int_{0^{-}}^{0^{+}} y(t') \, dt' = 0 \qquad ..(4.47b)$$

Therefore, we get

$$y(0^{+}) = a_{o} \int_{-\infty}^{0^{+}} \left[\int_{-\infty}^{0^{+}} g(t'') \, dt'' \right] dt' + a_{1} \int_{-\infty}^{0^{+}} g(t') \, dt'$$

$$..(4.48)$$

We shall consider the following cases:

(a) Let $g(t)$ be a doublet. Then $a_1 = 0$, since $\frac{dg(t)}{dt}$ does not have any physical meaning. It is known that

$$\int_{-\infty}^{0^{+}} g(t') \, dt' = U_{i}(t) \qquad ..(4.49a)$$

and

$$\int_{-\infty}^{0^{+}} \left[\int_{-\infty}^{0^{+}} g(t'') \, dt'' \right] dt' = U_{s}(0)^{+} = 1 \qquad ..(4.49b)$$

Hence, we get, from (4.48),

$$y(0^{+}) = a_{o} \qquad ..(4.50)$$

(b) Let $g(t)$ be an impulse. In this case, it is known that

$$\int_{-\infty}^{0^+} g(t') \, dt' = U_s(0^+) = 1 \qquad \ldots (4.51)$$

and

$$\int_{-\infty}^{0^+} \left[\int_{-\infty}^{0^+} g(t'') \, dt'' \right] dt' = U_r(0^+) = 0 \qquad \ldots (4.52)$$

Hence, we get from (4.48),

$$y(0^+) = a_1 \qquad \ldots (4.53)$$

If $g(t)$ is other than an impulse or a doublet (like the unit step, unit ramp or unit parabola), it is easily verified that $y(0^+) = 0$.

From (4.45), we have

$$\frac{dy}{dt}(0^+) = a_o \int_{-\infty}^{0^+} g(t') \, dt' + a_1 g(t) - b_1 y(0^+) \qquad \ldots (4.54)$$

(a) If $g(t)$ is a doublet, we have

$$\frac{dy}{dt}(0^+) = - a_o b_1 \qquad \ldots (4.55)$$

(b) If $g(t)$ is an impulse, we get

$$\frac{dy}{dt}(0^+) = a_o - a_1 b_1 \qquad \ldots (4.56)$$

238   Modeling and Analysis of Linear Physical Systems

Let us try out this procedure on a few simple circuits.

Example 4.11

Fig. 4.19(a) shows the schematic drawing of a pneumatic circuit in which a pump supplies a tank through a length of line. When the pump is suddenly turned on, it develops the pressure function p(t) $U_s(t)$. If the line can be modeled by a pure resistance, find the flow into the tank immediately after the pump is turned on.

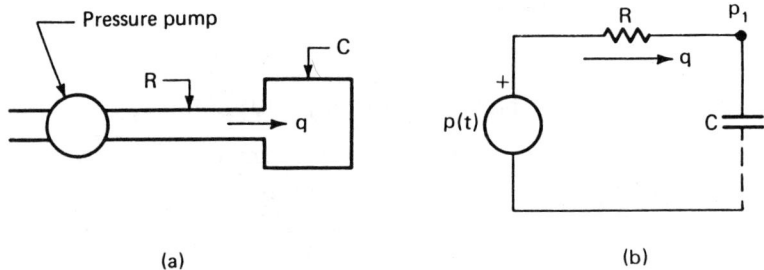

Figure 4.19 Fluid circuit (example 4.11)

The first step of the mathematical approach is to determine the differential equation that describes the flow, q(t). To do this, consider the analogous circuit diagram shown in Fig. 4.19(b). By KVLA, we know that:

$$\left[p(t) - p_1\right] + \left[p_1\right] = p(t) \qquad \ldots (4.57)$$

By substituting the circuit component equations, we obtain:

$$Rq + \frac{1}{q} \int q \, dt = p(t) \qquad ..(4.58)$$

By differentiation of (4.58), we get:

$$RC \frac{dq(t)}{dt} + q(t) = C \frac{dp(t)}{dt} = C U_i(t) \qquad ..(4.59)$$

According to the mathematical procedure enunciated, we must use repeated integration until there is only one non-integral term on the left hand side. In this case, after one integration, we have:

$$RC \, q(t) + \int_{-\infty}^{t} q(t') \, dt' = C U_s(t) \qquad ..(4.60)$$

and we have already reached the condition of only one non-integral term on the left-hand side. We may, therefore, evaluate the equation at $t = 0^+$ with the result that:

$$RC \, q(0^+) = C \text{ and } q(0^+) = \frac{1}{R} \qquad ..(4.61a)$$

since

$$\int_{-\infty}^{t} q(t') \, dt' = 0 \text{ when } t = 0^+ \qquad ..(4.61b)$$

Of course, we would have reached this same result by simply replacing the capacitor in Fig. 4.19(b) by a short-circuit. Then the flow through the resulting initial condition circuit would have been $q(0^+) = 1/R$

directly.

Example 4.12

The mechanical mass-spring-damper model shown in Fig. 4.20(a) is subjected to a velocity input, $v_s(t) = U_s(t)$. Find the initial value of the force taken by the spring $[f_k(0^+)]$ and the initial value of the time derivative $\dfrac{df_k}{dt}(0^+)$.

Fig. 4.20(b) shows the analogous circuit representation of the model. Since we are interested in the spring force, we must derive the differential equation for this variable. By use of the circuit laws and some algebra, we may express the spring force as

$$\frac{d^2 f_k}{dt^2} + \frac{b}{m}\frac{df_k}{dt} + \frac{k}{m}f_k = k\frac{dv_s}{dt} + \frac{bk}{m}v_s \qquad ..(4.62)$$

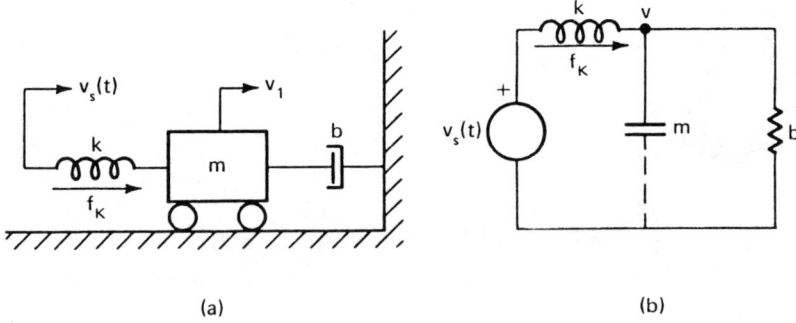

Figure 4.20 Mechanical circuit (example 4.12)

Basic Network Models

If we substitute $v_s = U_s(t)$ into (4.62), the result is:

$$\frac{d^2 f_k}{dt^2} + \frac{b}{m}\frac{df_k}{dt} + \frac{k}{m} f_k = k U_i(t) + \frac{bk}{m} U_s(t) \quad ..(4.63)$$

The mathematical approach requires integration until only one non-integral term remains on the left hand side. One integration produces:

$$\frac{df_k}{dt} + \frac{b}{m} f_k + \frac{k}{m} \int_{-\infty}^{t} f_k \, dt' = k U_s(t) + \frac{bk}{m} U_r(t)$$

$$..(4.64)$$

A second integration leads to:

$$f_k + \frac{b}{m}\int_{-\infty}^{t} f_k \, dt' + \frac{k}{m}\int_{-\infty}^{t}\left[\int_{-\infty}^{t'} f_k \, dt''\right] dt'$$

$$= k U_r(t) + \frac{bk}{m} U_p(t) \quad ..(4.65)$$

Now evaluation of the above result at $t = 0^+$ yields:

$$f_k(0^+) = 0 \quad ..(4.66)$$

since the integral terms are zero and the ramp and the parabolic terms are also zero at $t = 0^+$. To find $\frac{df_k}{dt}(0^+)$, we return to (4.64) after one integration and evaluate it at $t = 0^+$. The result is:

$$\frac{df_k}{dt}(0^+) = k \qquad \qquad ..(4.67a)$$

since we know that

$$f_k(0^+) = 0 \qquad \qquad ..(4.67b)$$

and the ramp function is zero at $t = 0^+$.

In the initial condition circuit method, the spring behaves like an open circuit. Thus $f_k(0^+)$ is immediately known to be zero. The equation of the spring is:

$$\frac{df_k}{dt} = k(v_s - v_1) \qquad \qquad ..(4.68)$$

At $t = 0^+$, $v_1(0^+) = 0$, because of the shorting of the mass. Thus:

$$\frac{df_k}{dt}(0^+) = k \qquad \qquad ..(4.69)$$

by this method also.

If impulsive functions remain on the right hand side after the appropriate number of integrations, the evaluation at $t = 0^+$ cannot be performed, since the left side integrals are not zero. In this circumstance, additional integrations are required to remove the impulsive functions. The subsequent evaluation at t =

Basic Network Models   243

0^+ then defines an integral of the variable, that is

$$\int_{-\infty}^{0^+} y(t') \, dt', \quad \int_{-\infty}^{0^+} \left[\int_{-\infty}^{0^+} y(t'') \, dt''\right] dt'$$

The initial conditions are then determined from this value by working back through successive integrations.

4.5 CIRCUIT MODELS FOR INITIAL ENERGY STORAGE

In Section 4.3, the energy storage elements had no stored energy prior to the application of the step function. With this restriction, the A-type storage element behaves initially as a short-circuit and the T-type storage element behaves initially as an open circuit.

The simplest way to treat storage elements with initial energy storage is to introduce a special circuit representation for such cases. Since these circuits must be based on physical equations, let us begin with the mathematical formulation of energy storage.

4.5.1 A-type Storage

Consider the elemental equation of the A-type storage device in similar form to (4.22):

$$av(t) = \frac{1}{C} \int_{-\infty}^{t} tv(t') \, dt' \qquad \ldots (4.70)$$

The use of the lower limit of integration as $t = -\infty$ includes the entire previous history of the storage device. We can separate the integral in (4.70) into two integrals. Thus:

$$av(t) = \frac{1}{C} \int_{-\infty}^{0} tv(t') \, dt' + \frac{1}{C} \int_{0}^{t} tv(t') \, dt' \quad \ldots (4.71)$$

We can interpret the first integral as the value of the across variable at $t = 0$ [that is, $av(0)$]. Hence, for all values of time greater than zero, the elemental equation is:

$$av(t) = av(0) \, U_s(t) + \frac{1}{C} \int_{0}^{t} tv(t') \, dt' \quad \ldots (4.72)$$

The first term on the right-hand side represents the value of the across variable due to initial energy storage. The second term is recognized as the standard expression for an A-type storage device. Now a circuit model or equivalent circuit must yield the same relation between $av(t)$ and $tv(t)$ as indicated in (4.72). This relation suggests a series connection of a source [to represent $av(0).U_s(t)$] and an A-type storage element. These components may be placed in either order. The corresponding circuit models are shown in Fig. 4.21. Each model has identically the same equilibrium equation as (4.72) provided that $C = C_e$. We, therefore, conclude that they are equivalent circuits that may be used to replace A-type elements with initial storage. The reason for designating the component in Fig. 4.21 as C_e

Basic Network Models

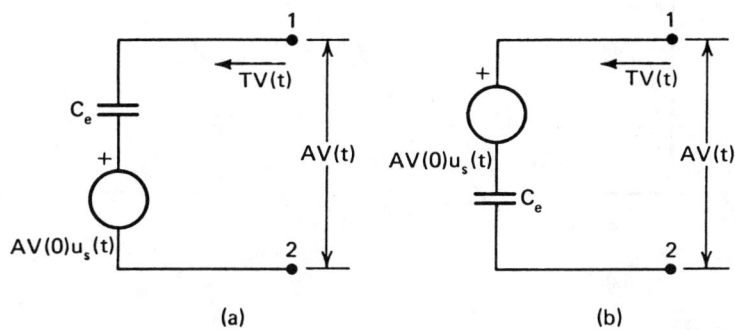

Figure 4.21 Circuit models for A-Type storage

is to emphasize that this component alone does not correspond to the original component. Since this is a true equivalent circuit, the terminals representing the actual component are marked 1 and 2 on Fig. 4.21.

Example 4.13

A tank filled with water is emptied through a long pipe as indicated in Fig. 4.22(a). If the initial height of the water is h_0, draw an equivalent circuit after the valve is opened. Locate the terminal on the circuit drawing that represents the pressure at the bottom of the tank.

The initially filled tank is a fluid capacitor with initial storage. We may choose either circuit model shown in Fig. 4.21 to represent this component. However, do not forget that one terminal of the equivalent circuit for a tank capacitor must always go

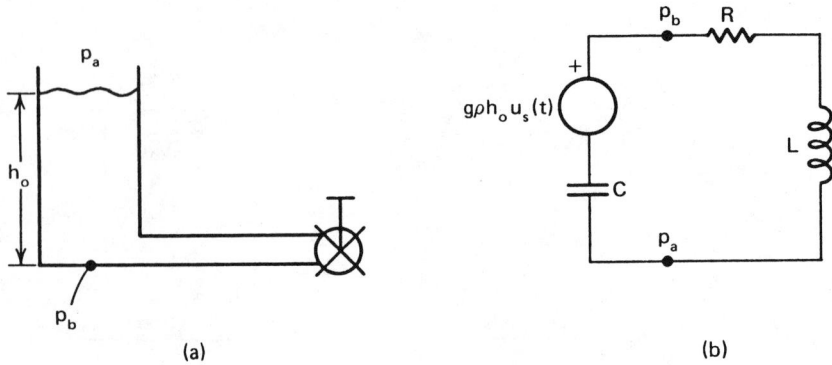

Figure 4.22 Fluid circuit (example 4.13)

to ground. The initial storage pressure is $\rho g h_o$ (where ρ is the water density and g is the acceleration due to gravity). Thus, the tank model has a source, $\rho g h_o \cdot U_s(t)$, in series with a capacitance. The general line model is a series combination of R and L. Fig. 4.22(b) shows the complete analogous circuit for the fluid arrangement in Fig. 4.22(a). The pressure at the bottom of the tank, p_b, is shown on the analogous circuit. Note that this terminal does not come between the source and the capacitor. There is no available terminal between these components.

4.5.2 T-type Storage

The corresponding model for the T-type storage element is obtained by performing the dual analysis. We begin with the basic elemental equation [Eq. (4.16)] in integral form. Thus:

Basic Network Models   247

$$tv(t) = \frac{1}{L} \int_{-\infty}^{t} av(t') \, dt' \qquad \ldots (4.73)$$

When expanded in the same manner as for the A-type element, (4.73) becomes:

$$tv(t) = \frac{1}{L} \int_{-\infty}^{0} av(t') \, dt' + \frac{1}{L} \int_{0}^{t} av(t') \, dt' \qquad \ldots (4.74)$$

The first integral is the value of the through variable at t = 0 [that is tv(0)]. For times greater than zero, we may express the elemental equation given in (4.74) as:

$$tv(t) = tv(0) \cdot U_s(t) + \frac{1}{L} \int_{0}^{t} av(t') \, dt' \qquad \ldots (4.75)$$

Now the first term on the right-hand side of (4.75) is the value of the through variable due to initial storage. The integral term is the standard relation for a T-type storage element. To form a circuit model from (4.75), note that each term represents a through variable. The sum of through variable terms must result in a parallel connection according to our basic circuit laws. Fig. 4.23 shows the circuit that produces the relation between tv(t) and av(t) given in (4.75) provided that L_e = L. Thus, a T-type storage element is modeled by an equivalent circuit in which there is a T-type element (having the same numerical parameter value) in parallel with a step

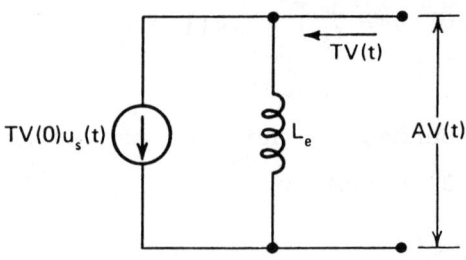

Figure 4.23 Circuit model for T-Type storage flow source.

Example 4.14

The spring-damper system shown in Fig. 4.24(a) is given an initial displacement, X_o, and then released. Obtain an equivalent circuit for the system and indicate the branch that carried the spring force, f_k.

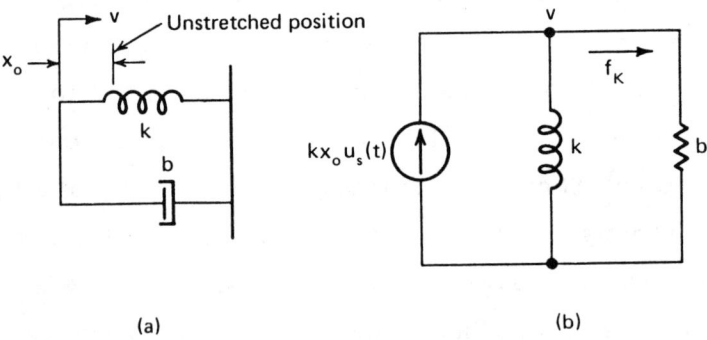

Figure 4.24 Mechanical circuit (example 4.14)

The spring in the mechanical arrangement is modeled by the parallel combination of a force source,

Basic Network Models   249

$kX_o \cdot U_s(t)$, and a spring component. The damper is also connected between the velocity terminal, v, and ground. Fig. 4.24(b) shows the analogous circuit diagram. Note that the initial displacement produces a reaction force that is in the direction of the assumed velocity. This force is modeled, therefore, as directed towards the node v. The spring force, f_k, is located also on Fig. 4.24(b). Note that this is not the through variable passing through the k element in the circuit. The spring force is the sum of the two force terms (the source contribution and the element contribution). This is a consequence of the circuit model for T-type storage.

The complete mathematical solutions for physical circuits with or without initial storage are discussed in Chapters V, VI and IX.

4.6 REALIZATION OF MATHEMATICAL OPERATIONS ON SINGULARITY FUNCTIONS BY THE CIRCUIT MODELS

In Chapter II, composite functions have been obtained as combinations of singularity functions which are, in turn, made possible by mathematical operations on such singularity functions. In this section, we shall show how such composite functions can actually be generated in the physical world using the basic circuit components presented so far.

It is shown so far in this chapter and in Chapter III that an Across Variable (Through Variable) is either a proportional or time-integral or time-derivative of a Through Variable (Across Variable). Therefore, the combinations of these mathematical operations have to be considered. Such circuits are very large in number and hence all of them cannot be covered here. Only representative circuits will be considered in order to illustrate the principles involved.

4.6.1 Realization of Proportional plus Integral

From this mathematical operation, one obtains the output $g_o(t)$ as

$$g_o(t) = A_1 g_i(t) + A_2 \int g_i(t) \, dt \qquad ..(4.76)$$

where $g_i(t)$ is the input,

and A_1 and A_2 are real constants.

If $g_i(t)$ is chosen as the Through Variable (that is, current), A_2 shall obviously correspond to a capacitor so that the output is an Across Variable (that is, voltage). Instead, if $g_i(t)$ is chosen as Across Variable (that is, voltage), A_2 shall obviously correspond to an inductor so that the output is Through Variable (that is, current). For both the cases, A_1 obviously corresponds to a resistor.

It is clear from the foregoing discussion that initially one has to make the choice regarding the input variable, which is dependent on the circuit. Then the

Basic Network Models

other variable is chosen as the output and the circuit is analyzed using the equilibrium equations. Illustrative examples are now given.

Example 4.15

Consider the circuit shown in Fig. 4.25, which is a series combination of a resistor, a capacitor and an ideal current source.

From KVL, we have

$e_{13}(t)$ = voltage across the current generator $i(t)$

$$= e_R(t) + e_C(t) \qquad \ldots (4.77a)$$

$$= R\, i(t) + \frac{1}{C} \int i(t)\, dt \qquad \ldots (4.77b)$$

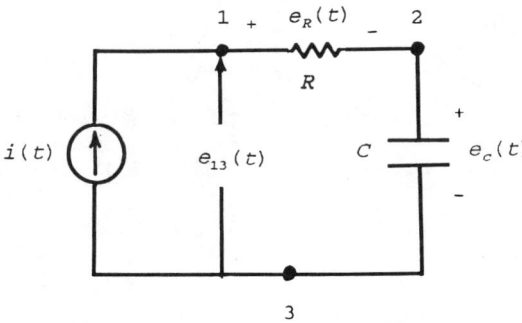

Figure 4.25 The circuit for example 4.15.

It is obvious that this combination realizes the operation of proportional plus integral, where the input is the current $i(t)$ and the output is the voltage $e_{13}(t)$.

At this point, if we consider Example 2.10, it is seen that

$$g_o(t) = e_{13}(t)$$
$$g_i(t) = i(t)$$
$$R = 2 \text{ ohms}$$
$$C = 1 \text{ farad}$$

..(4.78)

Considering the individual variables, it is observed that different components can have different waveforms even though these components are contained in the same network. The waveform across R will be $g_{op}(t)$; the waveform across C will be $g_{oi}(t)$. Also, the current generator waveform is $g_i(t)$, while the voltage developed across it has the waveform $g_o(t)$.

Example 4.16

Consider the circuit shown in Fig. 4.26, which is a parallel combination of a resistor, an inductor and a voltage generator.

In this circuit, $e(t)$ is the voltage of node (1) with respect to node (2) which is considered as the reference.

From KCL, we have

Basic Network Models

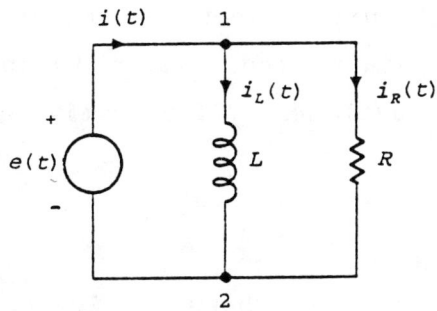

Figure 4.26 The circuit for example 4.16

$$i(t) = i_R(t) + i_L(t) \qquad ..(4.79a)$$

$$= \frac{e(t)}{R} + \frac{1}{L}\int e(t)\, dt \qquad .(4.79b)$$

This combination also realizes the operation of proportional plus integral, where the input is the voltage e(t) and the output is the current i(t).

4.6.2 Realization of Proportional plus Derivative

From this mathematical operation, one obtains the output $g_o(t)$ as

$$g_o(t) = A_1\, g_i(t) + A_2\, \frac{dg_i(t)}{dt} \qquad ..(4.80)$$

where $g_i(t)$ is the input,

and A_1 and A_2 are real constants.

254   Modeling and Analysis of Linear Physical Systems

Just as in the previous case, one has to make an initial choice regarding the input variable, which depends on the circuit. Then the other variable is chosen as the output and the circuit is analyzed using the equilibrium equations. Illustrative examples are now given.

Example 4.17

Consider the circuit shown in Fig. 4.27, which is a series combination of a resistor, an inductor and an ideal current source.

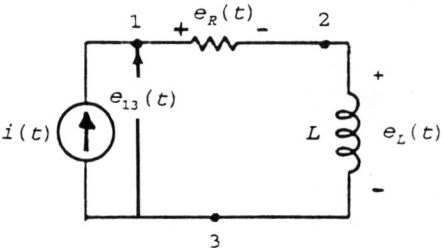

Figure 4.27. The circuit for example 4.17.

From KVL, we have

$e_{13}(t)$ = voltage across the current generator $i(t)$

$$= e_R(t) + e_L(t) \qquad \qquad ..(4.81a)$$

Basic Network Models

$$= R\, i(t) + L\, \frac{di(t)}{dt} \qquad \qquad ..(4.81b)$$

It is obvious that this combination realizes the operation of proportional plus derivative, where the input is the current $i(t)$ and the output is $e_{13}(t)$.

At this point, if we consider Example 2.11, it is seen that

$$e_{13}(t) = g_o(t)$$
$$i(t) = g_i(t) \qquad \qquad ..(4.82)$$
$$R = 1.5 \text{ ohms}$$
$$L = 1 \text{ henry}$$

In this case also, considering the individual variables, different components have different waveforms. The waveform across R will be $g_{op}(t)$; the waveform across L will be $g_{od}(t)$. Also the current generator has the waveform $g_i(t)$, while the voltage across it has the waveform $g_o(t)$.

Example 4.18

Consider the circuit shown in Fig. 4.28, which is a parallel combination of a resistor, a capacitor and a voltage generator.

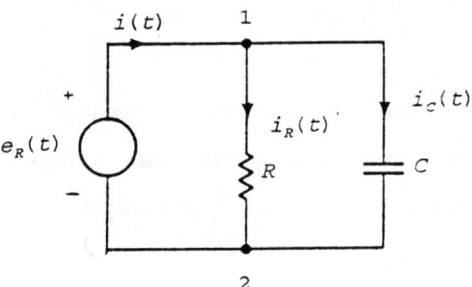

Figure 4.28. The circuit for example 4.18.

In this circuit, e(t) is the voltage at node (1) with respect to node (2) which is considered as the reference.

From KCL, we have

$$i(t) = i_R(t) + i_C(t) \qquad ..(4.83a)$$

$$= \frac{e(t)}{R} + C\frac{de(t)}{dt} \qquad ..(4.83b)$$

This combination also realizes the operation of proportional plus the derivative, where the input is the voltage e(t) and the output is the current i(t).

4.6.3 Realization of Proportional plus Derivative plus Integral

From this mathematical operation, one obtains the output as

Basic Network Models

$$g_o(t) = A_1 g_i(t) + A_2 \frac{dg_i(t)}{dt} + A_3 \int g_i(t) \quad \ldots (4.84)$$

where $g_i(t)$ is the input,

and A_1, A_2 and A_3 are real constants.

Here also, the initial choice regarding the input variable has to be made first. Then the other variable is chosen as the output and the circuit is analyzed using the equilibrium equations. Illustrative examples are now given.

Example 4.19

Consider the circuit shown in Fig. 4.29 which is a series combination of a resistor, an inductor, a capacitor and a current generator.

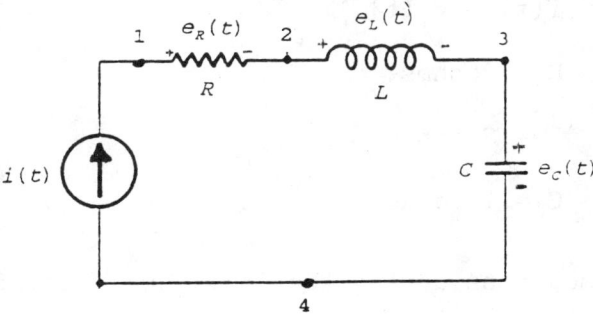

Figure 4.29. The circuit for example 4.19.

From KVL, we have

$e_{14}(t)$ = voltage across the current generator

$$= e_R(t) + e_L(t) + e_C(t) \qquad ..(4.85a)$$

$$= R\,i(t) + L\frac{di(t)}{dt} + \frac{1}{C}\int i(t)\,dt \qquad ..(4.85b)$$

It is obvious that this combination realizes the operation of proportional plus derivative plus integral, where the input is the current $i(t)$ and the output is $e_{14}(t)$.

At this point, if we consider Example 2.12, it is seen that

$$e_{14}(t) = g_o(t)$$
$$i(t) = g_i(t)$$
$$R = 2 \text{ ohms} \qquad ..(4.86)$$
$$L = 3 \text{ henrys}$$
$$C = 1 \text{ farad}$$

In this case also, considering the individual variables, different components have different waveforms. The waveform across R will be $g_{op}(t)$; the waveform across L will be $g_{od}(t)$ and that across C will be $g_{oi}(t)$. Also, the current generator has the waveform $g_i(t)$, while the voltage $e_{14}(t)$ has the waveform $g_o(t)$.

Example 4.20

Consider the circuit shown in Fig. 4.30, which is a parallel combination of a resistor, an inductor, a capacitor and a voltage generator.

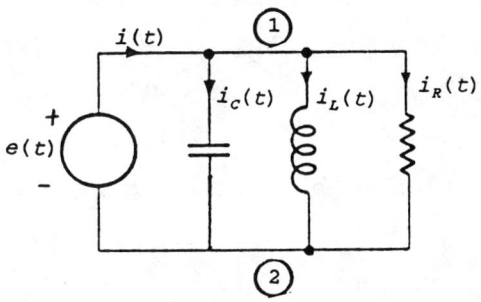

Figure 4.30. The circuit for example 4.20.

Here, e(t) is the voltage at node (1) with respect to node (2), which is the reference.

From KCL, we have

$$i(t) = i_R(t) + i_L(t) + i_C(t) \qquad ..(4.87a)$$

$$= \frac{e(t)}{R} + \frac{1}{L}\int e(t)\,dt + C\frac{de(t)}{dt} \qquad ..(4.87b)$$

This combination also realizes the operation of proportional plus derivative plus integral, where the input is the voltage e(t) and the output is the current i(t).

4.7 SUMMARY AND DISCUSSIONS

In this chapter, it is shown how a given system can be represented by an analogous electrical circuit diagram. With this knowledge, it is possible to formulate the mathematical equilibrium equations, starting from the two basic circuit laws KCL and KVL. Further, properties of storage elements have been discussed. Circuit models have been obtained when the storage elements contain initial stored energy. Some circuit realizations of mathematical operations of singularity functions have been discussed. It is clearly demonstrated that waveforms across elements in the same circuit can be different. However, it has to be emphasized that the two basic laws KCL and KVL have to be obeyed at any given instant of time and hence at all times.

4.8 PROBLEMS

4.1 Find an analogous electrical circuit for the fluid circuit shown in Fig. P4.1. If the constant flow pump were replaced by a constant pressure pump, which element could be removed without changing the pressure gain P_o/P_i?

4.2 A fluid system is shown in Fig. P4.2. How would you represent the system by an analogous electrical circuit?

Basic Network Models

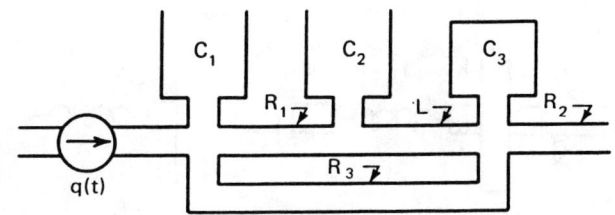

Figure P 4.1

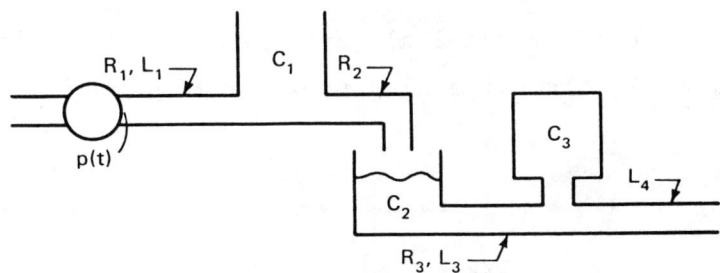

Figure P 4.2

4.3 Draw the electric circuit analog for the mechanical circuit shown in Fig. P4.3.

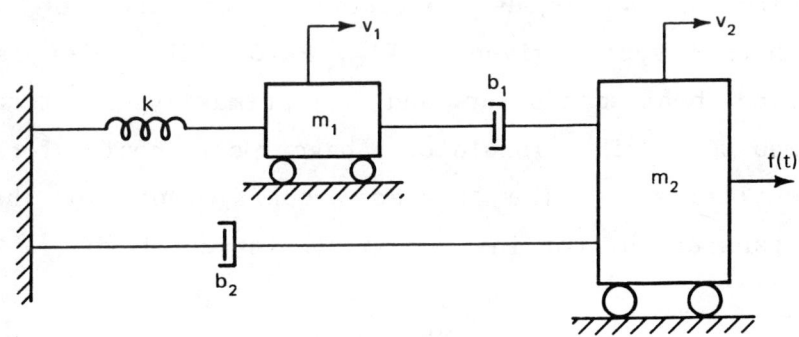

Figure P 4.3

4.4 Draw the electric circuit analog for the rotational mechanical circuit in Fig. P4.4.

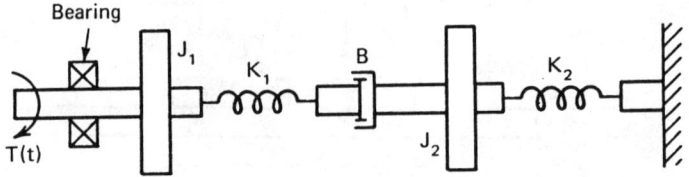

Figure P 4.4

4.5 Find the mechanical arrangement that has the electric circuit analog shown in Fig. P4.5.

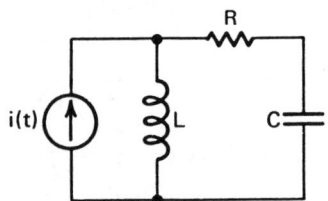

Figure P 4.5

4.6 Find an analogous electrical circuit for the thermal system given in Fig. P4.6. The water is a good heat conductor and is primarily a storage medium. The insulators have poor heat storage qualities. The surface coefficient of heat transfer, h, for insulation, I, is very high.

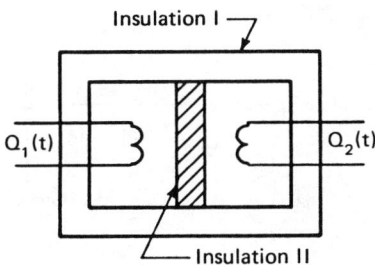

Figure P 4.6

4.7 For the electric circuit shown in Fig. P4.7, $R_1 = 2$ Ω, $R_2 = 5$ Ω, $C = 5$ μF and $L = 0.1$ H. The current generator produces the step input $i(t) = 5\,U_s(t)$, A. Find the following initial and final conditions:

(a) $i_{R_1}(0^+)$, $e_2(0^+)$, $i_{R_1}(\infty)$, $e_2(\infty)$.

(b) $\dfrac{de_2}{dt}(0^+)$, $\dfrac{di_L}{dt}(0^+)$

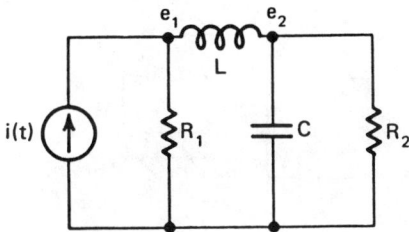

Figure P 4.7

4.8 Some initial and final conditions are desired for the electric circuit shown in Fig. P4.8. Find:

(a) $e_1(0^+)$, $e_2(\infty)$, $i_{R_1}(0^+)$, $i_{R_1}(\infty)$, $i_{R_3}(\infty)$

(b) $\dfrac{de_2}{dt}(0^+)$, $\dfrac{de_1}{dt}(0^+)$.

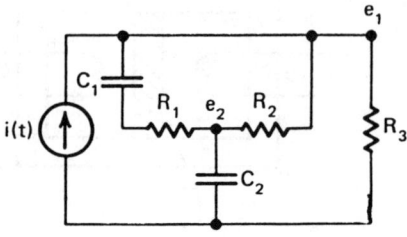

Figure P 4.8

4.9 In the mechanical circuit shown in Fig. P4.9, $b_1 = 1$ N.s/m, $b_2 = 3$ N.s/m, $k = 5$ N/m and $m = 20$ kg. The input velocity, $v(t) = 20\, U_s(t)$, m/s. Find:

(a) $f_m(0^+)$, $f_k(\infty)$, $v_m(\infty)$

(b) $\dfrac{dv_m}{dt}(0^+)$, $\dfrac{df_k}{dt}(0^+)$

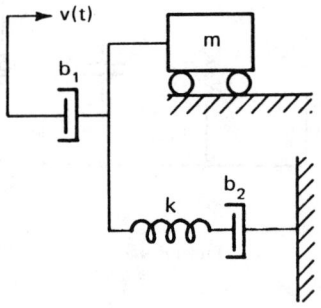

Figure P 4.9

4.10 The rotational mechanical circuit shown in Fig. P4.10 has $J = 100$ kg.m^2, $K = 20$ N.m and $B = 5$

Basic Network Models   265

N.m.s.   If the input torque $T(t) = 500\, U_s(t)$, N.m, find:

(a) $\omega_J(0^+)$, $\omega_J(\infty)$, $T_B(\infty)$

(b) $\dfrac{d\omega_J}{dt}(0^+)$

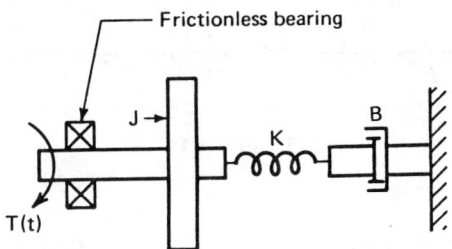

Figure P 4.10

4.11 The mechanical system shown in Fig. P4.11 is acted upon by a force, $f(t) = 10\, U_s(t)$, N. Initially, the 10 kg mass is moving away from the wall at a velocity of 1 m/s and initially there is no force in the spring. The damper $b = 4$ N.s/m and the spring $k = 1$ N/m. Find:

(a) The initial value of $\dfrac{df_k}{dt}$ and $\dfrac{df_b}{dt}$

(b) The final value of f_k and v_1.

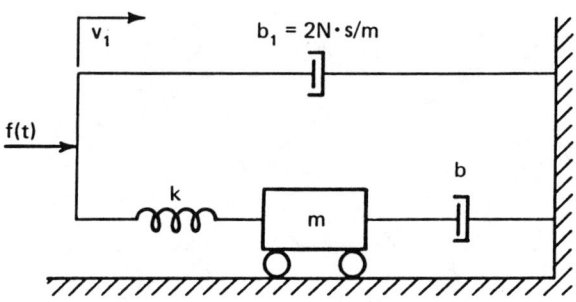

Figure P 4.11

4.12 An electric circuit is shown in Fig. P4.12. The current flowing through the 2-H inductor, $i_L(t)$, is known to be that shown in Fig. P4.12(b). Determine the source voltage, $e(t)$, that can produce the given $i_L(t)$. Sketch $e(t)$, and label all relevant voltages and times.

4.13 In the circuit shown in Fig. P4.13(a), $i_C(t)$, the current through C, is required to be as shown in Fig. P4.13(b). Determine $e(t)$ that will produce the required $i_C(t)$. Sketch the waveform and label the relevant voltages and times.

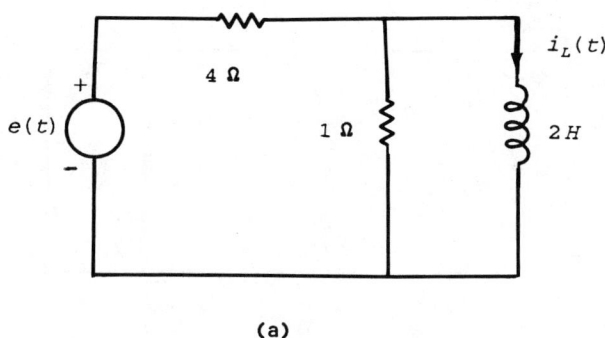

(a)

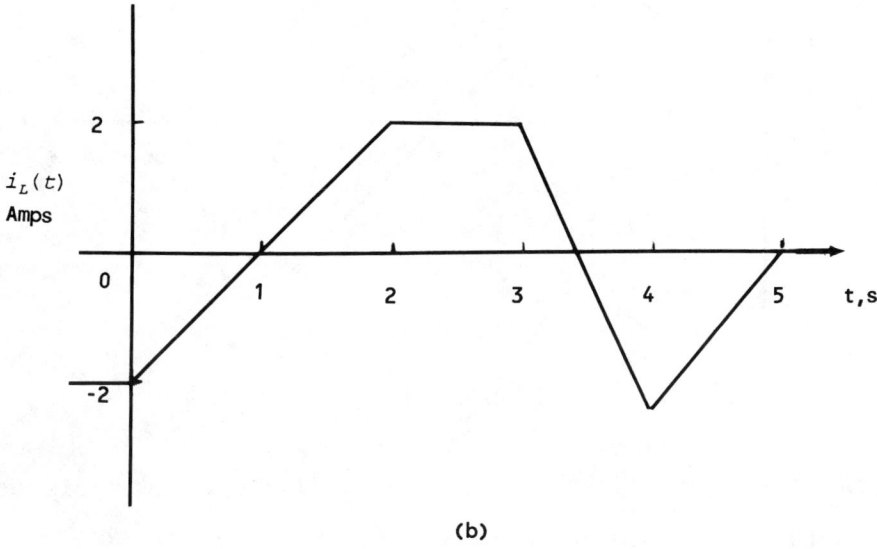

(b)

Figure P 4.12

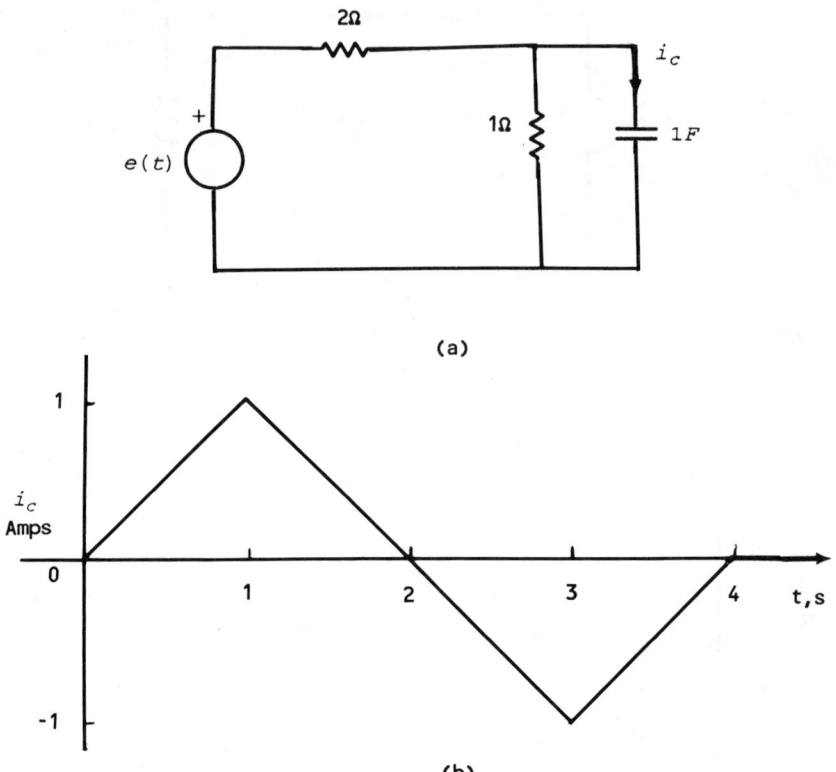

Figure P 4.13

4.14 If $i_R(t)$, the current flowing through the resistor in the network of Fig. P4.14(a), is as shown in Fig. P4.14(b), sketch:
(a) the voltage, $e(t)$,
(b) the current through the inductor, $i_L(t)$,
(c) the current through the capacitor, $i_C(t)$,
(d) the current drawn from the voltage source, $i(t)$.

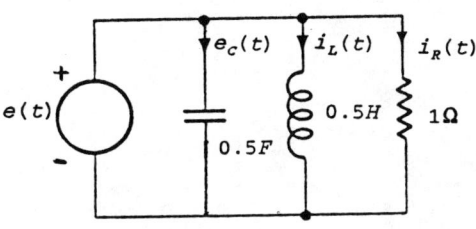

(a)

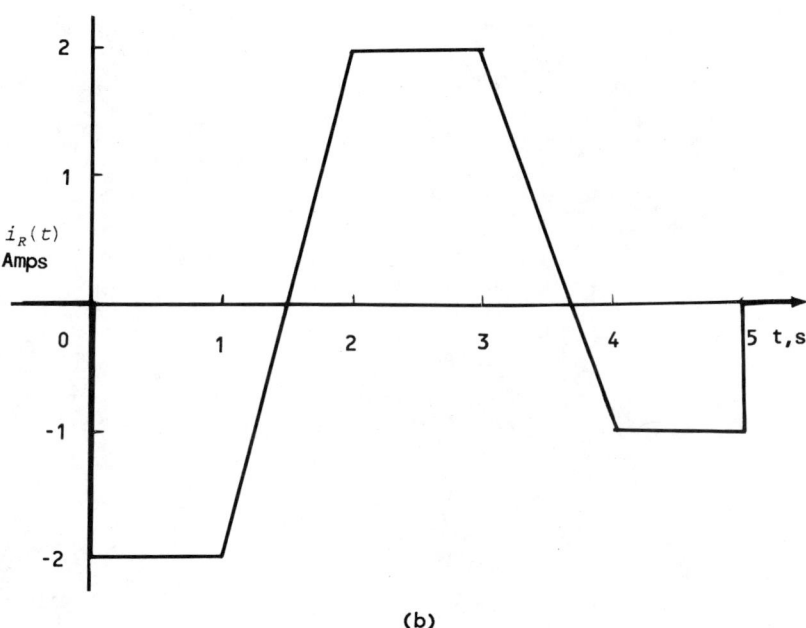

(b)

Figure P 4.14

CHAPTER 5

RESPONSE OF FIRST ORDER CIRCUITS

First order circuits normally contain only one storage element and may contain one or more dissipative elements. The storage element may be A-type or T-type. Irrespective of the total number of circuit elements, the circuit with one storage element will always be described by a first order differential equation.

In the first three sections of this Chapter, we consider circuits in which the storage element has no initial energy storage. This condition is sometimes referred to as **initially at rest**. In terms of the signal variables initially at rest means that the across variables of A-type storage elements and the through variables of T-type elements are equal to zero before the circuit is activated, that is, at t = 0⁻. Circuits with initial storage will be treated in Section 5.4.

The response of circuits is the time variation that occurs in a specific signal variable as the result of a time variation in another variable. Fig. 5.1 shows this concept in block diagram form. When the general input g(t) is operated on by a circuit, an output function y(t) is produced. The output is the response of the circuit to a particular input. The most general differential equation for first order circuits is:

$$\tau \frac{dy}{dt} + y = a_1 g(t) + a_2 \frac{dg(t)}{dt} \qquad ..(5.1)$$

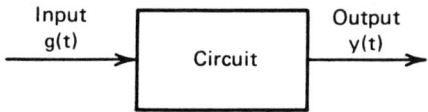

Figure 5.1 Block diagram of circuit

where τ, a_1 and a_2 are constants that depend on the values of the circuit elements. The constant, τ, is usually referred to as the **time constant** of the circuit. The physical significance of the time constant will become evident when we obtain solutions for (5.1). For circuits that have more than one dissipative element a_1 or a_2 may be zero. However, in the case of a circuit with precisely one storage element and one dissipative

Response of First Order Circuits 273

element, either a_1 or a_2 must be zero.

The following sections of this Chapter will develop the circuit response, $y(t)$, for a variety of input functions, $g(t)$.

5.1 STEP RESPONSE

Let us begin with a two-element circuit - one storage element and one dissipative element. If the storage element and the output variable associated with it are of the same type (e.g., a capacitor and the voltage across it or an inductor and the current through it), then $a_2 = 0$ in (5.1). In addition, the output variable in this type of circuit cannot change suddenly, when a step change is made in the input and therefore $y(0^+) = y(0^-) = 0$. We will call this a **Type 1** circuit. From (5.1) with $a_2 = 0$, the differential equation describing a Type 1 circuit is:

$$\tau \frac{dy}{dt} + y = a_1 \, g(t)$$
$$y(0^+) = 0$$
.. (5.2)

The input function $g(t)$ equals $G \, U_s(t)$, where G is the amplitude of the step change. Thus, (5.2) becomes

$$\tau \frac{dy}{dt} + y = a_1 \, G \, U_s(t) \qquad ..(5.3)$$

We will solve (5.3) for y in the classical manner as the sum of a homogeneous solution sometimes called complementary function), y_H and a particular solution (or particular integral), y_P. In engineering, we refer to the homogeneous solution as the natural response and the particular solution as the forced response. Thus:

$$y = y_H + y_P$$
$$\updownarrow \qquad \updownarrow$$
$$= y_{NATURAL} + y_{FORCED} \qquad ..(5.4)$$

The homogeneous solution is sometimes also called the unforced solution. The homogeneous differential equation is, therefore, obtained from the complete equation by setting all input functions equal to zero. The homogeneous and particular forms of (5.3) are:

$$\tau \frac{dy_H}{dt} + y_H = 0 \qquad ..(5.5a)$$

$$\tau \frac{dy_P}{dt} + y_P = a_1 \, G \, U_s(t) \qquad ..(5.5b)$$

Note that the addition of (5.5a) and (5.5b) yield (5.3). The homogeneous equation (5.5a) has variables which are separable and may be integrated directly to obtain the homogeneous solution. However, we do not use this method of solution here. We prefer rather a method that is applicable to this and more complicated circuits. In this method, we assume a trial solution for y_H in the form $A.\varepsilon^{rt}$ where A and r are constants to be determined. The rationale for this assumption is that the form $A.\varepsilon^{rt}$ can be made to represent all the expected response functions. Thus, if $y_H = A.\varepsilon^{rt}$ is substituted into (5.5a), the result is:

$$(\tau r + 1) A \varepsilon^{rt} = 0 \qquad ..(5.6)$$

Since the trial solution $A.\varepsilon^{rt} \neq 0$, (5.6) yields the condition that:

$$\tau r + 1 = 0 \qquad ..(5.7)$$

Eq. (5.7) is known as the **characteristic equation** of the circuit. The equation is satisfied, if

$$r = -1/\tau \qquad ..(5.8)$$

Thus, the homogeneous solution becomes:

$$y_H = A \varepsilon^{-t/\tau} \qquad ..(5.9)$$

To obtain the particular solution, let us consider (5.5b). When the input is a step function, the right-hand side is a constant for t > 0. Thus, the sum of the left hand terms must also be a constant. In effect, we need a function which added to its derivative yields a constant.

The particular solution that satisfies (5.5b) is:

$$y_P = a_1 \, G \, U_s(t) \qquad \qquad ..(5.10)$$

The correctness of this solution is readily verified by substituting it back into (5.5b). Another way of approaching the particular solution is to note that it is the same as the value of the variable after all the transients have disappeared. Remember that in physical circuits, the output is not changing a long time after a step has been applied. This means that dy/dt should be zero, when t approaches infinity. As a result, the particular solution for a circuit with a step input is equal to the value of the output variable at $t = \infty$. Thus we may also write:

$$y_P = y(\infty) \qquad \qquad ..(5.11)$$

The complete solution of (5.3) is the sum of the homogeneous and particular solutions given in (5.9) and (5.10) and is:

$$y(t) = A \, \varepsilon^{-t/\tau} + a_1 \, G \, U_s(t) \qquad \qquad ..(5.12)$$

Now that the complete solution has been obtained, we may evaluate the constant A from the initial condition $y(0^+) = 0$. Thus:

$$y(0) = A + a_1 \, G = 0 \qquad \qquad ..(5.13a)$$

from which we find that

$$A = -a_1 G \qquad \text{..(5.13b)}$$

The complete response of this first-order circuit to a step input is therefore:

$$y(t) = a_1 G [1 - \varepsilon^{-t/\tau}] U_s(t) \qquad \text{..(5.14)}$$

Fig. 5.2 shows the response described by (5.14). The ordinate is the normalized output $[y(t)/a_1G]$ and the abscissa is the number of time constants. At $t = \tau$ (one time constant), the normalized output is 0.632 of its final value. At two-time constants ($t = 2\tau$), the output is 0.865 and at three-time constants ($t = 3\tau$), it is 0.950.

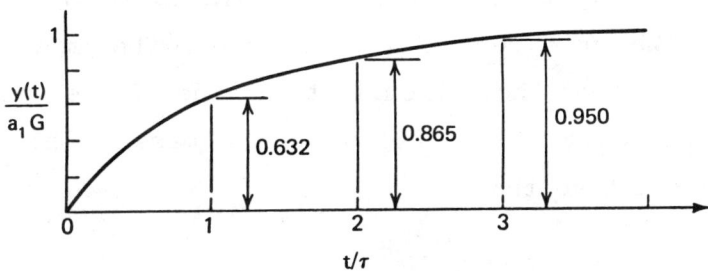

Figure 5.2 Normalized step response of type 1 circuit

We have presented a mathematical treatment of a Type 1 circuit. Now let us reinforce some of these ideas with an example of a particular physical circuit.

Example 5.1

A velocity step input is applied to the mechanical damper and mass arrangement shown in Fig. 5.3(a). If $v(t) = 2 U_s(t)$ m/s, $b = 5.0$ N.s/m and $m = 50$ kg, find $v_m(t)$.

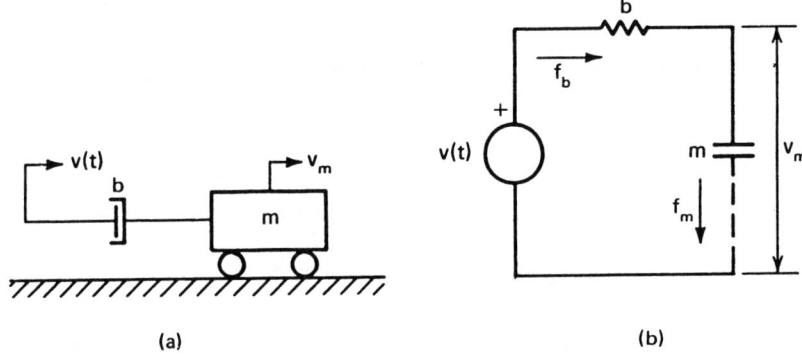

(a)   (b)

Figure 5.3 Mechanical circuit (example 5.1)

The first step is to place the physical arrangement into a circuit form. This is shown in Fig. 5.3(b). We may recognize from both the mechanical arrangement and the circuit that the force passing through the damper is applied to the mass. This is a series circuit so that:

$$f_m = f_b \qquad \ldots (5.15)$$

Now, from the elemental equations, we know that

$$f_m = m \frac{dv_m}{dt} \qquad \ldots (5.16a)$$

and

$$f_m = b(v - v_m) \qquad \ldots (5.16b)$$

If (5.16a) and (5.16b) are substituted in (5.15), we obtain

$$m \frac{dv_m}{dt} = b(v - v_m) \qquad \ldots (5.17)$$

which may be arranged in the form given in (5.2) for a Type 1 circuit as

$$(\frac{m}{b}) \frac{dv_m}{dt} + v_m = v(t) \qquad \ldots (5.18)$$

where (m/b) is the circuit time constant and $a_1 = 1$. We might have expected that this circuit would be of Type 1, since the output variable v_m is an across variable and the mass is an across storage element. For the particular component values given, $\tau = m/b = 50/5 = 10$ seconds, G = 2.0 m/s. The desired solution follows directly from (5.14) as:

$$v_m(t) = 2(1 - \varepsilon^{-t/10}) U_s(t), \text{ m/s} \qquad \ldots (5.19)$$

The mass starts from rest and ultimately reaches the velocity of the input. After 10 seconds (1 time constant), the mass is moving at 0.632 of the final velocity or 1.264 m/s.

The solution presented above is based on the mathematical development. Let us try the more physical approach of adding the natural and forced responses to obtain the same solution. The physical approach may be summarized in four steps:

(1) Find the homogeneous solution. For the first order circuit, this solution is always

$$y_H = A\,\varepsilon^{-t/\tau} \qquad \ldots(5.20)$$

For the mass-damper system,

$$v_{mH} = A\,\varepsilon^{-bt/m} \qquad \ldots(5.21)$$

(2) Find the particular solution. This is the value of the variable as t approaches infinity and can be found from the circuit by shorting T-type storage element and opening A-type storage element. For the mass-damper, the mass acts as an open-circuit and therefore

$$v_m(\infty) = v(\infty) = 2 \text{ m/s} \qquad \ldots(5.22)$$

(3) Find the initial condition of the variable. Immediately upon application of a step, the T-type application of a step, the T-type elements are open and A-type elements are short. The mass thus behaves initially as a short-circuit and $v_m(0^+) = 0$.

(4) Add the homogeneous and particular solutions together to get the complete solution

$$v_m(t) = A\,\varepsilon^{-bt/m} + 2 \qquad \ldots(5.23)$$

Response of First Order Circuits

and then evaluate A from the initial condition $v_m(0^+) = 0$. Thus

$$v_m(t) = 2(1 - \varepsilon^{-t/10}) U_s(t) \qquad ..(5.24)$$

represents the complete solution as before.

In this example, the utility and advantage of the physical approach is not obvious. When we deal with responses that depend on the derivatives of inputs, the benefits of the physical approach will become clear.

Let us continue with another two element circuit. However, in this case, the storage element and the output variable associated with it are of different types (e.g., a capacitor and the current through it or an inductor and the voltage across it). As a result, $a_1 = 0$ and (5.1) may be written as:

$$\tau \frac{dy}{dt} + y = a_2 \frac{dg(t)}{dt} \qquad ..(5.25)$$

We call this a **Type 2** circuit. In this circuit, the output variable, $y(t)$, can change suddenly so that in general $y(0^+) \neq 0$. If

$$g(t) = G U_s(t) \qquad ..(5.26)$$

then (5.25) becomes

$$\tau \frac{dy}{dt} + y = a_2 G U_i(t) \qquad ..(5.27)$$

Thus the application of a step input to the Type 2 circuit leads to a differential equation with an impulse function on the right hand side. We will first obtain a mathematical solution of (5.27). Then, in Example 5.2 which follows, we will use the physical approach.

If we integrate (5.27) with respect to time, the result is:

$$\tau \frac{d}{dt}\left[\int_0^t y\, dt'\right] + \left[\int_0^t y\, dt'\right] = a_2\, G\, U_s(t)$$

..(5.28)

Suppose we now define the new variable, $z(t)$, where

$$z(t) = \int_0^t y\, dt' \qquad ..(5.29)$$

When (5.28) is expressed in terms of z, we obtain:

$$\tau \frac{dz}{dt} + z = a_2\, G\, U_s(t) \qquad ..(5.30)$$

Furthermore, we recognize that from (5.29), $z(0^+) = 0$, since the upper and the lower limits of integration have the same value. Eq. (5.30) is now precisely the same form as (5.3). Therefore, the solution for z is the same as that given in (5.14) for y, namely:

Response of First Order Circuits

$$z(t) = a_2 G (1 - \varepsilon^{-t/\tau}) U_s(t) \qquad ..(5.31)$$

This is a solution for $z(t)$ and we seek a solution for $y(t)$. The relation between the functions is given in (5.29). Therefore, from (5.29) and (5.31), we find that:

$$y(t) = \frac{dz(t)}{dt}$$

$$= \frac{a_2 G}{\tau} \varepsilon^{-t/\tau} U_s(t) + a_2 G \left[1 - \varepsilon^{-t/\tau} \right] U_i(t)$$

$$..(5.32)$$

The last term on the right-hand side of (5.32) is identically equal to zero. This may be explained by noting that any function $g(t)$ multiplied by an impulse $U_i(t - t_o)$ is:

$$g(t) \cdot U_i(t - t_o) = g(t_o) \cdot U_i(t - t_o) \qquad ..(5.33)$$

Thus:

$$(1 - \varepsilon^{-t/\tau}) U_i(t) = U_i(t) - \varepsilon^{-t/\tau} u_i(t)$$
$$= U_i(t) - (1) U_i(t) = 0 \quad ..(5.34)$$

Thus the solution for the step input in a Type 2 circuit is:

$$y(t) = \frac{a_2 G}{\tau} \varepsilon^{-t/\tau} U_s(t) \qquad ..(5.35)$$

Fig. 5.4 shows the normalized step response described by (5.35). The output variable changes suddenly at t = 0 and thereafter has an exponential decay. The following example pertains to a Type 2 circuit.

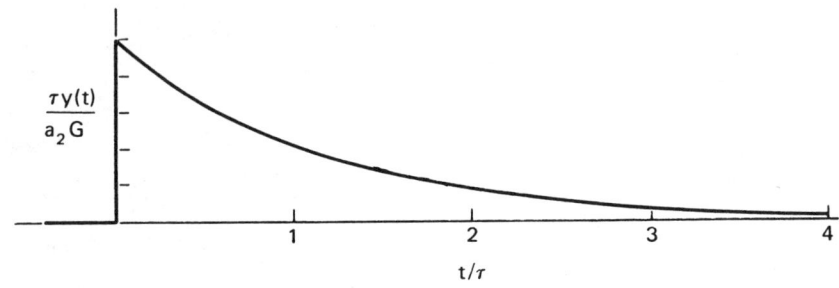

Figure 5.4 Normalized step response of type 2 circuit

Example 5.2

A pump supplies water to a drum through a long line as shown in Fig. 5.5(a). The pump is suddenly turned on and may be modeled as $p(t) = 2(10^4) \cdot U_s(t)$, Pa. The line has a fluid resistance $R = 500$ N·s/m^5 and the drum has a fluid capacitance $C = 0.8 \cdot (10^{-2})$ m^5/N. Find an expression for the flow rate into the drum $q(t)$.

Fig. 5.5(b) shows the circuit that represents the physical configuration of the problem. The drum is a fluid capacitance and therefore an A-type storage device. The output variable, $q(t)$, on the other hand, is a through variable. We expect, then, a governing equation of the form given in (5.25) for a Type 2 circuit. From the compatibility relations, the circuit in Fig. 5.5(b) yields:

Response of First Order Circuits

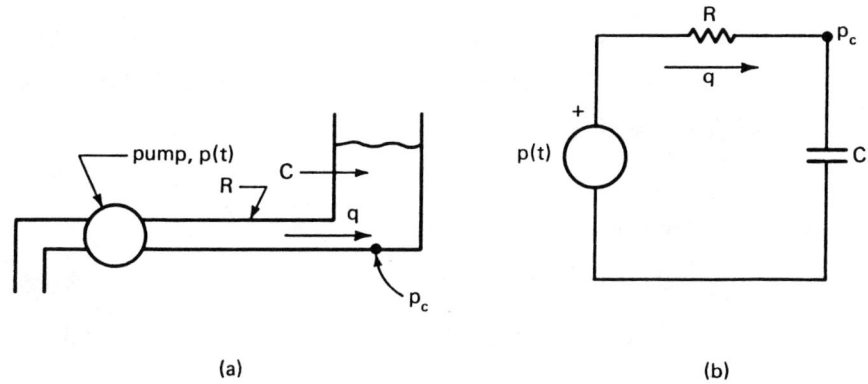

Figure 5.5 Fluid circuit (example 5.2)

$$\left[p(t) - p_c\right] + [p_c] = p(t) \qquad \ldots (5.36)$$

From the elemental equations

$$p(t) - p_c = Rq \qquad \ldots (5.36a)$$

and

$$p_c = \frac{1}{C}\int q\, dt \qquad \ldots (5.36b)$$

we have

$$Rq + \frac{1}{C}\int q\, dt = p(t) \qquad \ldots (5.37)$$

If we take the derivative of the above equation, the result is:

$$RC\frac{dq}{dt} + q = C\frac{dp(t)}{dt} \qquad \ldots (5.38)$$

This equation is in the form of (5.25) with $\tau = RC$

$= (500)(0.8)(10^{-2}) = 4$ seconds and $a_2 = C = 0.8(10^{-2})$ m^5/N. Thus we could write the expression for $q(t)$ directly from (5.35). However, let us follow the steps of the physical approach.

(1) The homogeneous solution is

$$q_H = A\ \varepsilon^{-t/RC} \qquad ..(5.39)$$

(2) The particular solution corresponds to

$$q(\infty) = 0 \qquad ..(5.40)$$

After a long time, the drum capacitance becomes an open circuit. From a physical viewpoint, the pressure at the bottom of the drum increases as the drum fills. Ultimately, a pressure is reached that is sufficiently high to prevent further flow passing from pump to drum.

(3) We have to determine the initial condition. From the circuit [Fig. 5.5(b)], we recognize that the drum acts as a short circuit for sudden changes. Thus instantaneously the circuit flow depends only on the resistance and pump pressure. Thus

$$q(0^+) = \frac{p(0^+)}{R} = \frac{2(10)^4}{500} = 40\ m^3/s \qquad ..(5.41)$$

Response of First Order Circuits

(4) The complete solution is the sum of the homogeneous and particular solutions and hence

$$q(t) = A\varepsilon^{-t/RC} + 0 \qquad \qquad ..(5.42)$$

and A may be evaluated from the initial condition so that:

$$q(t) = 40\,\varepsilon^{-t/4}\,U_s(t) \qquad \qquad ..(5.43)$$

This is the same result that would be obtained from (5.35).

Let us return now to the general differential equation for a first-order circuit given in (5.1).

$$\tau \frac{dy}{dt} + y = a_1\,g(t) + a_2 \frac{dg(t)}{dt} \qquad \qquad ..(5.1)$$

We may decompose (5.1) into equations that have already been solved.

$$\tau \frac{dy_1}{dt} + y_1 = a_1\,g(t) \qquad \qquad ..(5.44a)$$

$$\tau \frac{dy_2}{dt} + y_2 = a_2 \frac{dg(t)}{dt} \qquad \qquad ..(5.44b)$$

$$y_1 + y_2 = y \qquad \qquad ..(5.44c)$$

The step response solutions for (5.44a) and

(5.44b) are given in (5.14) and (5.35) respectively as:

$$y_1(t) = a_1 G \left[1 - \varepsilon^{-t/\tau}\right] U_s(t) \quad \ldots (5.46a)$$

and
$$y_2(t) = \frac{a_2 G}{\tau} \varepsilon^{-t/\tau} U_s(t) \quad \ldots (5.46b)$$

The complete step response solution for the general first order circuit is, therefore, the sum of (5.46a) and (5.46b). Hence,

$$y(t) = \left[a_1 G + (\frac{a_2}{\tau} - a_1)G \, \varepsilon^{-t/\tau}\right] U_s(t) \quad \ldots (5.47)$$

We will make use of another example to show that the physical approach produces the same result.

Example 5.3

Fig. 5.6 shows an electrical circuit which has a voltage source $e(t) = E.U_s(t)$. Find the current $i(t)$ through resistor R_1.

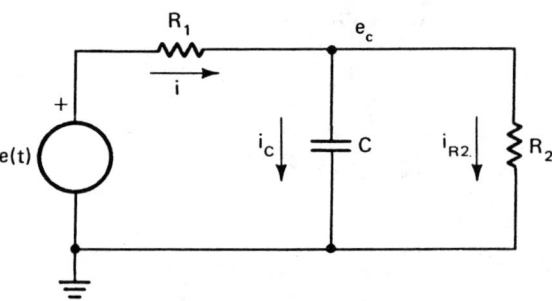

Figure 5.6 Electrical circuit (example 5.3)

Response of First Order Circuits

From the compatibility relation (KVL), we may write

$$\left[e(t) - e_c\right] + e_c = e(t) \qquad ..(5.48)$$

This expression may be applied for the loop 1 including R_1 and C and also for the loop 2 including R_1 and R_2. Substitution of the elemental equations then yields:

For loop 1 $\longrightarrow$ $R_1 i + \frac{1}{C} \int i_c \, dt = e(t)$

$$..(5.49a)$$

For loop 2 $\longrightarrow$ $R_1 i + R_2 i_{R_2} = e(t)$

$$..(5.49b)$$

If the equation from loop 1 is differentiated with respect to time and multiplied by C, we may rewrite the above as:

For loop 1 $\longrightarrow$ $R_1 C \frac{di}{dt} + i_c = C \frac{de(t)}{dt}$

$$..(5.50a)$$

Similarly,

For loop 2 $\longrightarrow$ $\frac{R_1}{R_2} i + i_{R_2} = \frac{e(t)}{R_2}$

$$..(5.50b)$$

If we add these equations and note that

$$i = i_c + i_{R_2} \qquad ..(5.51)$$

we obtain the differential equation for this circuit as:

$$\frac{R_1 R_2}{R_1 + R_2} C \frac{di}{dt} + i = \frac{e(t)}{R_1 + R_2} + \frac{R_2 C}{R_1 + R_2} \frac{de(t)}{dt}$$

$$..(5.52)$$

This is in the form of (5.1) with $\tau = R_1 R_2 C/(R_1 + R_2)$, $a_1 = 1/(R_1 + R_2)$ and $a_2 = R_2 C/(R_1 + R_2)$. The solution can be found directly from (5.47) with $G = E$. In the physical approach, however, we merely determine i_H, $i(0^+)$ and $i(\infty)$. The instantaneous value of the current is

$$i(0^+) = \frac{E}{R_1} \qquad ..(5.53)$$

This is a consequence of the short circuit behavior of the capacitance at $t = 0^+$. After a long time, the capacitance becomes an open circuit so that

$$i(\infty) = \frac{E}{(R_1 + R_2)} \qquad ..(5.54)$$

The homogeneous solution of all first-order circuits is of the same form, specifically

Response of First Order Circuits

$$i_H = A\,\varepsilon^{-t/\tau} \qquad \qquad ..(5.55)$$

Thus, the current i(t) may be expressed as:

$$i(t) = i_H + i(\infty)$$

$$= A\,\varepsilon^{-t/\tau} + \frac{E}{(R_1 + R_2)} \qquad ..(5.56)$$

Now, from the initial condition,

$$i(0^+) = \frac{E}{R_1} \qquad \qquad ..(5.57a)$$

we may evaluate A as:

$$i(0^+) = A + \frac{E}{(R_1 + R_2)} = \frac{E}{R_1} \qquad ..(5.57b)$$

and therefore

$$A = \frac{R_2 E}{R_1(R_1 + R_2)} \qquad \qquad ..(5.57c)$$

Thus the complete step response of the circuit in (5.52) is:

$$i(t) = \left[\frac{E}{R_1 + R_2} + \frac{R_2 E}{R_1(R_1 + R_2)}\,\varepsilon^{-t/\tau}\right] U_s(t) \quad ..(5.58a)$$

$$= \frac{E}{R_1 + R_2}\left[1 + \frac{R_2}{R_1}\,\varepsilon^{-t/\tau}\right] U_s(t) \qquad ..(5.58b)$$

Fig. 5.7 shows the normalized step response for this circuit. The magnitude of the current depends on the ratio R_2/R_1.

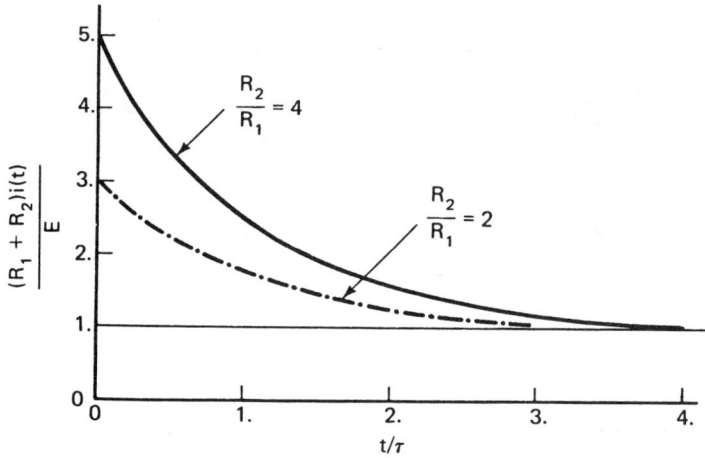

Figure 5.7 Normalized step response of electrical circuit (example 5.3)

5.2 RAMP, PARABOLIC AND IMPULSE RESPONSES

In the previous section, we have developed the response of a first-order circuit to a step input. Now, the response, $y(t)$, to any general input function $g(t)$ can also be obtained by direct mathematical solution of (5.1). However, when the input function is related to the step function by integration or differentiation, a simpler method is possible. In these cases, we may make use of a property of linear differential equations to determine the response without solving the equations. To illustrate the rationale behind this procedure, suppose we designate the unit step response for the general first-order circuit as y_s. Then we may rewrite (5.1) as:

Response of First Order Circuits

$$\tau \frac{dy_s}{dt} + y_s = a_1 U_s(t) + a_2 U_i(t) \quad \ldots (5.59)$$

If we subject the same circuit to a ramp input and call the response, y_r, then (5.1) becomes:

$$\tau \frac{dy_r}{dt} + y_r = a_1 U_r(t) + a_2 U_s(t) \quad \ldots (5.60)$$

Now, if (5.59) is integrated with respect to time, the result is:

$$\tau \frac{\left[\int_0^t y_s \, dt'\right]}{dt} + \left[\int_0^t y_s \, dt'\right] = a_1 U_r(t) + a_2 U_s(t)$$

$$\ldots (5.61)$$

Let us compare (5.60) and (5.61). The right hand sides are identical. Thus the left-hand sides must also be equal to each other. This condition is satisfied only if

$$y_r = \int_0^t y_s \, dt' \quad \ldots (5.62)$$

The result given in (5.62) shows that the ramp response of a linear circuit is equal to the integral of the circuit step response. We may generalize this concept. Fig. 5.8(a) shows a block diagram in which the input $U_s(t)$ produces the output response $y_s(t)$. Fig. 5.8(b) shows the circuit with ramp input $U_r(t)$ and output response $y_r(t)$. Since the input in Fig. 5.8(b) is the integral of the input in Fig. 5.8(a), the output of Fig. 5.8(b) is the integral of the output of Fig. 5.8(a). This is merely a restatement of what we have

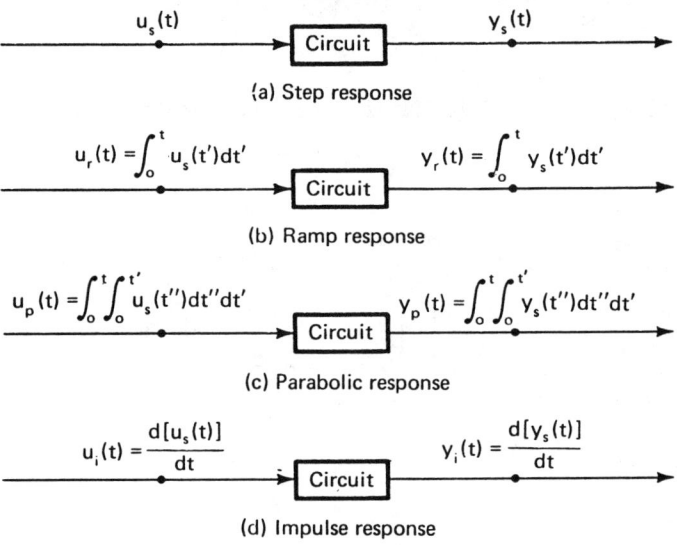

Figure 5.8 Integral and differential property of linear circuits

proved in (5.62). In a similar way, the parabolic and impulse responses may be related to the step response. Fig. 5.8(c) shows the circuit with an input $U_p(t)$. The output response in this case, y_p, is the double integral of the step response or the single integral of the ramp response. The circuit with impulse function input, $U_i(t)$, is shown in Fig. 5.8(d). The input is the derivative of a step input. Therefore, the corresponding impulse response, $y_i(t)$, is the derivative

Response of First Order Circuits

of the step response.

We may now apply this integral and differential property of linear differential equations to find the ramp, parabolic and impulse responses of the general first-order circuit. Eq. (5.47) gives the step response as:

$$y_s(t) = \left[a_1 + \left\{\frac{a_2}{\tau} - a_1\right\}\varepsilon^{-t/\tau}\right] U_s(t) \quad \ldots (5.63)$$

Thus, the general ramp response is:

$$y_r(t) = \int_0^t y_s(t')\, dt'$$

$$= \left[a_1 t + (a_2 - a_1\tau)(1 - \varepsilon^{-t/\tau})\right] G\, U_s(t) \quad \ldots (5.64)$$

The parabolic response is:

$$y_p(t) = \int_0^t y_r(t')\, dt' = \int_0^t \int_0^{t'} y_s(t'')\, dt''\, dt'$$

$$= \left[\frac{a_1 t^2}{2} + (a_2 - a_1\tau)(t + \tau - \tau\varepsilon^{-t/\tau})\right] G\, U_s(t)$$

$$\ldots (5.65)$$

and the impulse response is:

$$y_i(t) = \frac{d(y_s(t))}{dt}$$

$$= \frac{a_2 G}{\tau} U_i(t) - \frac{1}{\tau}\left\{\frac{a_2}{\tau} - a_1\right\} G \varepsilon^{-t/\tau} U_s(t)$$

..(5.66)

The solutions given in (5.63), (5.64), (5.65) and (5.66) are valid for the step, ramp, parabolic and impulse responses of all first-order circuits with initial rest conditions. In the following examples, however, we choose to extend the physical approach rather than to rely on these general formulas. When the input is not a step function, we add a fifth step to the physical approach. The first four steps remain the same and the complete step response is obtained. In step 5, we perform a mathematical operation on the step response to get the desired response.

Example 5.4

A mechanical spring and damper are connected together as shown in Fig. 5.9(a). The spring constant k = 25 N/m and the damping b = 75 N.s/m. Find the velocity of the connecting point, if the applied force, f(t) is (a) 50 $U_r(t)$, N and (b) 50 $U_i(t)$, N.

The analogous circuit for the mechanical arrangement is shown in Fig. 5.9(b). To find the differential equation, we apply the continuity equation at the node, v. Thus:

Response of First Order Circuits

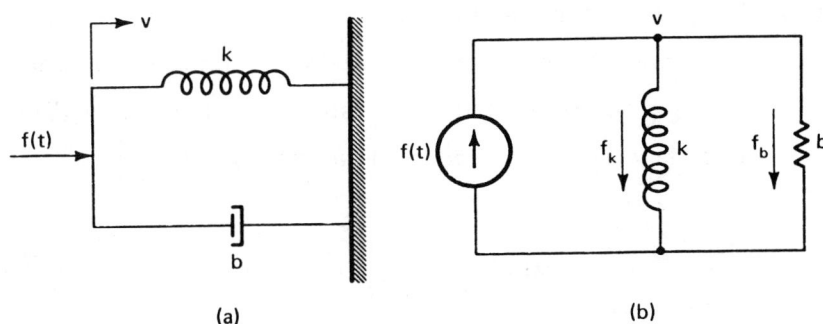

Figure 5.9 Mechanical circuit (example 5.4)

$$f_b + f_k = f(t) \qquad ..(5.67)$$

Now, by substituting the elemental equations into the above, we obtain:

$$bv + k \int v \, dt = f(t) \qquad ..(5.68)$$

This equation can be placed in the standard form by differentiating and dividing by k. We have:

$$\frac{b}{k}\frac{dv}{dt} + v = \frac{1}{k}\frac{df(t)}{dt} \qquad ..(5.69)$$

We will first find the response of the system to the step force, $f(t) = 50.U_s(t)$, N.

(1) The homogeneous solution

$$v_H = A\,\varepsilon^{-kt/b} = A\,\varepsilon^{-t/3} \qquad ..(5.70)$$

can be written by inspection.

(2) The particular solution:

A long time after a sudden change is made, the spring behaves like a short circuit. Thus the velocity is grounded and we may write:

$$v(\infty) = v_P = 0 \qquad \qquad ..(5.71)$$

(3) The initial condition:

The short-time performance of a spring is similar to an open-circuit. For a sudden change, the damper takes all the force instantaneously. We may, therefore, specialize the elemental equation of the damper at $t = 0^+$ as:

$$f_b(0^+) = f(0^+) = bv(0^+) \qquad ..(5.72a)$$

or $\qquad v(0^+) = 50/75 = 2/3 \qquad \qquad ..(5.72b)$

(4) The complete step response, v_s, is:

$$v_s(t) = A\,\varepsilon^{-t/3} + 0 \qquad \qquad ..(5.73)$$

When the initial condition is used, the value of A is determined as 2/3 and the complete solution is:

$$v_s(t) = (2/3)\,\varepsilon^{-t/3}\,U_s(t),\ m/s \qquad ..(5.74)$$

Response of First Order Circuits

(5) To find the ramp response, v_r, and the impulse response, v_i, recall that:

$$v_r = \int_0^t v_s(t') \, dt' \qquad \ldots (5.75a)$$

and

$$v_i = \frac{dv_s}{dt} \qquad \ldots (5.75b)$$

so that

$$v_r(t) = 2(1 - \varepsilon^{-t/3}) U_s(t), \text{ m/s} \qquad \ldots (5.76)$$

and $v_i(t) = (2/3) U_i(t) - (2/9) \varepsilon^{-t/3} U_s(t)$, m/s

$$\ldots (5.77)$$

These results agree with those that would be obtained from (5.64) and (5.66) with

$$\tau = 3, \; a_1 = 0, \; a_2 = 1/25 \text{ and } G = 50.$$

The responses are shown graphically in Figs. 5.10(a) and (b). For a ramp force input, the system approaches a constant velocity. When the force input is an impulse, the system moves instantaneously to some position where the spring is depressed. Thereafter, the system returns slowly to its static equilibrium position.

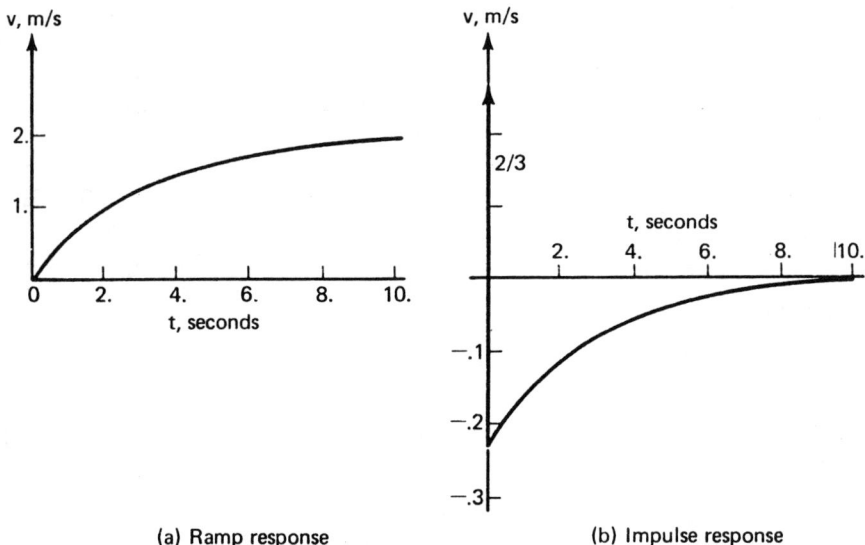

Figure 5.10 Responses for mechanical circuit in example 5.4

Example 5.5

Fig. 5.11 shows an RL series electric circuit. If R = 10 ohms and L = 1 H, find the current, i(t), for voltage inputs:

(a) $e(t) = 5\, U_r(t)$, volts, and
(b) $e(t) = 20\, U_p(t)$, volts.

To find the governing differential equation, we apply the compatibility condition around the circuit, which is

$$e_L + (e - e_L) = e \qquad \ldots (5.78)$$

Response of First Order Circuits

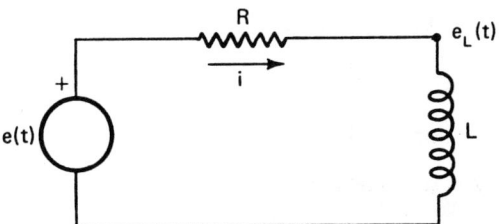

Figure 5.11 Electric circuit (example 5.5)

Substitution of the elemental relations leads directly to:

$$L \frac{di}{dt} + Ri = e(t) \qquad ..(5.79)$$

or in standard form:

$$\frac{L}{R} \frac{di}{dt} + i = \frac{1}{R} e(t) \qquad ..(5.80)$$

We follow the usual procedure now and find the solution of the circuit when $e(t) = U_s(t)$, volts.

(1) The homogeneous solution

$$i_H = A \varepsilon^{-Rt/L} = A \varepsilon^{-10t} \qquad ..(5.81)$$

is found directly since this is always the form for a first-order circuit.

(2) The particular solution:

After a long time following the step input, the inductor behaves as a short circuit. Thus:

$$i_p = i(\infty) = e(\infty)/R = 0.1 \qquad ..(5.82)$$

(3) The initial condition:

Initially the inductor behaves as an open circuit. Therefore,

$$i(0^+) = 0 \qquad ..(5.83)$$

(4) The complete step response is:

$$i_s(t) = A\,\varepsilon^{-10t} + 0.1 \qquad ..(5.84)$$

and with the initial condition used $A = -0.1$ so that we may write:

$$i_s(t) = 0.1(1 - \varepsilon^{-10t})\,U_s(t),\ A \qquad ..(5.85)$$

(5) (a) The response to $5U_r(t)$ is $i_r(t)$ and may be expressed as:

$$i_r(t) = 5\int_0^t i_s(t')\,dt' \qquad ..(5.86)$$

and when the integration operation on the step response is performed, we get

$$i_r(t) = 0.5(t - 0.1 + 0.1\,\varepsilon^{-10t})\,U_s(t) \qquad ..(5.87)$$

Response of First Order Circuits

This response is shown plotted in Fig. 5.12(a). Note that the asymptote in this form of ramp response always intersects the time axis at $t = \tau$.

(b) The response to $20\, U_p(t)$ is $i_p(t)$ and is defined as:

$$i_p(t) = 20 \int_0^t \int_0^{t'} i_s(t'')\, dt''\, dt' \qquad ..(5.88)$$

If the double integration is performed, the result is:

$$i_p(t) = 2\left[\frac{t^2}{2} - 0.1\,t + 0.01 - 0.01\,\varepsilon^{-10t}\right] U_s(t)$$

$$..(5.89)$$

The parabolic response is shown plotted in Fig. 5.12(b). The response may be checked from (5.64) and (5.66).

5.3 RESPONSE TO COMPOSITE FUNCTIONS

In the previous section, we observed one of the important properties of a linear system. Specifically if an input, $g(t)$, produces a response, $y(t)$, then other inputs that are related mathematically to the original input (that is, $dg(t)/dt$ or $\int g(t)\, dt$) produce responses with the same mathematical operation (that is, $dy(t)/dt$

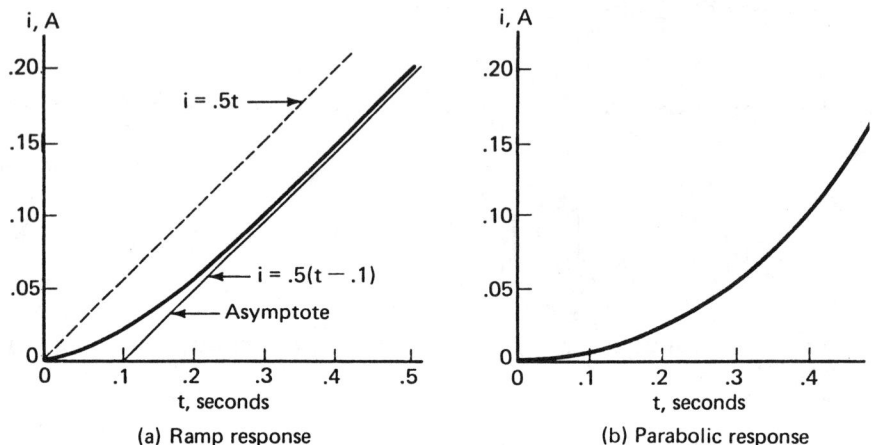

Figure 5.12 Responses for electric circuit in example 5.5

or $\int y(t)\, dt$).

Another important property of linear systems is superposition. Consider that the response $y_1(t)$ to an input $g_1(t)$ for the general first-order circuit is described by the differential equation:

$$\tau \frac{dy_1}{dt} + y_1 = a_1 g_1 + a_1 \frac{dg_1}{dt} \qquad \ldots (5.90)$$

Now the response $y_2(t)$ of this same system to another input $g_2(t)$ is formulated in a similar way as:

$$\tau \frac{dy_2}{dt} + y_2 = a_2 g_2 + a_2 \frac{dg_2}{dt} \qquad \ldots (5.91)$$

If we add (5.90) and (5.91), the result is:

$$\tau \frac{d(y_1 + y_2)}{dt} + (y_1 + y_2) = a_1(g_1 + g_2) + a_2 \frac{d(g_1 + g_2)}{dt}$$

..(5.92)

As a consequence of (5.92), note that if the input had been the function $\{g_1(t) + g_2(t)\}$, the corresponding output response would be $\{y_1(t) + y_2(t)\}$. Thus, if we apply a composite input to a linear circuit, we may use the component portions of the input separately to obtain an output for each. The complete response to a composite input is then the sum of the output component portions. In more general terms, we may state, for the above inputs and outputs, that an input $\{A_1 g_1(t) + A_2 g_2(t)\}$ will produce an output of $\{A_1 y_1(t) + A_2 y_2(t)\}$ where A_1 and A_2 are constants.

In addition to the above properties, we may also time shift the response of linear circuits. For example, if input g(t) causes output y(t), then input g(t-T) will cause output y(t-T). The shifting feature of linear circuits with zero initial conditions is one we will often use in conjunction with composite input functions. The following examples will clarify the superposition and shifting properties.

Example 5.6

Fig. 5.13(a) shows a first-order mass-damper system with the mass, m = 100 kg and the damper, b = 20 N.s/m. A force, f(t), of the form shown in Fig. 5.13(b) is applied to the system.

Determine the velocity of the mass at the end of 1 second, v(1), and at the end of 3 seconds, v(3).

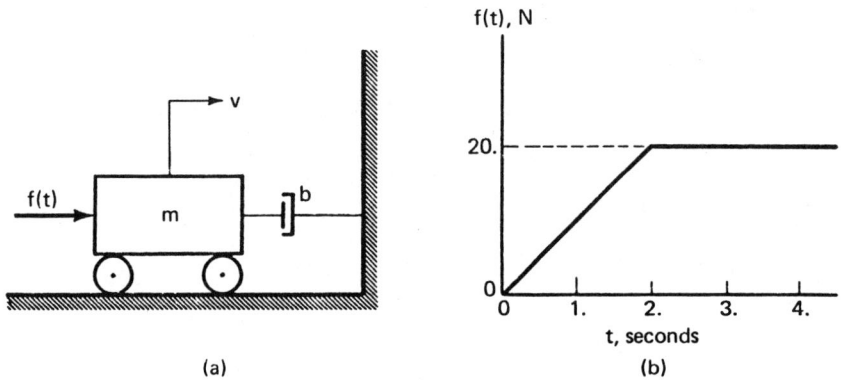

Figure 5.13 Mass-damper system (example 5.6)

The analogous circuit for this mechanical system is presented in Fig. 5.14. Remember that the mass always appears as a capacitance to ground.

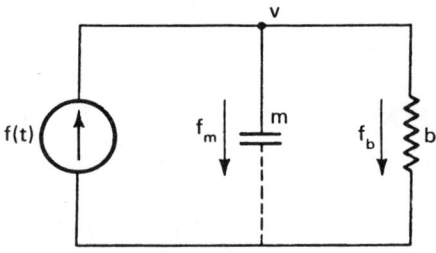

Figure 5.14 Analogous circuit for mass-damper system (example 5.6)

Response of First Order Circuits

The continuity condition (KCLA) at either node yields:

$$f_m + f_b = f(t) \qquad ..(5.93)$$

From the elemental relations, we know immediately that

$$f_m = 100 \, dv/dt \qquad ..(5.94a)$$

and $\qquad f_b = 20 \, v. \qquad ..(5.94b)$

The input function, $f(t)$, may be separated into the sum of two ramp functions,

$$f(t) = 10 \, U_r(t) - 10 \, U_r(t-2) \qquad ..(5.95)$$

Thus the system equation is:

$$5 \frac{dv}{dt} + v = \frac{f(t)}{20} \qquad ..(5.96)$$

As before, we begin by finding the response of the system to a step input, $f(t) = U_s(t)$. For the step input, the system equation is:

$$5 \frac{dv_s}{dt} + v_s = \frac{1}{20} U_s(t) \qquad ..(5.97)$$

Now with the above equation, the regular procedure will yield $v_s(t)$.

(1) The homogeneous solution is

$$v_H = A\,\varepsilon^{-0.2t} \qquad \qquad ..(5.98)$$

(2) The particular solution:

A capacitor does not pass steady current. It behaves as an open circuit, when the voltage across it is steady. The analogous mechanical component, mass, takes no force a long time after a sudden change. In this system, all the force supplied after a long time is used to overcome damping. As a result.

$$v_P = f(\infty)/b = 1/20 \qquad \qquad ..(5.99)$$

(3) The initial condition:

A mass initially at rest cannot suddenly have an initial velocity with a finite driving force and hence

$$v_s(0^+) = 0 \qquad \qquad ..(5.100)$$

(4) The complete step response is the sum of the homogeneous and particular solutions. Hence,

$$v_s(t) = A\,\varepsilon^{-0.2t} + \frac{1}{20} \qquad \qquad ..(5.101)$$

From the initial condition (5.100), we get

$$A = -1/20 \qquad \qquad ..(5.101)$$

Therefore, we have

$$v_s(t) = \frac{1}{20}(1 - \varepsilon^{-0.2t})\, U_s(t),\ \text{m/s} \qquad \ldots (5.102)$$

(5) (a) The response to $10\, U_r(t)$ is $v_1(t)$ where

$$v_1(t) = 10\int v_s(t')\, dt' \qquad \ldots (5.103a)$$

$$v_1(t) = 0.5\left[t + 5\varepsilon^{-0.2t} - 5\right] U_s(t) \qquad \ldots (5.103b)$$

(b) The response to $-10\, U_r(t-2)$ is $-v_1(t-2)$ by the shifting property of linear equations. Hence

$$-v_1(t-2) = -0.5\left[(t-2) + 5\varepsilon^{-0.2(t-2)} - 5\right] U_s(t-2)$$
$$\ldots (5.104)$$

This is the same function as $v(t)$ with all the t arguments replaced by $(t-2)$. The complete response to the given function, $f(t)$, is therefore the superposition of parts (a) and (b) or

$$v(t) = 0.5\left[t + 5\varepsilon^{-0.2t} - 5\right] U_s(t)$$
$$- 0.5\left[(t-2) + 5\varepsilon^{-0.2(t-2)} - 5\right] U_s(t-2)$$
$$\ldots (5.105)$$

(6) Evaluation of the response at the desired times of 1 and 3 seconds:

From the complete response, we obtain:

$$v(1) = 0.5\left[1 + 5\varepsilon^{-0.2} - 5\right] - 0 = 0.047 \text{ m/s}$$

..(5.106)

Note that the second term in the response does not contribute to v(1), because the output response can never come before the input which produced it.

$$v(3) = 0.5\left[3 + 5\varepsilon^{-0.6} - 5\right] - 0.5\left[1 + 5\varepsilon^{-0.2} - 5\right]$$

$$= 0.325 \text{ m/s} \qquad ..(5.107)$$

Example 5.7

In the series RC electric circuit shown in Fig. 5.15(a), the value of R is 100 ohms and the value of C is 0.1 F. Find the voltage drop across the resistor $e_R(t)$, when the voltage source e(t) is the rectangular pulse shown in Fig. 5.15(b). Compare this response to the response when $e(t) = 20\,U_i(t)$, volts.

From the continuity condition at the node, e_R, and the elemental equations, the differential equation for this circuit is:

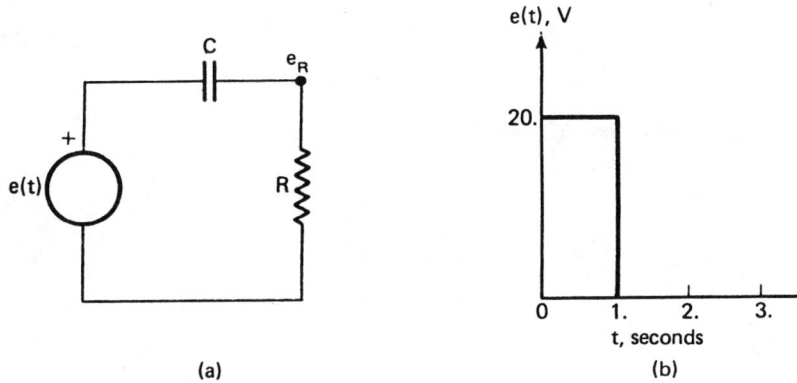

Figure 5.15 Electrical circuit (example 5.7)

$$10 \frac{de_R}{dt} + e_R = \frac{de(t)}{dt} \quad \quad ..(5.108)$$

The rectangular pulse shown in Fig. 5.15(b) may be described in terms of singularity functions as:

$$e(t) = 20\, U_s(t) - 20\, U_s(t-1) \quad \quad ..(5.109)$$

We start by finding the response of the circuit to the unit step function $e(t) = U_s(t)$.

(1) The homogeneous solution,

$$e_H = A\, \varepsilon^{-t/10} \quad \quad ..(5.110a)$$

(2) The particular solution,

$$e_P = 0 \quad \quad ..(5.110b)$$

(3) The initial condition,

$$e_R(0^+) = e(0^+) = 1 \qquad \ldots (5.110c)$$

(4) The complete solution,

$$e_R(t) = \varepsilon^{-t/10} U_s(t) \qquad \ldots (5.110d)$$

This solution is the response to a unit step function. The response to the rectangular pulse function given is, therefore, by superposition and shifting:

$$e_{RP}(t) = 20\, \varepsilon^{-t/10} U_s(t) - 20\, \varepsilon^{-(t-1)/10} U_s(t-1)$$
$$\ldots (5.111)$$

The response to the impulse function is:

$$e_{Ri}(t) = 20\, \frac{de_R}{dt} = 20\, U_i(t) - 2\, \varepsilon^{-t/10} U_s(t)$$
$$\ldots (5.112)$$

The rectangular pulse and impulse responses are shown plotted in Fig. 5.16. Both responses consist of two terms. The impulse response (solid line) has a positive impulse at t = 0 and a negative decaying exponential.

The rectangular pulse response (dashed line)

represents the difference between an exponential function and an exponential function delayed. At t = 1, the rectangular pulse changes discontinuously. Note that as time increases, the two responses approach each other. At t = 10 seconds, the difference between the

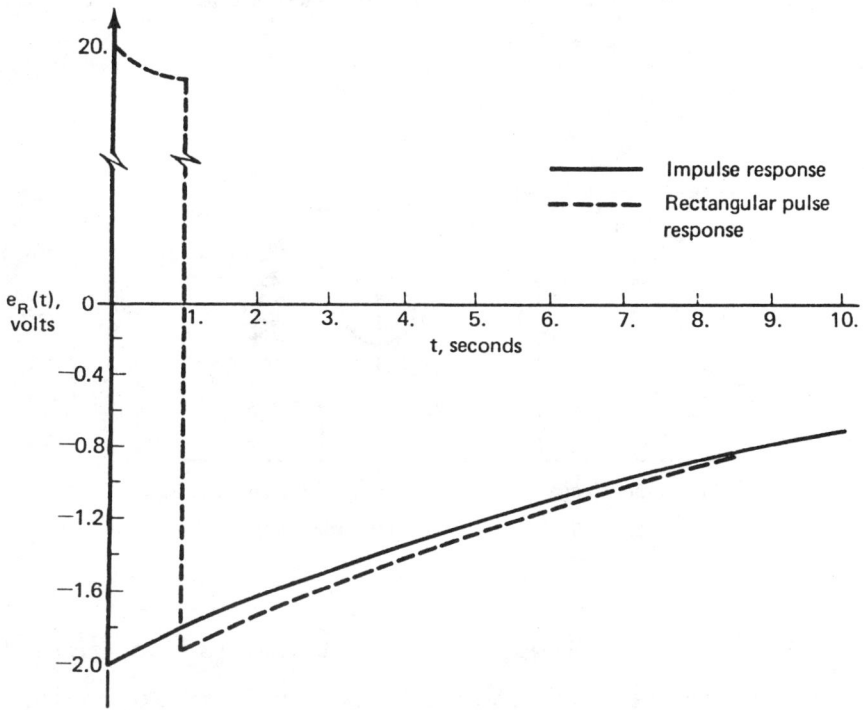

Figure 5.16 Comparison of impulse and rectangular pulse response (example 5.7)

responses is 0.038 volts. We may state, in general, that a rectangular pulse of short duration compared to the circuit time constant produces a response similar to an impulse of the same strength.

5.4 RESPONSE WITH INITIAL STORAGE

In the circuits previously considered, the storage elements had no initial energy storage. We now turn our attention to first-order circuits with energy storage. Although the treatment of initial storage is identical for all physical circuits, we will illustrate the principles involved by referring to the electrical signal variables, e and i.

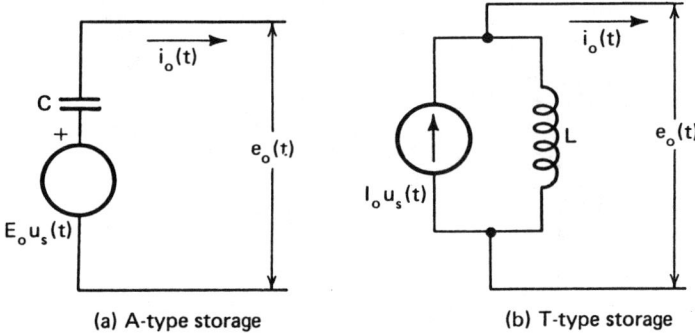

(a) A-type storage   (b) T-type storage

Figure 5.17 Circuit models for initial storage

Fig. 5.17 reviews the circuit models for initial storage that were developed in Chapter IV. For A-type storage, a potential source, $E_0 U_s(t)$, is placed in series with the capacitive component. For T-type storage, a flow source, $I_0 U_s(t)$, is placed in parallel with the inductive component. E_0 represents the magnitude of the initial voltage across the capacitor and I_0 represents the magnitude of the initial current through the inductor. When a component has initial storage, we replace the component by one of the circuit

Response of First Order Circuits 315

models for initial storage shown in Fig. 5.17.

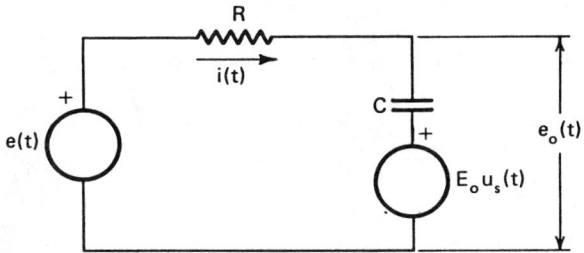

Figure 5.18 Series RC circuit with initial storage

Fig. 5.18 shows the equivalent circuit for a series RC circuit with initial storage. Note that the storage element (the capacitor) is replaced by a storage element and a source in series. Suppose we wish to determine the current response i(t), and the voltage response across the capacitor, $e_o(t)$, for a source input, e(t). To obtain these responses, we make use of the principle of superposition of two sources. The circuit is treated as if it had two separate sources:
(1) the input source, e(t), and

(2) the storage source, $E_0 U_s(t)$.
The effect of each source is then added to form the complete response.

Fig. 5.19(a) shows the input source acting alone. In this diagram, the A-storage source is zero or a short-circuit. The resulting circuit is then the circuit without initial storage that we have already analyzed. The circuit equations without storage are:

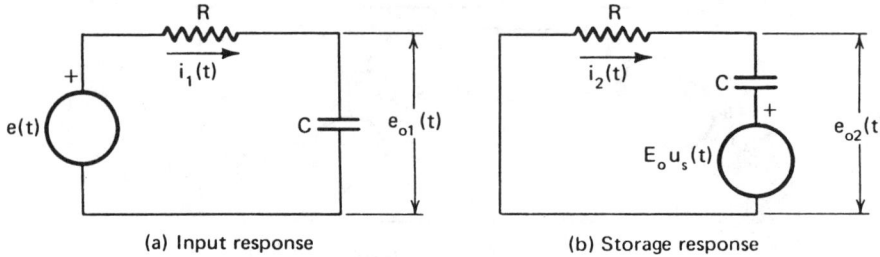

Figure 5.19 Separation of sources in series RC circuit with storage

$$RC \frac{de_{o_1}}{dt} + e_{o_1} = e(t) \qquad ..(5.113a)$$

$$RC \frac{di_1}{dt} + i_1 = C \frac{de(t)}{dt} \qquad ..(5.113b)$$

Fig. 5.19(b) shows the storage source acting alone. The input voltage source is zero in this case and, therefore, it has been replaced by a short circuit. The circuit equations without an input source are:

$$RC \frac{de_{o_2}}{dt} + e_{o_2} = RC \left[E_0 U_s(t) \right] = RC\, E_0\, U_i(t)$$

$$..(5.114a)$$

$$RC \frac{di_2}{dt} + i_2 = -C \frac{d}{dt}\left[E_0 U_s(t) \right] = -C\, E_0\, U_i(t)$$

$$..(5.114b)$$

Response of First Order Circuits

Now, by the superposition of sources, we have the complete response for the voltage as:

$$e_o = e_{o_1} + e_{o_2} \qquad \ldots (5.115a)$$

and for the current as:

$$i = i_1 + i_2 \qquad \ldots (5.115b)$$

The input response portion of the complete response has been dealt with in the previous sections and remains unchanged. It will, of course, depend on the input function applied. The storage portion of the response, on the other hand, can be obtained by solving (5.114). The result is:

$$e_{o_2} = E_0 \, \varepsilon^{-t/RC} \, U_s(t) \qquad \ldots (5.116a)$$

$$i_2 = -\left(\frac{E_0}{R}\right) \varepsilon^{-t/RC} \, U_s(t) \qquad \ldots (5.116b)$$

This portion of the response does not depend on the input function and is, therefore, the same for any input function. The storage response is always in the characteristic form $A \, \varepsilon^{-t/\tau}$ where A is the initial value of the variable due to storage. Let us demonstrate that this result holds also for T-type storage.

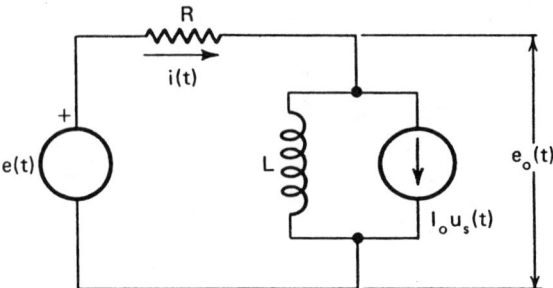

Figure 5.20 Series RL circuit with initial storage

Fig. 5.20 shows a series RL circuit with initial storage. The inductor has been replaced by an inductor and flow source in parallel. To find the complete responses, e_o and i, we may treat each source separately and then superimpose the results. Fig. 5.21(a) shows the circuit with the flow storage source equal to zero or open-circuited. This circuit is an RL series circuit without storage. The responses due to the input source depend on solutions to the equations:

$$\frac{L}{R}\frac{de_{o_1}}{dt} + e_{o_1} = \frac{L}{R}\frac{de(t)}{dt} \qquad ..(5.117a)$$

$$\frac{L}{R}\frac{di_1}{dt} + i_1 = \frac{1}{R} e(t) \qquad ..(5.117b)$$

This portion of the response has already been found in the previous sections. It depends on the type of input function, $e(t)$, that is applied.

Response of First Order Circuits

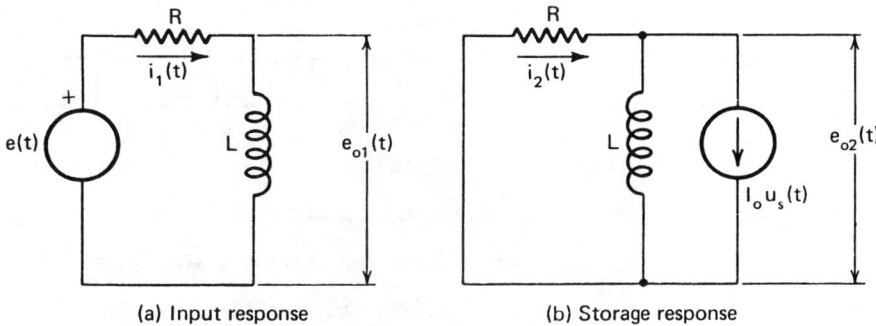

Figure 5.21 Separation of sources in series RL circuit with storage

Fig. 5.21(b) shows the series RL circuit with the input voltage source set equal to zero. The circuit differential equations for this portion of the response are:

$$\frac{L}{R}\frac{de_{o_2}}{dt} + e_{o_2} = L\frac{d}{dt}\left[I_0 U_s(t)\right] = L I_0 U_i(t)$$

$$..(5.118a)$$

$$\frac{L}{R}\frac{di_2}{dt} + i_2 = -\frac{L}{R}\frac{d}{dt}\left[I_0 U_s(t)\right] = -\frac{L}{R} I_0 U_i(t)$$

$$..(5.118b)$$

The storage portion of the response for the RL circuit can be determined by solving (5.117) and is:

$$e_{o_2} = R I_0 \varepsilon^{-Rt/L} U_s(t) \qquad ..(5.118a)$$

$$i_2 = - I_0 \varepsilon^{-Rt/L} U_s(t) \qquad ..(5.118b)$$

Thus, as in the RC circuit, the response due to storage in the RL circuit has the characteristic form $A\varepsilon^{-t/\tau}$ where A is the initial value of the variable due to storage. When the output current response is desired, care must be exercised to make sure that the sign of the storage contribution is correctly handled. The following examples will clarify the solutions of first-order circuits with storage.

Example 5.8

A leaky water tank is being filled from an overhead pipe [Fig. 5.22(a)]. The tank has a fluid capacitance $C = 0.01 \text{ m}^5/\text{N}$ and the leak may be modeled as a linear fluid resistance $R = 1500 \text{ N.s/m}^5$. The flow into the tank, q(t), has the function shown in Fig. 5.22(b). At the instant the flow begins (t = 0), the height of the water in the tank is 10 m. Find

(a) the pressure at the bottom of the tank, p(t),
(b) the height of the water after 20 seconds.

The analogous circuit is shown in Fig. 5.23. As previously indicated, we obtain separately the responses due to the flow and storage sources.

Response of First Order Circuits

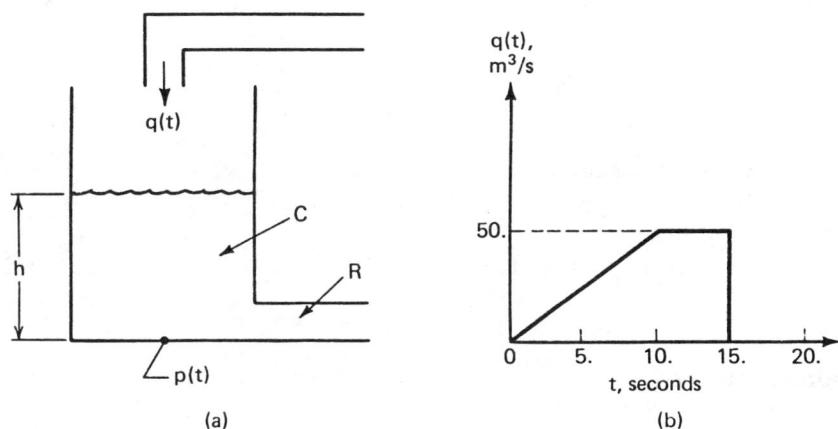

Figure 5.22 Fluid circuit (example 5.8)

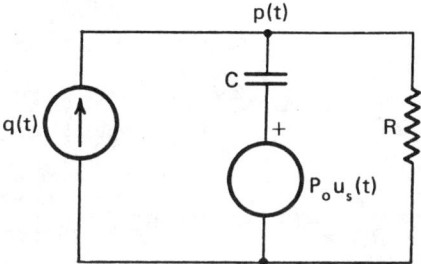

Figure 5.23 Analogous circuit for example 5.8

Step Response due to $q(t) = U_s(t)$ - no storage

(1) The homogeneous solution,

$$p_H = A \, \varepsilon^{-t/RC} \quad\quad \text{..(5.119a)}$$

(2) The particular solution,

$$p_P = R(1) \quad\quad \text{..(5.119b)}$$

(3) The initial condition - no storage,

$$p(0^+) = 0 \qquad \qquad ..(5.119c)$$

(4) Unit step response,

$$p_s(t) = R(1 - \varepsilon^{-t/RC}) \qquad \qquad ..(5.119d)$$

Response due to

$$q(t) = 5 U_r(t) - 5 U_r(t - 10) - 50 U_s(t - 15)$$
$$..(5.119e)$$

(5) $p_1(t) = 5R \left[t + RC\, \varepsilon^{-t/RC} - RC \right] U_s(t)$

$\qquad - 5R \left[(t - 10) + RC\, \varepsilon^{-(t - 10)/RC} - RC \right] U_s(t - 10)$

$\qquad - 50R \left[1 - \varepsilon^{-(t - 15)/RC} \right] U_s(t - 15) \qquad ..(5.119f)$

Since $R = 1500$ N.s/m^5, $C = 0.01$ m^5/N, $\qquad ..(5.119g)$

$p_1(t) = 7500 \left[t + 15\, \varepsilon^{-t/15} - 15 \right] U_s(t)$

$\qquad - 7500 \left[(t - 10) + 15\, \varepsilon^{-(t - 10)/15} - 15 \right] U_s(t - 10)$

$\qquad - 75,000 \left[1 - \varepsilon^{-(t - 15)/15} \right] U_s(t - 15) \quad ..(5.119h)$

Response of First Order Circuits

Response due to storage - no input.

(6) $p_2(t) = \rho g h_o \, \varepsilon^{-t/RC} = (1000)(9.81)(10) \, \varepsilon^{-t/15} U_s(t)$

$= 98,100 \, \varepsilon^{-t/15} U_s(t)$, Pa ..(5.119i)

Complete Response

(7) $p(t) = p_1(t) + p_2(t)$ ..(5.119j)

$= 7500 \left[t + 28.1 \, \varepsilon^{-t/15} - 15 \right] U_s(t)$

$-7500 \left[(t - 10) + 15 \, \varepsilon^{-(t-10)/15} - 15 \right] U_s(t - 10)$

$-75,000 \left[1 - \varepsilon^{-(t-15)/15} \right] U_s(t - 15)$ Pa ..(5.119k)

This is the solution to part (a).

$p(20) = 7500 \left[5 + 7.4 \right] - 7500 \left[-5 + 7.7001 \right]$

$\quad - 75,000 \left[1 - 0.717 \right]$

$= 51,475$ Pa ..(5.120a)

$h(20) = (51,475/9,810) = 5.25$ m ..(5.120b)

This is the solution to part (b).

Example 5.9

The mechanical arrangement shown in Fig. 5.24(a) consists of a spring and damper. The spring has a constant of k = 20 N/m and is initially stretched a

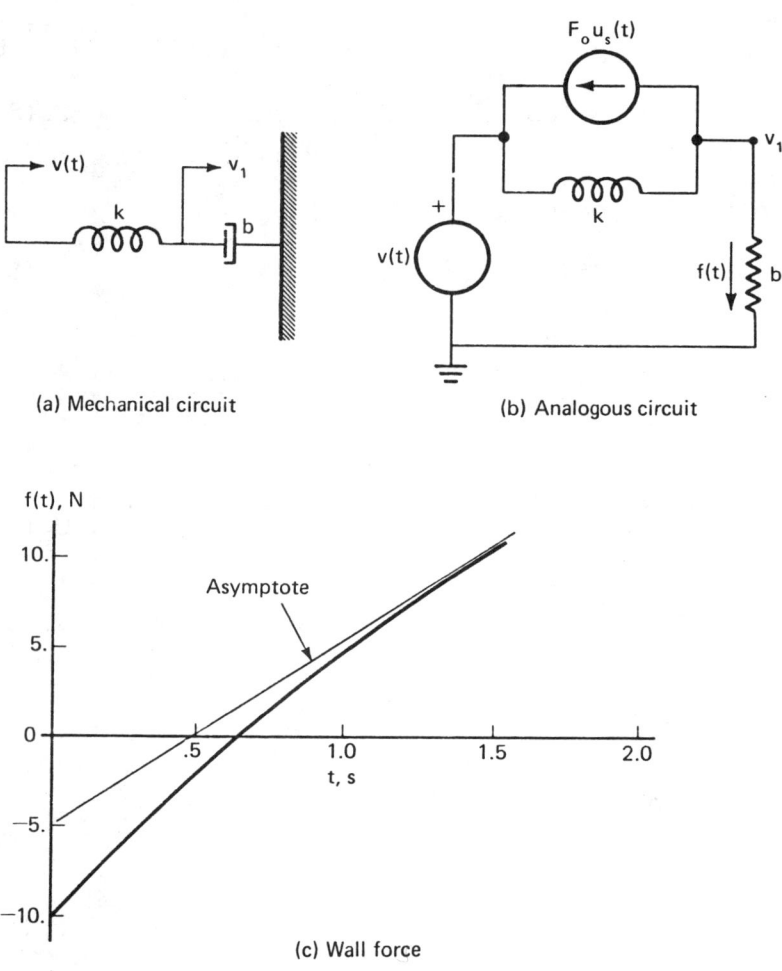

Figure 5.24 Circuit for example 5.9

distance of 0.5 m. The damper has a value, b = 10 N.s/m. A velocity input $v(t) = U_r(t)$ is applied to the free end of the stretched spring. Find the reaction force that the mechanical circuit exerts on the wall [f(t)].

Fig. 5.24(b) shows the analogous circuit diagram. From the initial stretch of the spring,

$$F_o = k x = (20)(0.5) = 10 \text{ N}. \quad \quad ..(5.121)$$

Response to a unit step input - no storage.

(1) The homogeneous solution, $f_H = A \varepsilon^{-kt/b}$ ..(5.122a)

(2) The particular solution, $f_p = b(1)$ ..(5.122b)

(3) The initial condition, $f(0) = 0$ ..(5.122c)

(4) The complete solution,

$$f_s(t) = b(1 - \varepsilon^{-kt/b}) U_s(t) \quad \quad ..(5.122d)$$

Response to ramp input - no storage

(5) $f_1(t) = b \left[t + \dfrac{b}{k} \varepsilon^{-kt/b} - \dfrac{b}{k} \right] U_s(t)$ ..(5.122e)

Storage Response - no input

(6) $f_2(t) = - F_o \varepsilon^{-kt/b} U_s(t)$ ..(5.122f)

Superposition of Responses

(7) $f(t) = f_1(t) + f_2(t)$

$$= 10 \left[t + (1/2) \varepsilon^{-2t} - (1/2) \right] U_s(t) - 10 \varepsilon^{-2t} U_s(t)$$

..(5.122g)

$$f(t) = \left[10 t - 5 \varepsilon^{-2t} - 5 \right] U_s(t), \text{ N} \qquad ..(5.122h)$$

Initially, the force on the wall acts to the left. When t = 0.64, there is no force on the wall and thereafter (t > 0.64), the force acts to the right. The complete response is shown in Fig. 5.24(c).

5.5 SUMMARY AND DISCUSSIONS

In this chapter, we have discussed the time-domain solution of a first-order network, which consists of a series or a parallel combination of a resistor and either a A-type or a T-type storage element. It is shown that the solution in both cases consists of two parts, namely (i) the homogeneous solution and (ii) the particular solution. The response due to the homogeneous part decays exponentially with time. The particular solution is dependent on the type of input to the system. For a step input, the particular solution can be obtained either by solving the first-order differential equation or by physical considerations, namely by the behavior of the storage element under steady-state conditions. The initial condition is obtained by the behavior of the storage element at the time of application of the unit step. This is required

Response of First Order Circuits

to obtain the complete solution of the differential equation governing the network.

Starting from the solution of the network for a unit step input, the complete solution for a composite input function can be obtained by the application of the the principle of superposition, which has the following properties:

(a) Let $y_1(t)$ be the response to an input $x_1(t)$ and let $y_2(t)$ be the response to an input $x_2(t)$. Then, the response to an input $[a_1 \cdot x_1(t) + a_2 \cdot x_2(t)]$ will be $[a_1 \cdot y_1(t) + a_2 \cdot y_2(t)]$. This property permits us to do the following: If more than one input is involved, we can obtain the response to one input at a time making all other inputs equal to zero. Then the complete response to a composite input is the sum of individual responses so obtained. Also, if a particular input is multiplied by a constant a, the corresponding output is also multiplied by the same constant a.

(b) The response of the circuit may also be shifted in time. That is, if the input $x_1(t)$ is shifted in time to $x_1(t - T)$, the output is also shifted in time to $y_1(t - T)$. Of course, this requires that the initial conditions shall be zero at time T also.

(c) If the input is subjected to a mathematical operation, namely integration or differentiation, the output is also subjected to the same mathematical operation.

With suitable combination of these three properties, it is possible to obtain the response of a system for a composite function which can be resolved into a sum of singularity functions, as discussed in Chapter II.

5.6 PROBLEMS

5.1 In the RC network shown in Fig. P5.1, $R_1 = 4,000\ \Omega$, $R_2 = 6,000\ \Omega$ and $C = 50\ \mu F$. If $e_i(t) = 200\ U_s(t)$, find:

(a) the current through the capacitor, $i_c(t)$,
(b) $e_o(t)$.

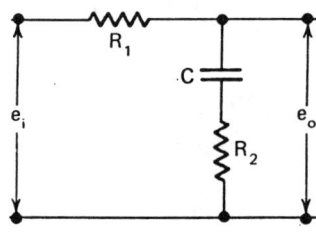

Figure P 5.1

5.2 Consider the two RC circuits shown in Fig. P5.2 connected to an ideal source. Find for each circuit:

(a) $i(t)$ and $e_1(t)$, if the source is a voltage source $e(t) = 4.0\ U_s(t)$, V.
(b) $i(t)$ and $e_1(t)$, if the source is a current source $i(t) = 20\ U_s(t)$, A.

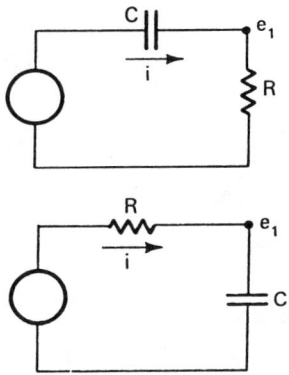

Figure P 5.2

5.3 A step input $e(t) = 5\, U_s(t)$, V is applied to the electric circuit shown in Fig. P5.3. If $R_1 = 1,000$ Ω and $R_2 = 4,000$ Ω, what is the value of the capacitance so that $e_c(2) = 2.4$ V.

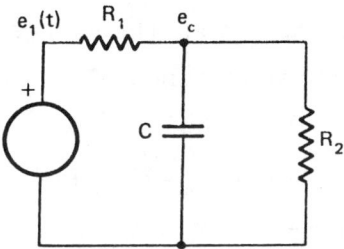

Figure P 5.3

5.4 A series RL electric circuit has R = 50 Ω and L = 2 H. If the voltage input given in Fig. P5.4 is applied as a source, find the current and the voltage drop across the inductance as functions of time. (Assume that $i_L(0^-) = 0$ Amps).

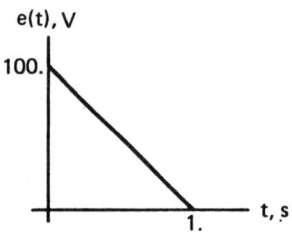

Figure P 5.4

5.5 An electric circuit contains a parallel combination of R and L (R = 3.75 Ω and L = 1.5 H). The circuit receives its input from a current source 2 $U_r(t)$, A. However, the initial current in the inductor is 1.0 A. Find the voltage drop across the resistance as a function of time.

5.6 A water tank is drained through a long pipeline. The tank has a capacitance C = (10^{-4}) m^5/N and initially has 5.0 m height of water. If the pipeline can be modeled as a pure resistance R = $4.0(10^5)$ $N \cdot s/m^2$, find:
   (a) The pressure at the bottom of the tank as a function of time after the tank valve is suddenly opened.
   (b) The height of the water in the tank at t = 25 s.

5.7 A pump supplies water to the bottom of a large tank (C = 25 m^5/N) through a long line (R = 2.0 $N \cdot s/m^5$). When there is initially no water in the tank, the system requires that the pump be started and be capable of causing the tank level to reach 3.0 m in 50 s.

(a) The chief engineer has selected a pump that develops a pressure $39.2(10^3) U_s(t)$, Pa. Will this meet the requirements? What height is reached in 50 s?

(b) An engineering student claims that the pump must deliver the pressure function, $p(t) = A U_s(t) - (A - 39.2(10^3)) U_s(t - 10)$. What should be the value of A to satisfy the system requirements?

5.8 A pump delivering water to a tank through a resistive line ($R_1 = 2,000$ N.s/m^5) is suddenly turned on. Water flows out of the tank through another resistance ($R_2 = 8,000$ N.s/m^5). The capacitance of the tank $C = 9(10^{-3})$ m^5/N. The output of the pump may be modeled as $P U_s(t)$. Pa.

(a) What is the maximum value of P so that the water height in the tank never exceeds 5.0 m?

(b) Repeat part 'a', if $C = 20(10^{-3})$ m^5/N.

5.9 An open water tank can be filled in two different ways: from the top and from the bottom. In both schemes, the source is a large reservoir of constant water level 8.16 m. The resistance $R = 200,000$ N.s/m^5, the capacitance $C = 0.001$ m^5/N and the density of water is 1,000 kg/m^3. Find, for each method of filling, the time for the water to reach a height of 4.0 m in an initially empty tank.

5.10 A pipe carrying water is 10 m long and has a diameter of 2 cm. The pipe has a pump at one end and is connected to the atmosphere at the other end. The pipe must be modeled with both resistance and inertance. If the pump delivers $P\, U_s(t)$, Pa, find the value of P that makes the Reynolds number in the pipe 2,000 at 20 s.

5.11 The mass-damper system shown in Fig. P5.11 is initially at rest. If $b_1 = 60$ N.s/m, $b_2 = 120$ N.s/m and m = 175 kg, find:

(a) The velocity of the mass at t = 0.25 s, when $v(t) = 8\, U_s(t)$ m/s.
(b) The time required for the mass to reach the velocity attained in part 'a'.
(c) The effect of a change in b_1 on the time required in part 'b'.

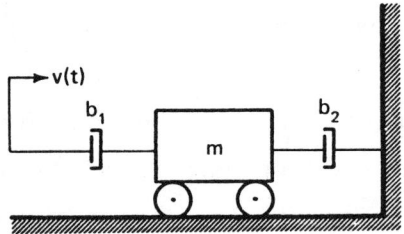

Figure P 5.11

5.12 In the mechanical system shown in Fig. P5.12, b = 260 N.s/m, k = 80 N/m, and the spring is initially compressed a distance of 0.25 m. If the input force $f(t) = 200\, U_s(t)$, N, find v(t) and x(t).

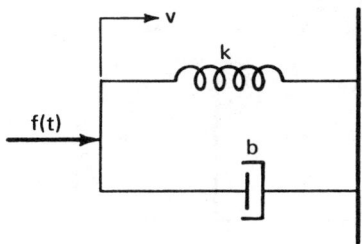

Figure P 5.12

5.13 In a carnival game, contestants try to hit a lever hard enough to cause a mass to move up a pole and strike a bell. The minimum force required for the mass to reach the ball is 70 $U_i(t)$, N. If the mass is 3.0 kg and we assume that the friction between mass and pole may be modeled as a linear damper of 5.0 N.s/m, find:
(a) The time for the mass to reach the ball when the minimum impulsive force is applied.
(b) The height of the bell, h.

5.14 The two horizontal force impulses, shown in Fig. P5.14, are applied to a mass of 20 kg that is resting on a table. The time period, T, is the time required for the first impulse to cause the velocity of the mass to reach 0.0062 m/s. If the friction between mass and table may be modeled as viscous damping, b = 25 N.s/m, how far will the mass move from its initial position?

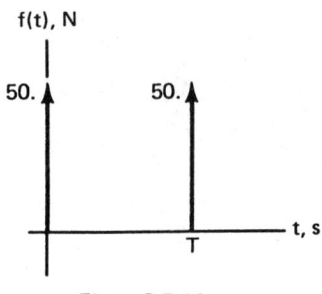

Figure P 5.14

5.15 A golf ball lies on the green 10 m from the hole. The ball has a mass of 0.1 kg and the grass provides a damping of 0.038 N.s/m. The golfer applies a force with his putter that can be modeled as $f(t) = 0.4\, U_i(t)$, N. If the ball moves in a straight line towards the hole, what is its velocity when it reaches the hole?

5.16 To get from the first floor to ground in an emergency, firepersons often slide down a metal pole. The average fireperson may be assumed to have a mass of 75 kg and requires 2 s to descend the distance, h, between first floor and ground. If the friction between the fireperson and pole can be modeled as damping b = 500 N.m/s, find:
(a) The velocity of the fireperson at the time of reaching the ground.
(b) The distance between first floor and ground.

5.17 A first-order mechanical system consists of two components in parallel connected to one end to a fixed support. If the input force to the free end is 600 $U_s(t)$, N and the resulting output velocity is shown in Fig. P5.17, identify the components and find their values.

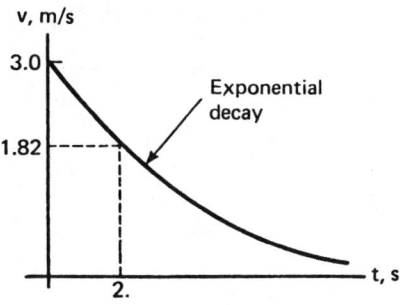

Figure P 5.17

5.18 For the rotary mechanical system shown in Fig. P5.18, angular velocity $\omega(t) = 5 U_r(t) - 5 U_r(t - 10)$, rad/s. If the damper B = 20 N.m.s and the moment of inertia of the mass is J = 10 kg.m^2, find:

(a) $\omega_1(t)$.
(b) The effect of a larger J on $\omega_1(t)$.

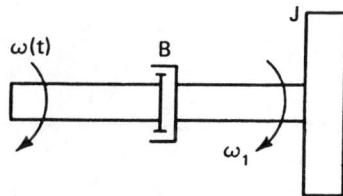

Figure P 5.18

5.19 In the rotational mechanical system shown in Fig. P5.19, the flywheel is initially moving with an angular velocity of ω_o rad/s clockwise. An external torque of $T\,U_s(t)$, N.m is applied to the flywheel in a counterclockwise direction. Find the resulting flywheel velocity in terms of J, B, T, ω_o and t.

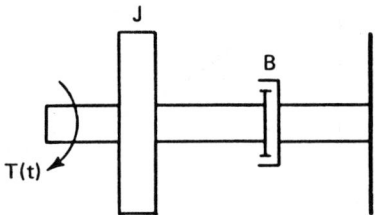

Figure P 5.19

5.20 A rotational spring-damper system shown in Fig. P5.20 has k = 2 N.m and B = 6 N.m.s. The applied angular velocity function is:
$$\omega(t) = 8\,U_s(t) - U_r(t-4) + U_r(t-20), \text{ rad/s}$$
Find:
(a) The angular velocity $\omega_1(t)$.
(b) The torque in the damper at 8 s [that is, $T_B(8)$].

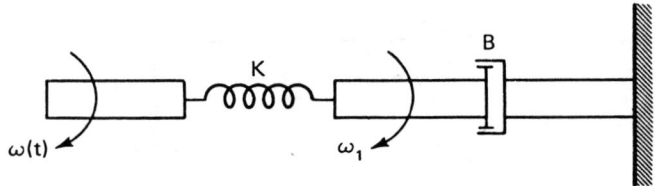

Figure P 5.20

5.21 In a test to find the moment of inertia of the rotor of a generator, the rotor is run up to a constant angular velocity and then the power source is turned off. The time for the rotor to reach a lower angular velocity is measured. Measurements show that the initial angular velocity is 500 rad/min and it takes 3.5 minutes for the rotor to slow down to 400 rad/min. If the retardation due to friction can be modeled as damping B = 110 N.m.s., what is the moment of inertia of the rotor?

5.22 The temperature inside a particular house is kept at 20°C. The heating system which maintains this temperature breaks down on a day when the outside temperature is a constant 5°C. After the breakdown, it takes 8 hours for the temperature inside to reduce to 15°C. The thermal resistance of the house may be represented by R = $5(10^{-8})\,^\circ$C.hr/J. Find:
  (a) The temperature in the house as a function of time.
  (b) The thermal capacitance of the house.

5.23 The water in an electric water heater is initially at 20°C. An immersion heater is suddenly turned on and supplies 4,000 $U_s(t)$, W. If the resistance of the insulation around the water tank is $0.02\,^\circ$C.s/J and the specific heat of water is 4,180 J/$^\circ$C.kg, find:

(a) The mass of water that can be heated to $55°C$ in one hour.

(b) The maximum temperature of the water after a long time.

5.24 The process of making a certain wine involves 'slow cooking' at $35°C$ above room temperature for several months. The wine has a density of $1,000$ kg/m^3 and a specific heat of $4,180$ W.s/kg.$°C$. To cook the wine, a cubic shaped container, 0.5 m on each side is fabricated with cork board 2 cm thick. The thermal conductivity of cork is 0.05 W/m.$°C$. An electric light bulb within the container acts as the heater. Find:

(a) The power required for the light bulb.

(b) The time constant of the system.

CHAPTER 6

RESPONSE OF SECOND ORDER CIRCUITS

In the previous chapter, we considered circuits with only one storage element. These circuits could all be described by a first-order differential equation and a first-order characteristic equation. Since the characteristic equation had a single root, the natural response of first-order circuits was limited to a single form, $A\varepsilon^{-t/\tau}$.

In this chapter, we consider circuits with two storage elements. Analysis of these circuits leads to second-order differential equations and a quadratic characteristic equation. Thus the characteristic equation always has two roots for second-order circuits. The natural response may now have any of three forms, one of which is generally similar to the natural response of first-order circuits, the other two being significantly different. The form that is appropriate for a particular circuit depends on the roots of the characteristic equation.

When the general input g(t) is operated on by a second-order circuit, the output function y(t) is produced. The most general differential equation for second-order circuits is:

$$\frac{d^2y}{dt^2} + 2\delta\omega_N \frac{dy}{dt} + \omega_N^2 y = a_1 g(t) + a_2 \frac{dg(t)}{dt} + a_3 \frac{d^2g(t)}{dt^2}$$

..(6.1)

where δ, ω_N, a_1, a_2 and a_3 are constants that depend on the values of the circuit elements. The constant, δ, is called the **damping ratio** and ω_N is called the **undamped natural frequency**.

The effect of δ and ω_N on the response will be made clear, when we derive the various forms of the natural response.

Let us begin our consideration of two storage element circuits, however, by suggesting a procedure for placing a desired output variable in the general form given in (6.1).

6.1 FORMULATION OF DIFFERENTIAL EQUATIONS

We will limit our discussion here to circuits with two storage elements and only one dissipative element. Of course, the circuit laws we use are general and could be applied to circuits of any complexity. However, we place this restriction now to reduce the algebraic manipulation required in the formulation of the circuit differential equations. Circuits with more components are more conveniently treated by the impedance method which will be presented in Chapter VII.

Consider, as a first illustration, the series RLC circuit shown in Fig. 6.1. There is one current (T-type) output variable, $i(t)$, and three potential (A-type) output variables, $e_R(t)$, $e_L(t)$ and $e_C(t)$. If

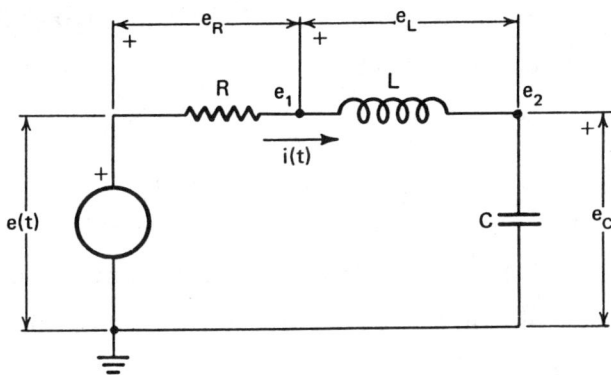

Figure 6.1 Series RLC circuit

we use the concept of nodal potential, we can reduce the number of unknown potentials to two, $e_1(t)$ and $e_2(t)$. Then, the three original potential differences can be expressed in terms of the nodal potentials and the source potential. Thus:

$$e_R = e - e_1, \quad e_L = e_1 - e_2, \quad \text{and} \quad e_C = e_2 \qquad ..(6.2)$$

We are left, therefore, with the variables $i(t)$, $e_1(t)$ and $e_2(t)$ which we wish to relate to the input, $e(t)$, through a differential equation. The best way to accomplish this is to follow these general steps:

(a) Select the type of variable that has the fewer unknowns. In this case, the variable is $i(t)$, the single T-type variable. [There are two A-type variables, $e_1(t)$ and $e_2(t)$].

(b) If the variable selected in part (a) is T-type, begin by writing a compatibility equation (KVL). If the variable is A-type, write a continuity equation (KCL). For the series RLC, we therefore write the compatibility equation

$$[e(t) - e_1] + [e_1 - e_2] + [e_2] = e(t) \qquad ..(6.3)$$

(c) Substitute the elemental equations into the equation obtained in part (b). In this case, it will be (6.3). Thus:

$$R\,i + L\frac{di}{dt} + \frac{1}{C}\int i\,dt = e(t) \qquad ..(6.4)$$

(d) Perform the mathematical operations required to put (6.4) into the form given in (6.1). Thus

Response of Second Order Circuits

$$\frac{d^2 i}{dt^2} + \frac{R}{L}\frac{di}{dt} + \frac{1}{LC} i = \frac{1}{L}\frac{de(t)}{dt} \qquad ..(6.5)$$

Now that we have found the differential equation that relates $i(t)$ to the input $e(t)$, we may easily express the other outputs $e_1(t)$ and $e_2(t)$ in differential equation form.

(e) Substitute a single elemental equation into (6.4). For example, to find e_2, use the elemental equation

$$e_2 = \frac{1}{C} \int i\, dt \qquad ..(6.6a)$$

or
$$i = C \frac{de_2}{dt} \qquad ..(6.6b)$$

The result is:

$$\frac{d^2 e_2}{dt^2} + \frac{R}{L}\frac{de_2}{dt} + \frac{1}{LC} e_2 = \frac{1}{LC} e(t) \qquad ..(6.7)$$

To find e_1, substitute

$$i = \frac{e(t) - e_1}{R} \qquad ..(6.8)$$

into (6.4) or (6.5) and obtain

$$\frac{d^2e_1}{dt^2} + \frac{R}{L}\frac{de_1}{dt} + \frac{1}{LC}e_1 = \frac{1}{LC}e(t) + \frac{d^2e(t)}{dt^2} \qquad ..(6.9)$$

Note that the coefficients of the terms on the left-hand side of the differential equations for the outputs i(t), $e_1(t)$ and $e_2(t)$ are the same. That is, the coefficient of the first derivative term is R/L and the coefficient of the third term is 1/LC. Other circuits, in general, will have other coefficients. However, for any given particular circuit, the differential equations for all the circuit variables will have the same coefficients. The coefficients are a characteristic of the circuit and do not depend on the output variable.

To further demonstrate the procedure for formulating the circuit differential equation, consider the circuit shown in Fig. 6.2. The circuit has an inductance branch in parallel with a series RC branch. The input signal is from a current source, i(t). Suppose we wish to find the differential equation that relates the potential $e_1(t)$ to the input source, i(t). The steps to follow are:

(a) Select the variable with fewer unknowns. There are two nodal variables e_1 and e_2. There is only one mesh variable $i_1(t)$. Thus, we select the variable, $i_1(t)$, and find its differential equation before the differential equation for $e_1(t)$.

Response of Second Order Circuits

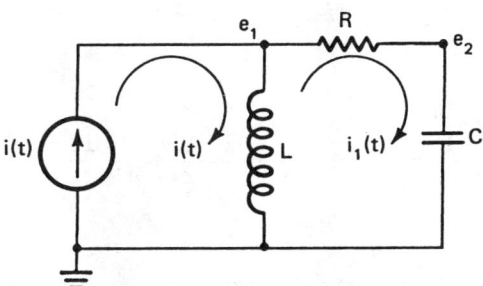

Figure 6.2 Parallel-Series RLC circuit

(b) Compatibility relation: Since the variable in part (a) is $i_1(t)$, we write the compatibility equation around the mesh

$$- e_1 + (e_1 - e_2) + e_2 = 0 \qquad ..(6.10)$$

(c) Substitute the elemental equations. We have

$$- L \frac{d}{dt}\left[i(t) - i_1\right] + R\, i_1 + \frac{1}{C}\int i_1\, dt = 0$$
$$..(6.11)$$

(d) Perform the mathematical operations. We get

$$\frac{d^2 i_1}{dt^2} + \frac{R}{L}\frac{di_1}{dt} + \frac{1}{LC} i_1 = \frac{d^2 i(t)}{dt^2} \qquad ..(6.12)$$

(e) Substitute the elemental equation for L in (6.12) and simplify. Thus:

$$i_1 = i(t) - \frac{1}{L}\int e_1\, dt \qquad \ldots (6.13)$$

and we obtain

$$\frac{d^2 e_1}{dt^2} + \frac{R}{L}\frac{de_1}{dt} + \frac{1}{LC} e_1 = R\frac{d^2 i}{dt^2} + \frac{1}{C}\frac{di}{dt} \qquad \ldots (6.14)$$

We will follow this formulation procedure in the examples throughout this chapter.

6.2 NATURAL RESPONSE OF SECOND ORDER CIRCUITS

The homogeneous form of the general differential equation for second-order circuits given in (6.1) is:

$$\frac{d^2 y_H}{dt^2} + 2\delta\omega_N \frac{dy_H}{dt} + \omega_N^2 y_H = 0 \qquad \ldots (6.15)$$

This is the general equation without the forcing functions and produces the unforced or natural response. As in the case of first-order circuits, we assume a trial solution of the form:

$$y_H = A\,\varepsilon^{rt} \qquad \ldots (6.16)$$

If this solution is substituted into (6.15), we obtain:

$$\left[r^2 + 2\delta\omega_N r + \omega_N^2\right] A\,\varepsilon^{rt} = 0 \qquad \ldots (6.17)$$

Once again, the trial solution is not equal to

Response of Second Order Circuits

zero, so that the bracketed term in (6.17) must be equal to zero, and

$$r^2 + 2\delta\omega_N r + \omega_N^2 = 0 \qquad ..(6.18)$$

Eq. (6.18) is the general characteristic equation of second-order circuits. If we use the quadratic formula to solve (6.18) for r, the result is:

$$r = -\delta\omega_N \pm \omega_N \sqrt{\delta^2 - 1} \qquad ..(6.19)$$

Eq. (6.19) represents the roots of the characteristic equation. The form of the natural response depends on whether the roots are real and distinct, real and equal, or a complex conjugate pair. This, in turn, depends on the value of δ, the damping ratio. From (6.19), observe that when:

(a) $\delta > 1$, the roots are real and distinct,
(b) $\delta < 1$, the roots are a complex conjugate pair,
(c) $\delta = 1$, the roots are real and equal.

Each of these conditions produces a different form for the natural response and is treated separately.

(a) $\delta > 1$:

The natural response under this condition is sometimes called the **overdamped** response, since large amounts of damping or resistance in circuits lead to large values of δ. In this case, the characteristic equation has two real and distinct roots, r_1 and r_2. The natural response is similar to that of a first-order circuit and has the form:

$$y_H(t) = A_1 \varepsilon^{r_1 t} + A_2 \varepsilon^{r_2 t} \qquad ..(6.20)$$

Note that the sum of the roots, $(r_1 + r_2)$, will always be equal to $-2\delta\omega_N$, the negative of the coefficient of r in the characteristic equation.

The constants A_1 and A_2 cannot be evaluated until we obtain the complete solution and use the initial conditions. However, (6.20) shows that the natural response will be the sum or difference of decaying exponentials.

(b) <u>δ < 1</u>:

When the damping ratio is less than unity, the response is termed **underdamped**. This condition occurs when a small amount of damping or resistance is present in the second-order circuit. The characteristic equation has complex conjugate roots that are designated from (6.19) as:

$$r_1 = -\alpha + j\omega_d \qquad ..(6.21a)$$
$$r_2 = -\alpha - j\omega_d \qquad ..(6.21b)$$

where α = the damping constant = $\delta\omega_N$ ..(6.22a)

Response of Second Order Circuits

and ω_d = the damped natural frequency = $\omega_N \sqrt{(1-\delta^2)}$

..(6.22b)

The natural response for the underdamped circuit is then:

$$y_H(t) = A_1 \varepsilon^{[-(\alpha - j\omega_d)t]} + A_2 \varepsilon^{[-(\alpha + j\omega_d)t]}$$

..(6.23)

Eq. (6.23) involving the complex exponentials is inconvenient to work with. We may place the equation in a more usable form by applying the Euler relation:

$$\varepsilon^{j\theta} = \cos \theta + j \sin \theta \qquad ..(6.24)$$

Thus, (6.23) may be reduced to:

$$y_H(t) = \varepsilon^{-\alpha t} \left[(A_1 + A_2) \cos \omega_d t + j(A_1 - A_2) \sin \omega_d t \right]$$

..(6.25)

The natural response $y_H(t)$ described by (6.25) must be a real quantity for all values of time. Remember that it represents the solution of a real physical problem. For this condition to be satisfied, the constants A_1 and A_2 must be complex conjugates. Then their sum will be a real number and their difference is imaginary. It is, therefore, convenient to express the natural response of an underdamped second order circuit as:

$$y_H(t) = \varepsilon^{-\alpha t} \left[A_3 \cos \omega_d t + A_4 \sin \omega_d t \right] \quad \text{..(6.26)}$$

Fig. 6.3 shows some typical underdamped natural responses for decaying sine and cosine functions. Note that the responses are contained between an envelope of decaying exponentials $\varepsilon^{-0.1t}$ and $-\varepsilon^{-0.1t}$. For the cases shown, $\alpha = 0.1 \text{ s}^{-1}$ and $\omega_d = 1.0$ rad/s. Thus the damping ratio here is $\delta = 0.0995$ and the natural

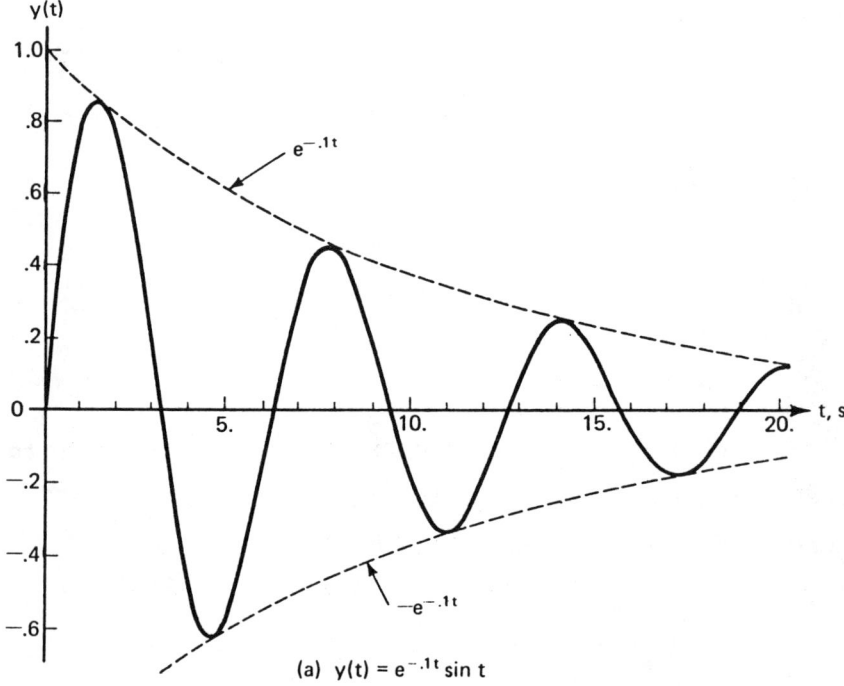

(a) $y(t) = e^{-.1t} \sin t$

Figure 6.3 Some typical underdamped natural responses ($\delta = 0.0995$, $\omega_N = 1.005$ rad/s)

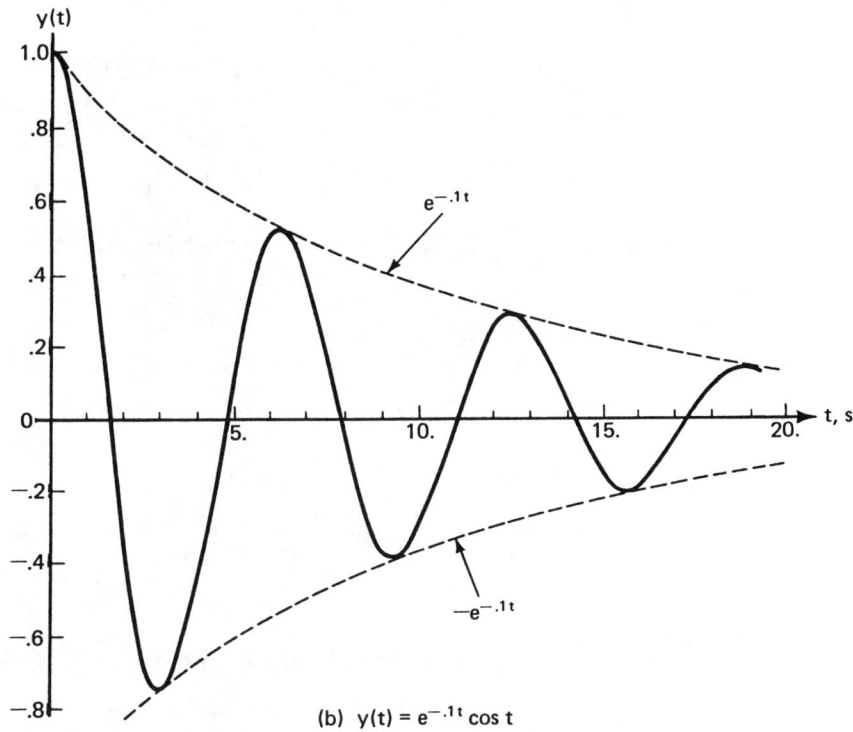

Figure 6.3 Some typical underdamped natural responses ($\delta = 0.0995$, $\omega_N = 1.005$ rad/s) (cont'd)

frequency $\omega_N = 1.005$ rad/s. When the damping ratio is larger (but still less than 1.0), the enveloping exponentials decay more rapidly and the magnitude of the peaks decreases.

(c) $\underline{\delta = 1}$:

When the roots are real and equal, the system is **critically damped**. We may treat this as the limiting case of the overdamped response in which the two real roots approach each other. Suppose we designate the roots as r_1 and $(r_1 + \Delta r)$. Then the natural response may be written as:

$$y_H(t) = \lim_{\Delta r \to 0} \left[A_1 \varepsilon^{r_1 t} + A_2 \varepsilon^{(r_1 + \Delta r)t} \right] \quad ..(6.27)$$

If we add and subtract $A_2 \varepsilon^{r_1 t}$ to the right-hand side of (6.27), we may express it as:

$$y_H(t) = (A_1 + A_2) \varepsilon^{r_1 t} +$$
$$\lim_{\Delta r \to 0} \left[A_2 \left\{ \varepsilon^{(r_1 + \Delta r)t} - \varepsilon^{r_1 t} \right\} \right] \quad ..(6.28)$$

The constants A_1 and A_2 are arbitrary. We may select

$$A_5 = A_1 + A_2 \quad ..(6.29a)$$

and $$A_6 = A_2 \Delta r \quad ..(6.29b)$$

When A_5 and A_6 are evaluated from the complete response, it can be shown that neither are functions of Δr. Thus, with this choice, we may express (6.28) as:

$$y_H(t) = A_5 \varepsilon^{r_1 t} +$$
$$A_6 \lim_{\Delta r \to 0} \left[\frac{\varepsilon^{(r_1 + \Delta r)t} - \varepsilon^{r_1 t}}{\Delta r} \right] \quad .(6.30)$$

Now, from the definition of a derivative, we recognize

that:

$$\frac{d(\varepsilon^{r_1 t})}{dr_1} = \lim_{\Delta r \to 0} \left[\frac{\varepsilon^{(r_1+\Delta r)t} - \varepsilon^{r_1 t}}{\Delta r} \right] = t\, \varepsilon^{r_1 t} \quad ..(6.31)$$

Thus, the homogeneous solution for the critically damped condition reduces to:

$$y_H(t) = (A_5 + A_6\, t)\, \varepsilon^{r_1 t} \quad ..(6.32)$$

Another way to approach the critically damped case is to assume a homogeneous solution of the form

$$y_H(t) = h(t)\, \varepsilon^{r_1 t} \quad ..(6.33)$$

Then the substitution of this form into (6.15) shows that

$$\frac{d^2 h(t)}{dt} = 0 \quad ..(6.34)$$

and therefore $\quad h(t) = A_5 + A_6\, t \quad ..(6.35)$

Fig. 6.4 compares the natural response of a typical critically damped circuit and an equivalent overdamped circuit in which the roots are close together. The responses are nearly the same. There is never a significant difference in appearance between an overdamped response with damping ratio near unity and a critically damped response.

Underdamped or critically damped responses may only occur in circuits that have one T-type storage device and one A-type storage device. The oscillatory or underdamped response is the result of an internal energy transfer between an A-type and a T-type element. Circuits that contain only one type of element (that is, two T-types or two A-types) must always be overdamped. For the damping ratio to be equal to or less than unity, elements of both types must be present. However, the presence of both element types is merely a necessary condition and does not ensure an underdamped or critically damped response.

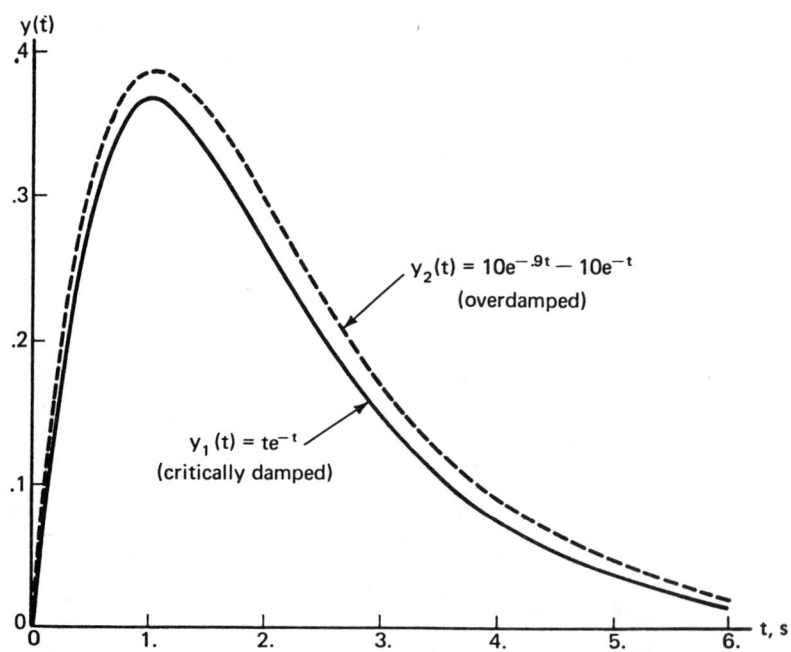

Figure 6.4 Typical critically damped natural response

6.3 STEP RESPONSE OF SECOND-ORDER CIRCUITS

As in the case of first order circuits, we can find the response of second-order circuits by a mathematical approach and a physical approach. The difference between the two approaches is more evident for second-order circuits. To obtain the step response, the input function is

$$g(t) = G\, U_s(t) \qquad \ldots (6.36)$$

(a) Mathematical Approach

In the mathematical approach, we first formulate the differential equation. The next step is to decompose the equation so that only one term appears as a forcing function on the right hand side. For the general second order circuit represented by (6.1), the decomposition results in:

$$\frac{d^2 y_1}{dt^2} + 2\delta\omega_N \frac{dy_1}{dt} + \omega_N^2 y_1 = G\, U_s(t) \qquad \ldots (6.37a)$$

$$\frac{d^2 y_2}{dt^2} + 2\delta\omega_N \frac{dy_2}{dt} + \omega_N^2 y_2 = G\, U_i(t) \qquad \ldots (6.37b)$$

$$\frac{d^2 y_3}{dt^2} + 2\delta\omega_N \frac{dy_3}{dt} + \omega_N^2 y_3 = G\, U_d(t) \qquad \ldots (6.37c)$$

$$y_s(t) = a_1 y_1 + a_2 y_2 + a_3 y_3 \qquad \ldots (6.37d)$$

From the principles of linear differential equations, we may relate the solutions for y_2 and y_3 to the solution of y_1. In effect, y_1 is the solution of a particular equation for a step input forcing function; y_2 and y_3 are solutions of the same equation for an impulse and doublet function respectively. This means that:

$$y_2 = \frac{dy_1}{dt} \qquad \ldots (6.38a)$$

$$y_3 = \frac{d^2 y_1}{dt^2} \qquad \ldots (6.38b)$$

and the complete solution for the general equation may be expressed solely in terms of y_1. From (6.38) and (6.37d), we obtain:

$$y_s(t) = a_1 y_1 + a_2 \frac{dy_1}{dt} + a_3 \frac{d^2 y_1}{dt^2} \qquad \ldots (6.39)$$

Thus, the complete step response, $y_s(t)$, depends only on y_1. The problem is reduced, therefore, to finding a solution for y_1 in (6.37a) and then applying (6.39). The solution for y_1 will always be the sum of a homogeneous and a particular solution. The homogeneous solution depends on the damping ratio, δ, as indicated in the previous section. The particular solution of (6.37a) is the value of y_1 after all the transients have disappeared (that is, $\frac{dy_1}{dt} = 0$, and $\frac{d^2 y_1}{dt^2} = 0$).

Response of Second Order Circuits 357

The solution for y_1 is then

$$y_1(t) = y_H + \frac{G}{\omega_N^2} \quad \quad ..(6.40)$$

where the particular solution is G/ω_N^2. The homogeneous portion of (6.40) contains two arbitrary constants which must be evaluated from the initial conditions, $y_1(0^+)$ and $\frac{dy_1}{dt}(0^+)$. At this point, we must remember that the decomposed equation (6.37a) alone does not represent the general second-order circuit. The initial conditions for the decomposed equation are always $y_1(0^+) = 0$ and $\frac{dy_1}{dt}(0^+) = 0$. The initial conditions for the actual circuit will generally be different and can be determined directly as indicated in Section 4.3.

The mathematical solution may be summarized in the following steps:

(1) **Formulation**: Find the complete circuit differential equation by the methods given in Section 6.1.

(2) **Decomposition**: Replace the right-hand side of the complete equation by the step magnitude, G. Determine the response $y_1(t)$, to the constant forcing function using $y_1(0^+) = 0$ and $\frac{dy_1}{dt}(0^+) = 0$.

(3) **Addition**: Apply (6.39) and obtain the complete step

response.

(b) Physical Approach:

In the physical approach, we also begin with the formulation of the differential equation. However, in this approach, we are really interested in finding the characteristic equation. From the characteristic equation, we determine the homogeneous solution. The procedure is identical to the physical approach used for first-order circuits and is itemized as:

(1) **Homogeneous Solution:** This solution depends on the value of the damping ratio, δ, and contains two arbitrary constants.

(2) **Particular Solution:** The particular solution for a step input is the value of the output variable at infinite time. Thus $y_p = y_s(\infty)$.

(3) **Initial Conditions:** These conditions are found from a consideration of initial circuit performance during a sudden change (Section 4.3). They are not necessarily equal to zero. For the second-order circuit, we need $y_s(0^+)$ and $\frac{dy_s}{dt}(0^+)$.

(4) **Complete Solution:** Combine the homogeneous and particular solutions to form the complete solution. Use the initial conditions to determine the arbitrary constants.

Response of Second Order Circuits

We will demonstrate the solution procedure for the step response of second-order circuits with a few examples.

Example 6.1

The mechanical arrangement shown in Fig. 6.5(a) is subjected to a driving force

$$f(t) = 10\, U_s(t),\, N \qquad ..(6.41)$$

If the mass, m = 1 kg, the damper, b = 3 N.s/m, and the spring, k = 2 N/m, find the velocity of the mass.

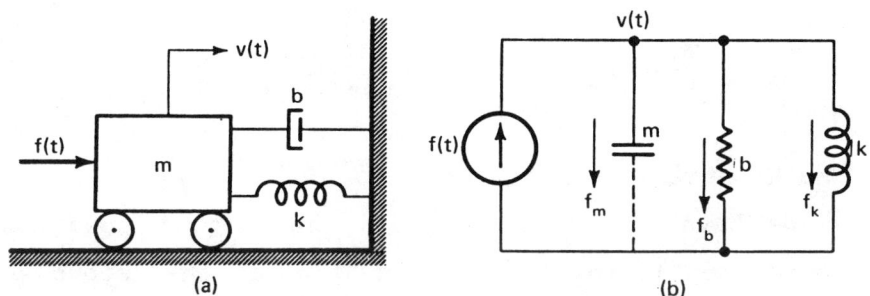

Figure 6.5 Mechanical circuit (example 6.1)

Fig. 6.5(b) shows the analogous circuit for the mechanical arrangement. The circuit consists of three elements in parallel and a T-type source (force source). To formulate the differential equation, we use the steps given in Section 6.1. The variable, v, (A-type), has the fewest unknowns and so we begin with a continuity equation:

$$f_m + f_b + f_K = f(t) \quad \quad ..(6.42)$$

The substitution of the elemental equation yields

$$m \frac{dv}{dt} + bv + k \int v \, dt = f(t) \quad \quad ..(6.43)$$

and by differentiation, the differential equation for this circuit is:

$$\frac{d^2v}{dt^2} + \frac{b}{m}\frac{dv}{dt} + \frac{k}{m}v = \frac{1}{m}\frac{d}{dt}[f(t)] \quad \quad ..(6.44)$$

We replace the components by their given values to obtain:

$$\frac{d^2v}{dt^2} + 3\frac{dv}{dt} + 2v = 10 \, U_i(t) \quad \quad ..(6.45)$$

(a) Mathematical Approach

We have already formulated the circuit equation. The next step is to place the equation in the decomposed form

$$\frac{d^2v_1}{dt^2} + 3\frac{dv_1}{dt} + 2v_1 = 10 \quad \quad ..(6.46)$$

with $\quad v_1(0^+) = 0 \quad\quad ..(6.47a)$

and $\quad \frac{dv_1}{dt}(0^+) = 0 \quad$ (by definition) $\quad ..(6.47b)$

If we compare the coefficients of this equation with the standard form of (6.1), we find that:

$$2\delta\omega_N = 3 \quad \text{and} \quad \omega_N^2 = 2 \quad \quad ..(6.48)$$

Thus, the damping ratio, $\delta = 1.06$, and the circuit is overdamped. The characteristic equation is:

$$r^2 + 3r + 2 = 0 \quad \quad ..(6.49)$$

and the roots of the equation are $r_1 = -1$ and $r_2 = -2$. The homogeneous solution is:

$$v_H = A_1 \varepsilon^{-t} + A_2 \varepsilon^{-2t} \quad \quad ..(6.50)$$

The particular solution, $v_{1p} = 10/2 = 5$. The complete solution for $v(t)$ is:

$$v_1(t) = A_1 \varepsilon^{-t} + a_2 \varepsilon^{-2t} + 5 \quad \quad ..(6.51)$$

We now evaluate the arbitrary constants from the zero initial conditions and find that:

$$v_1(0^+) = A_1 + A_2 + 5 = 0 \quad \quad ..(6.52a)$$

$$\frac{dv_1}{dt}(0^+) = -A_1 - 2A_2 = 0 \quad \quad ..(6.52b)$$

with the result that $A_1 = -10$ and $A_2 = 5$. Thus, the complete solution for $v_1(t)$ is:

$$v_1(t) = \left[-10\,\varepsilon^{-t} + 5\,\varepsilon^{-2t} + 5\right] U_s(t) \quad \ldots (6.53)$$

This is not the velocity of the mass. To obtain the velocity of the mass, we must use (6.39) which in this case reduces to:

$$v(t) = \frac{d}{dt}\left[v_1(t)\right] \quad \ldots (6.54)$$

If we perform the differentiation, we get

$$v(t) = (10\,\varepsilon^{-t} - 10\,\varepsilon^{-2t})\,U_s(t)$$
$$+ (-10\,\varepsilon^{-t} + 5\,\varepsilon^{-2t} + 5)\,U_i(t) \quad \ldots (6.55)$$

The second term on the right (the impulse term) is equal to zero for all values of time. Thus, the velocity of the mass is:

$$v(t) = \left[10\,\varepsilon^{-t} - 10\,\varepsilon^{-2t}\right] U_s(t) \quad \ldots (6.56)$$

(b) Physical Approach:

The homogeneous solution is in the same form as in the mathematical approach.

$$v_H = A_3\,\varepsilon^{-t} + a_4\,\varepsilon^{-2t} \quad \ldots (6.57)$$

We find the particular solution by inspection of the analogous circuit diagram [Fig. 6.5(b)]. After a long time, the mass behaves like an open circuit and the spring like a short circuit. Thus,

$$v_p = v(\infty) = 0 \qquad ..(6.58)$$

The initial conditions are also determined from the circuit diagram noting that the mass is now a short and the spring is open. Thus, we immediately know that

$$v(0^+) = 0 \qquad ..(6.59)$$

To determine $\frac{dv}{dt}(0^+)$, we look for an elemental equation that contains $\frac{dv}{dt}$. For example,

$$f_m(t) = m\frac{dv}{dt} \qquad ..(6.60)$$

This equation holds for all values of time. If we specialize the equation for $t = 0^+$, the result is:

$$\frac{dv}{dt}(0^+) = \frac{f_m(0^+)}{m} = \frac{f(0^+)}{m} = 10 \qquad ..(6.61)$$

Since the mass behaves as short, all the force passes through this component at $t = 0^+$, and $f_m(0^+) = f(0^+)$.

The physical approach thus yields a complete solution

$$v(t) = A_3 \varepsilon^{-t} + A_4 \varepsilon^{-2t} + 0 \qquad ..(6.62)$$

and the arbitrary constants are determined from the initial conditions derived from the circuit performance. Thus,

$$v(0^+) = A_3 + A_4 = 0 \qquad ..(6.63a)$$

and
$$\frac{dv}{dt}(0^+) = -A_3 - 2A_4 = 10 \qquad ..(6.63b)$$

from which $A_3 = +10$ and $A_4 = -10$. The complete step response for the velocity of the mass is:

$$v(t) = \left[10\, \varepsilon^{-t} - 10\, \varepsilon^{-2t}\right] U_s(t) \qquad ..(6.64)$$

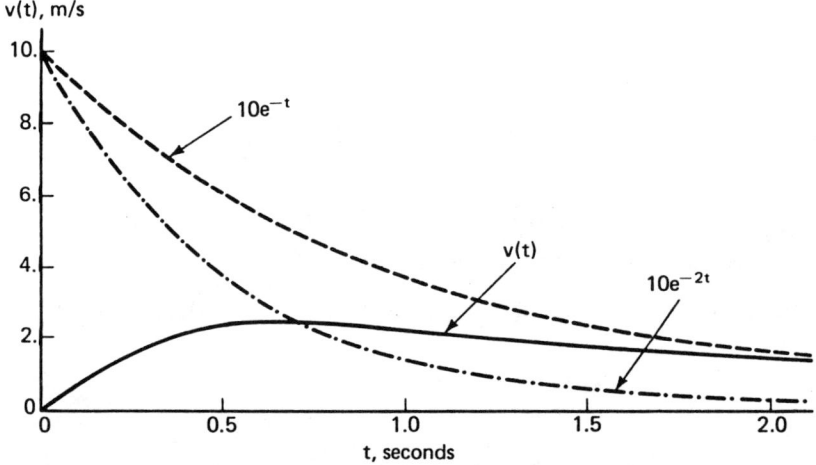

Figure 6.6 Velocity of mass (example 6.1)

Response of Second Order Circuits

which is the same result that was obtained by the mathematical approach. Fig. 6.6 shows the velocity of the mass. The velocity reaches a peak of 2.39 m/s when t = 0.7 s.

Example 6.2

The electric circuit shown in Fig. 6.7 has an input function

$$e_i(t) = U_s(t), \text{ v} \qquad ..(6.65)$$

The component values are R = 10 Ω, L = 0.592 H, and C = 0.01 F. Find the value of the output voltage, $e_o(t)$.

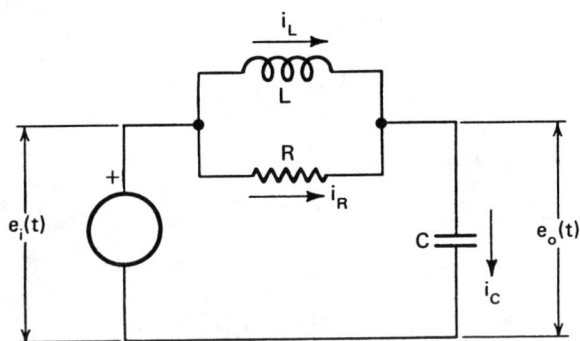

Figure 6.7 Electric circuit (example 6.2)

From the continuity equation at the output node:

$$i_C = i_R + i_L \qquad ..(6.66)$$

If the elemental equations are used to replace the currents above, the result is:

$$C \frac{de_o}{dt} = \frac{e_i - e_o}{R} + \frac{1}{L} \int (e_i - e_o) \, dt \qquad ..(6.67)$$

By differentiation and rearrangement, we have:

$$\frac{d^2 e_o}{dt^2} + \frac{1}{RC} \frac{de_o}{dt} + \frac{1}{LC} e_o = \frac{1}{LC} e_i(t) + \frac{1}{RC} \frac{d}{dt} [e_i(t)]$$
$$..(6.68)$$

With the component values, this becomes

$$\frac{d^2 e_o}{dt^2} + 10 \frac{de_o}{dt} + 169 e_o = 169 e_i(t) + 10 U_i(t)$$
$$..(6.69)$$

(a) Mathematical Approach:

In the decomposed form, we have

$$\frac{d^2 e_{o1}}{dt^2} + 10 \frac{de_{o1}}{dt} + 169 e_{o1} = 1 \qquad ..(6.70)$$

with $\quad e_{o1}(0^+) = 0 \qquad ..(6.71a)$

and $\quad \dfrac{de_{o1}}{dt}(0^+) = 0 \quad$ (by definition) $\qquad ..(6.71b)$

By comparison with the general second-order circuit differential equation, the coefficients of the second and third terms on the left-hand side yield:

Response of Second Order Circuits

$$2\delta\omega_N = 10 \quad \text{and} \quad \omega_N^2 = 169 \quad \quad ..(6.72)$$

From these relations, we find that:

$$\omega_N = 13 \text{ rad/s and } \delta = 0.385 \quad \quad ..(6.73)$$

The circuit is underdamped and hence

$$\alpha = \delta\omega_N = 5 \text{ s}^{-1} \quad \quad ..(6.74a)$$

and

$$\omega_d = \omega_N(1 - \delta^2)^{1/2} = 12 \text{ rad/s} \quad \quad ..(6.74b)$$

Therefore, the homogeneous solution is:

$$e_H = \varepsilon^{-5t}\left[A_1 \sin 12t + A_2 \cos 12t\right] \quad \quad ..(6.75)$$

The particular solution is:

$$e_p = 1/169 \quad \quad ..(6.76)$$

This will be the value of e_{o1}, when all the derivatives are set equal to zero indicating that the variable is no longer changing.

The form of the complete solution for e_{o1} is:

$$e_{o1}(t) = \varepsilon^{-5t}\left[A_1 \sin 12t + a_2 \cos 12t\right] + 1/169$$

$$..(6.77)$$

From the initial condition

$$e_{o1}(0^+) = 0 \qquad \qquad ..(6.78a)$$

we find that $\qquad A_2 = -1/169 = -0.00591 \qquad ..(6.78b)$

From the condition

$$\frac{de_{o1}}{dt}(0^+) = 0, \qquad \qquad ..(6.79a)$$

we find that $\qquad A_1 = (5/12) \, A_2 = -0.00246 \qquad ..(6.79b)$

The complete solution for e_{o1} is:

$$e_{o1}(t) = \varepsilon^{-5t}\left[-0.00246 \sin 12t - 0.00591 \cos 12t\right] + 0.00591 \qquad ..(6.80)$$

From (6.39), the output voltage, $e_o(t)$, is related to $e_{o1}(t)$ by:

$$e_o(t) = 169 \, e_{o1}(t) + 10 \, \frac{de_{o1}(t)}{dt} \qquad ..(6.81)$$

By substituting the expression for $e_{o1}(t)$ in the above equation, we find that the output voltage is:

$$e_o(t) = \left[\varepsilon^{-5t}(0.417 \sin 12t - \cos 12t) + 1\right] U_s(t) \qquad ..(6.82)$$

Response of Second Order Circuits

This is the complete step response of the circuit.

(b) Physical Approach:

The homogeneous equation for this circuit is already known as:

$$e_H = \varepsilon^{-5t} \left[A_3 \sin 12t + A_4 \cos 12t \right] \qquad ..(6.83)$$

The particular solution for a step input is the final value of the variable. From the circuit diagram of Fig. 6.7, the longtime performance after a step causes the inductor to act as a short and the capacitor to act as an open circuit. The final output voltage, therefore, equals the input voltage

$$e_o(\infty) = e_i(\infty) = 1 \text{ V} \qquad ..(6.84)$$

The complete solution has the form:

$$e_o(t) = \varepsilon^{-5t} \left[A_3 \sin 12t + A_4 \cos 12t \right] + 1 \qquad ..(6.85)$$

The arbitrary constants A_3 and A_4, must be evaluated from the initial conditions in the circuit subject to a step input. For this circuit,

$$e_o(0^+) = 0 \qquad ..(6.86)$$

because the capacitor is a short circuit to a sudden change. The value of other initial conditions is obtained by evaluating the elemental equation for the capacitance at $t = 0^+$. Thus:

$$\frac{de_o}{dt}(0^+) = \frac{1}{C} i(0^+) = \frac{1}{RC} e_i(0^+) = 10 \qquad ..(6.87)$$

From these two initial conditions, we find that

$$A_4 = -1 \text{ and } A_3 = 0.417 \qquad ..(6.88)$$

so that the complete solution is:

$$e_o(t) = \left[\varepsilon^{-5t}(0.417 \sin 12t - \cos 12t) + 1\right] U_s(t) \qquad ..(6.89)$$

which agrees with the previous result. Fig. 6.8 shows

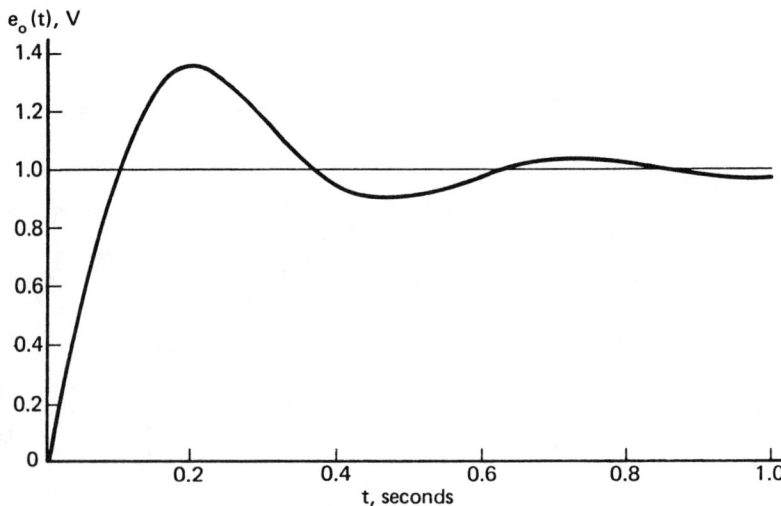

Figure 6.8 Output voltage (example 6.2)

the step response. The output voltage starts from zero and reaches 1.375 volts at t = 0.2 s. Thereafter, the voltage approaches the step value of 1 v in an oscillatory manner.

6.4 RESPONSE OF SECOND-ORDER CIRCUITS TO OTHER INPUTS

We have developed the response of second-order circuits to excitation by a step function. Thus, as we have shown in Chapter V, for first-order circuits, we may use the properties of linear differential equations to determine the response for other inputs. The response of a linear circuit to inputs that are related to the step function by integration or differentiation can be obtained from the step response by integration or differentiation. To find the ramp or impulse response, we may rewrite from (5.62) and (5.66)

$$y_r(t) = \int_0^t y_s(t') \, dt' \qquad ..(6.90a)$$

$$y_i(t) = \frac{d}{dt}\left[y_s(t)\right] \qquad ..(6.90b)$$

When the input is a composite function, we may separate it into a sum of basic functions and find the composite response by superposition of the basic function response. As a result, if we desire the response of a second-order circuit to any specific input, we first determine its step response. Then we use the relation between the specific input and the step input to operate appropriately on the step response.

Example 6.3

Water flows into a tank from above and then out through a fluid line [Fig. 6.9(a)]. The capacitance of the tank, $C = 0.1 \text{ m}^5/\text{N}$, and the line has resistance, $R = 120 \text{ N·s/m}^5$ and inertance, $L = 1000 \text{ N·s}^2/\text{m}^5$. Fig. 6.9(b) shows the flow rate $q(t)$ into the tank. Find an expression for the flow rate out of the tank, $q_o(t)$.

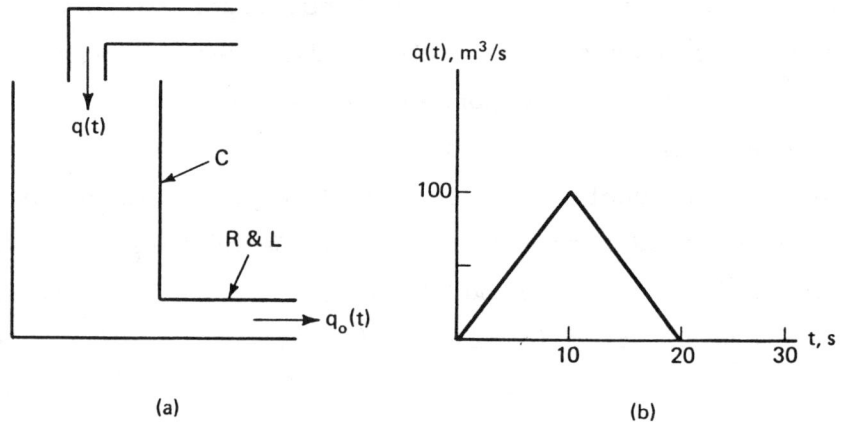

Figure 6.9 Fluid circuit (example 6.3)

The analogous circuit for the tank and line combination is shown in Fig. 6.10.

The inlet flow rate is a flow source. The tank represents a capacitance to ground and the line in this case is a series RL to ground. We designate the pressure at the bottom of the tank as p_1. The terminal, p_2, between R and L is fictitious in the sense that we cannot locate it physically. However, it is a convenient concept for the circuit analysis. A

Response of Second Order Circuits

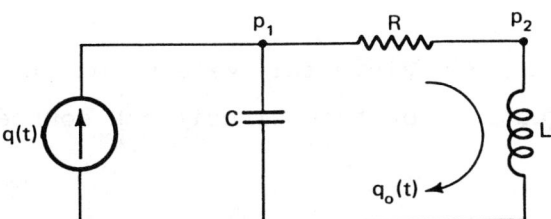

Figure 6.10 Analogous circuit (example 6.3)

compatibility relation around the $q_o(t)$ mesh yields:

$$P_1 = (P_1 - P_2) + P_2 \quad \text{(KVLA)} \qquad ..(6.91)$$

If the elemental equations are substituted into the compatibility relations, the result is:

$$\frac{1}{C} \int \left[q(t) - q_o(t) \right] dt = R\, q_o + L \frac{dq_o}{dt} \qquad ..(6.92)$$

By differentiation and rearrangement, we obtain the standard form:

$$\frac{d^2 q_o}{dt^2} + \frac{R}{L} \frac{dq_o}{dt} + \frac{1}{LC} q_o = \frac{1}{LC} q(t) \qquad ..(6.93)$$

The flow source, $q(t)$, may be described in terms of input functions as:

$$q(t) = 10\ U_r(t) - 20\ U_r(t - 10) + 10\ U_r(t - 20)$$
$$..(6.94)$$

Thus, when the elemental values are used, the differential equation for this circuit and source is:

$$\frac{d^2 q_o}{dt^2} + 0.12\ \frac{dq_o}{dt} + 0.01\ q_o = 0.1\ U_r(t) - 0.2\ U_r(t - 10)$$
$$+ 0.1\ U_r(t - 20) \qquad ..(6.95)$$

Comparison of the coefficients of this circuit equation with the standard form of (6.15) gives

$$2\delta\omega_N = 0.12 \quad \text{and} \quad \omega_N^2 = 0.01 \qquad ..(6.96)$$

From this, we find that

$$\omega_N = 0.1\ \text{rad/s} \quad \text{and} \quad \delta = 0.6 \qquad ..(6.97)$$

Thus, the circuit is underdamped.

We proceed now through the physical approach to find the step response. Since

$$\alpha = \delta\omega_N = 0.06\ \text{s}^{-1} \qquad ..(6.98a)$$

and $\qquad \omega_d = \omega_N(1 - \delta^2)^{1/2} = 0.08\ \text{rad/s} \qquad ..(6.98b)$

the homogeneous solution is:

$$q_H = \varepsilon^{-0.06t}\left[A_1 \sin 0.08t + A_2 \cos 0.08t\right]$$
..(6.99)

A long time after a unit step change, the inertance is shorted and the capacitance is open. Under these circumstances, all the source flow passes directly to the output and

$$q_p = q_o(\infty) = 1 \qquad ..(6.100)$$

The complete solution for the step input is:

$$q_{os}(t) = \varepsilon^{-0.06t}\left[A_1 \sin 0.08t + A_2 \cos 0.08t\right] + 1$$
..(6.101)

The initial conditions are now required to evaluate the arbitrary constants A_1 and A_2. Instantaneously, the capacitance is short and the inertance is open. Thus:

$$q_o(0^+) = 0 \qquad ..(6.102a)$$

$$\frac{dq_o}{dt}(0^+) = \frac{p_2(0^+)}{L} = 0 \qquad ..(6.102b)$$

With the use of these conditions,

$$A_1 = -0.75 \quad \text{and} \quad A_2 = -1 \qquad ..(6.103)$$

so that the step response is:

$$q_{os}(t) = 1 - \varepsilon^{-0.06t}\left[0.75 \sin 0.08t + \cos 0.08t\right]$$
$$..(6.104)$$

To obtain the ramp response [Eq. (6.90a)], we must integrate the step response between the limits t' = 0 and t' = t. Integration yields the unit ramp response as:

$$q_{or}(t) = t - 1.2 + 1.25\, \varepsilon^{-0.06t} \cos(0.08t + \phi)$$
$$..(6.105a)$$

where $\quad \phi = 16.3^\circ$ or 0.28 rad $\qquad ..(6.105b)$

This is the unit ramp response. However, recall that the input function is:

$$q(t) = 10\, U_r(t) - 20\, U_r(t - 10) + 10\, U_r(t - 20)$$
$$..(6.106)$$

Thus, by superposition and multiplication, the response for the given input flow may be expressed as:

$q_o(t) =$

$$10\left[t - 1.2 + 1.25\, \varepsilon^{-0.06t} \cos(0.08t + 0.28)\right] U_s(t)$$

$$- 20\left[t - 11.2 + 1.25\, \varepsilon^{-0.06(t-10)} \cos\left\{0.08(t-10) + 0.28\right\}\right] U_s(t-10)$$

$$+ 10\left[t - 21.2 + 1.25\, \varepsilon^{-0.06(t-20)} \cos\left\{0.08(t-20) + 0.28\right\}\right] U_s(t-20)$$

..(6.107)

Fig. 6.11 shows the output flow response and the input flow for this circuit. When the circuit is

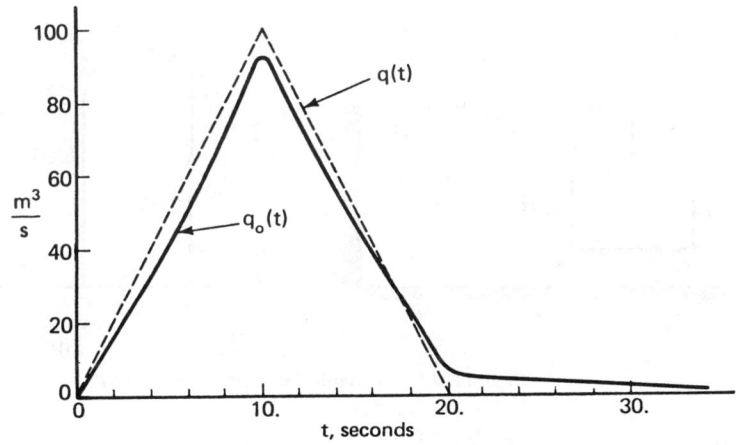

Figure 6.11 Output flow (example 6.3)

subjected to this input, the output flow initially lags behind the input flow. This fluid is stored in the tank. When the input is turned off, (t = 20), there is still some fluid left in the tank. This fluid runs out at a slow rate, as shown.

Example 6.4

Fig. 6.12 shows a simple mechanical circuit that consists of a mass, a spring and a damper. The applied force, f(t), is a series of two impulses delivered by a hydraulic ram. In terms of input functions, the applied force is:

$$f(t) = 1000\ U_i(t) + 1500\ U_i(t - 10),\ N. \quad ..(6.108)$$

If the mass m = 1000 kg, the spring, k = 7.5 N/m and the damper, b = 37.5 N.s/m, find the velocity of the mass as a function of time.

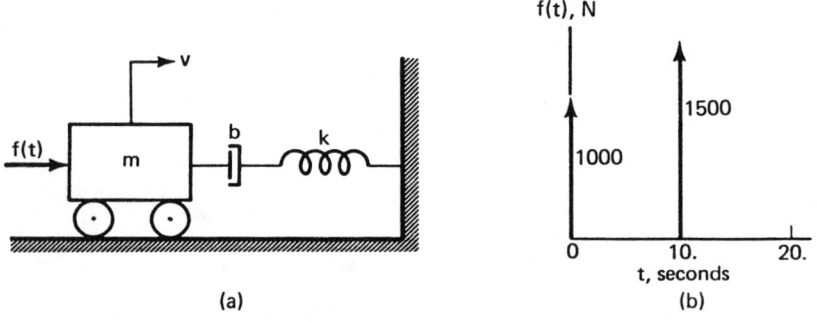

Figure 6.12 Mechanical circuit (example 6.4)

The analogous circuit for this example is the same

as the circuit in Example 6.3 (Fig. 6.10). The usual procedure for formulating the circuit equation yields:

$$\frac{d^2v}{dt^2} + \frac{k}{b}\frac{dv}{dt} + \frac{k}{m}v = \frac{k}{mb}f(t) + \frac{1}{m}\frac{df(t)}{dt} \qquad ..(6.109)$$

For the particular elements given

$$\frac{d^2v}{dt^2} + 0.2\frac{dv}{dt} + 0.0075\,v = 0.2(10^{-3})f(t) + 10^{-3}\frac{df(t)}{dt}$$
$$..(6.110)$$

A comparison of the coefficients with (6.15) indicates that the damping ratio, $\delta = 1.155$, and the circuit is therefore overdamped. The characteristic equation of the circuit is:

$$r^2 + 0.2\,r + 0.0075 = 0 \qquad ..(6.111)$$

which has two real roots. The roots are $r_1 = -0.05$ and $r_2 = -0.15$. The homogeneous solution has the form:

$$v_H = A_1\,\varepsilon^{-0.05t} + A_2\,\varepsilon^{-0.15t} \qquad ..(6.112)$$

If we use the physical approach, we must now find the final and the initial conditions for a step input. The final velocity depends only on the damper and the unit step force and hence

$$v_P = v(\infty) = \frac{f(\infty)}{b} = 0.0267 \quad \quad ..(6.113)$$

The initial conditions are:

$$v(0^+) = 0, \quad \frac{dv}{dt}(0^+) = \frac{1}{m} = 0.001 \quad \quad ..(6.114)$$

The complete response with arbitrary constants evaluated from the initial conditions is:

$$v_s(t) = 0.0033\,\varepsilon^{-0.15t} - 0.03\,\varepsilon^{-0.05t} + 0.0267$$
$$..(6.115)$$

We use (6.90b) and superposition to determine the response to the two impulses as:

$$v(t) = \left[1.5\,\varepsilon^{-0.05t} - 0.5\,\varepsilon^{-0.15t}\right]U_s(t)$$
$$+ \left[2.25\,\varepsilon^{-0.05(t-10)} - 0.75\,\varepsilon^{-0.15(t-10)}\right]U_s(t-10)$$
$$..(6.116)$$

The response is shown plotted in Fig. 6.13. The effects of the first impulse are still appreciable when the second impulse is applied. To find the distance moved by the mass, we would integrate the velocity response.

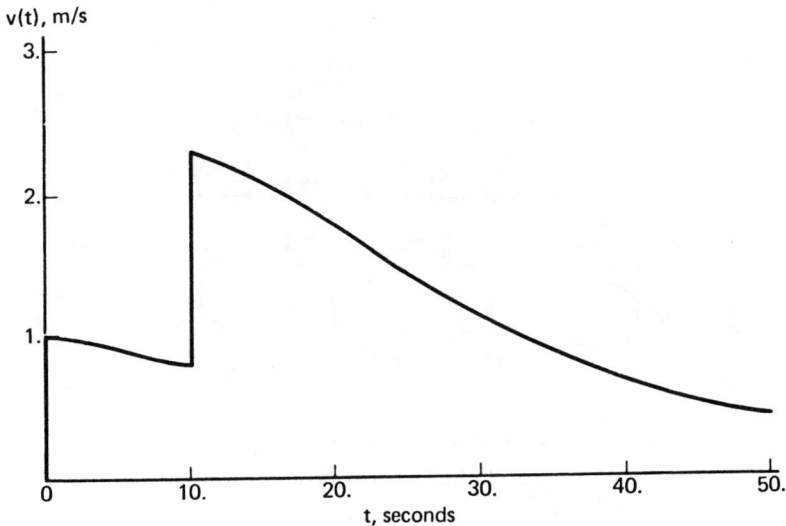

Figure 6.13 Velocity of mass (example 6.4)

6.5 RESPONSE OF SECOND-ORDER CIRCUITS WITH INITIAL STORAGE

Recall that when an A-type or T-type component has initial storage, we replace the component by a special circuit model. For A-type storage, a potential source, $E_o \cdot U_s(t)$, is placed in series with an A-type component having the same value. For T-type storage, a flow source, $I_o \cdot U_s(t)$, is placed in parallel with a T-type component having the same value.

Fig. 6.14 shows a series RLC circuit with initial storage in both capacitor and inductor. The resulting circuit has three sources. To obtain, for example, the current $i(t)$, we use the superposition of sources. We

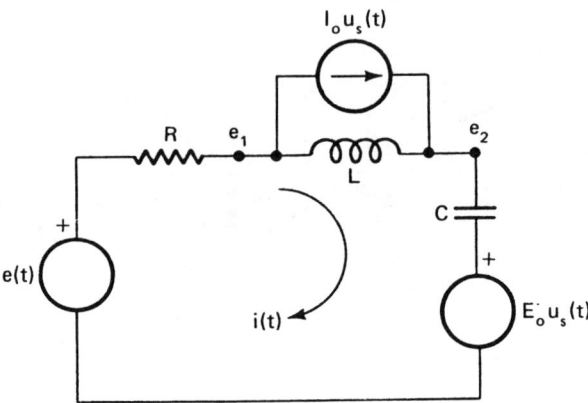

Figure 6.14 RLC circuit with initial storage

must, therefore, determine the effects of each source acting separately. Then

$$i(t) = i_1(t) + i_2(t) + i_3(t) \qquad ..(6.117)$$

where $i_1(t)$, $i_2(t)$ and $i_3(t)$ are the responses due to $e(t)$, $I_o \cdot U_s(t)$ and $E_o \cdot U_s(t)$ respectively. Fig. 6.1 shows the appropriate circuit diagram to determine $i_1(t)$. Both storage sources have been suppressed. The differential equation for $i_1(t)$ has been derived in Section 6.1, and is:

$$\frac{d^2 i_1}{dt^2} + \frac{R}{L}\frac{di_1}{dt} + \frac{1}{LC} i_1 = \frac{1}{L}\frac{de(t)}{dt} \qquad ..(6.118)$$

We must solve this equation by one of the methods

Response of Second Order Circuits

previously given. The form of e(t), of course, must be specified.

To find $i_2(t)$, the flow due to T-type storage, the input source, e(t), and the A-type storage source, $E_o \cdot U_s(t)$, are suppressed. The circuit acting with only the current storage source, $I_o \cdot U_s(t)$, is shown in Fig. 6.15(a). The differential equation for this circuit is found by inserting the component equations into the compatibility relation

$$- e_1 + (e_1 - e_2) + e_2 = 0 \qquad ..(6.119)$$

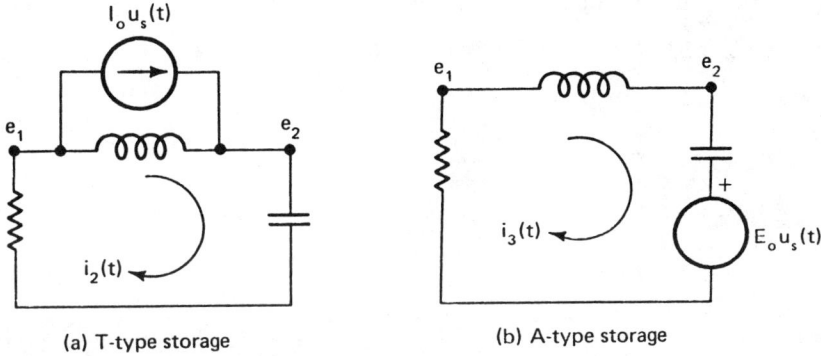

(a) T-type storage (b) A-type storage
Figure 6.15 Storage sources acting alone in RLC circuit

The result is:

$$Ri_2 + L\frac{d}{dt}\left[i_2 - I_o \cdot U_s(t)\right] + \frac{1}{C}\int i_2 \, dt = 0$$

$$..(6.120)$$

Now, differentiation and rearrangement leads to:

$$\frac{d^2 i_2}{dt^2} + \frac{R}{L}\frac{di_2}{dt} + \frac{1}{LC} i_2 = I_o \cdot U_d(t) \qquad ..(6.121)$$

The equation representing $i_2(t)$ has the same characteristic form as the equation for $i_1(t)$. In all cases, the characteristic equation of the circuit will not depend on the storage. However, when the storage is present, the input or the driving function will be specified by the product of the magnitude of the storage and some input function or functions.

The circuit acting with only the voltage storage source, $E_o \cdot U_s(t)$, is shown in Fig. 6.15(b). From the compatibility relation (6.119) and the component relations, we have:

$$Ri_3 + L\frac{di_3}{dt} + \frac{1}{C}\int i_3\, dt + E_o \cdot U_s(t) = 0 \qquad ..(6.122)$$

or

$$\frac{d^2 i_3}{dt^2} + \frac{R}{L}\frac{di_3}{dt} + \frac{1}{LC} i_3 = -\frac{1}{L} E_o \cdot U_i(t) \qquad ..(6.123)$$

The input in this case is specified as $-E_o \cdot U_i(t)/L$. Now, we could solve the differential equations for i_1, i_2 and i_3 by the mathematical approach. However, it is much more convenient to use the physical approach. This requires the determination of the initial and the final conditions. For storage type sources, the final condition is always zero. We might expect this from a practical viewpoint, since the stored energy is limited and eventually must run out. The following examples will demonstrate the treatment of second order circuits with storage.

Example 6.5

Fig. 6.16(a) shows a pump, p(t), supplying water to a tank through a fluid line. The tank has a capacitance $C = 0.167 \text{ m}^5/\text{N}$ and initially is partially full of water so that the initial pressure at the bottom of the tank is 5000 Pa. The line has a resistance $R = 50 \text{ N.s/m}^5$ and an inertance $L = 100 \text{ N.s}^2/\text{m}^5$. When the pump is turned on, it takes 10 seconds to develop its full pressure. The pump pressure-time relation [p(t) vs t] is shown in Fig. 6.16(b). Find the pressure at the bottom of the tank as a function of time [$p_c(t)$].

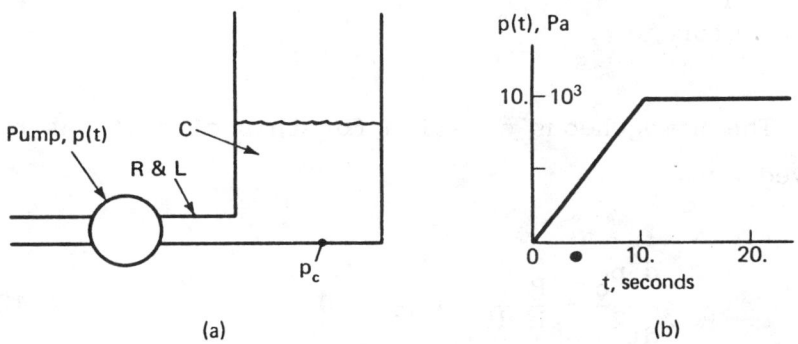

Figure 6.16 Fluid circuit (example 6.5)

Fig. 6.17 shows the analogous circuit. There are two sources, (i) the pump and (ii) the initial A-type storage.

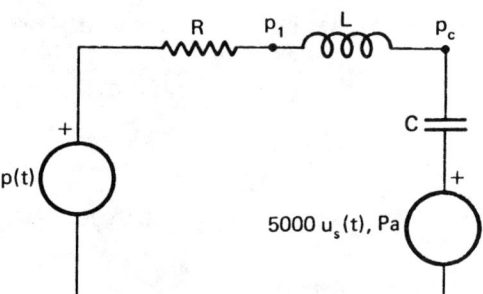

Figure 6.17 Analogous circuit (example 6.5)

To find $p_c(t)$, we must sum the responses of the sources acting alone. Thus:

$$p_c(t) = p_{c_1}(t) + p_{c_2}(t) \qquad ..(6.124)$$

where p_{c_1} is due to the pump and p_{c_2} is due to the initial storage.

The homogeneous equation for this circuit has been derived as:

$$\frac{d^2 p_c}{dt^2} + \frac{R}{L}\frac{dp_c}{dt} + \frac{1}{LC} = 0 \qquad ..(6.125)$$

and for the components given, the characteristic equation is:

$$r^2 + 0.5\, r + 0.06 = 0 \qquad ..(6.126)$$

Response of Second Order Circuits 387

The circuit is overdamped with roots of -0.2 and -0.3.

Pump Contribution (Storage source suppressed)
(a) Unit Step Response
(i) Homogeneous Solution,
$$p_H = A_1 \varepsilon^{-0.2t} + A_2 \varepsilon^{-0.3t} \qquad ..(6.127a)$$
(ii) Final Conditions, $p(\infty) = 1$ ..(6.127b)
(iii) Initial Conditions,
$$p(0^+) = 0, \quad \frac{dp}{dt}(0^+) = 0 \qquad ..(6.127c)$$
(iv) Step Response,
$$p_{1s}(t) = -3\varepsilon^{-0.2t} + 2\varepsilon^{-0.3t} + 1 \qquad ..(6.127d)$$

(b) Response to $p(t)$

Now $\quad p(t) = 1000\, U_r(t) - 1000\, U_r(t - 10) \qquad ..(6.128)$

Thus, we must integrate the step response and multiply it by 1000. The result is:

$$\begin{aligned}p_{c1}(t) = &\left[1000\, t - 8333 + 15000\, \varepsilon^{-0.2t} - 6670\, \varepsilon^{-0.3t}\right] U_s(t) \\ &- \left[1000\, (t-10) - 8333 + 15000\, \varepsilon^{-0.2(t-10)} - 6670\, \varepsilon^{-0.3(t-10)}\right] U_s(t-10)\end{aligned}$$
$$..(6.129)$$

Storage Source Contribution (Pump source suppressed)

(a) Unit Step Response

(i) Homogeneous solution,
$$p_H = A_3 \varepsilon^{-0.2t} + A_4 \varepsilon^{-0.3t} \qquad ..(6.130a)$$

(ii) Final condition, $p(\infty) = 0$ ..(6.130b)

This is always zero for storage purposes.

(iii) Initial Conditions,
$$p(0^+) = 1, \quad \frac{dp}{dt}(0^+) = 0 \qquad ..(6.130c)$$

(iv) Step response,
$$p_{2s}(t) = 3\varepsilon^{-0.2t} - 2\varepsilon^{-0.3t} \qquad ..(6.130d)$$

(b) Response to $5000.U_s(t)$

$$p_{C_2}(t) = \left[15000\,\varepsilon^{-0.2t} - 10000\,\varepsilon^{-0.3t}\right] U_s(t) \qquad ..(6.131)$$

Tank pressure

We may sum the contributions from the pump and the storage source to obtain:

$$p_C(t) = \left[1000t - 8333 + 30000\varepsilon^{-0.2t} - 16670\varepsilon^{-0.3t}\right] U_s(t)$$

$$- \left[1000(t-10) - 8333 + 15000\,\varepsilon^{-0.2(t-10)}\right.$$

$$-6670 \, \varepsilon^{-0.3(t - 10)}\Big] \, U_s(t - 10) \quad \quad ..(6.132)$$

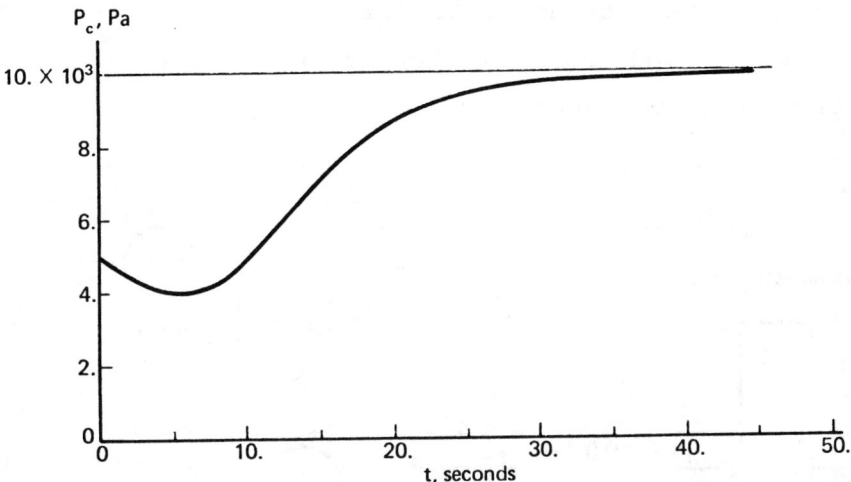

Figure 6.18 Tank pressure (example 6.5)

Fig. 6.18 shows the tank pressure $p_c(t)$. The initial pressure is 5000 Pa. Immediately after the pump is turned on, the pressure decreases. This is a consequence of the pump characteristic which does not come up to the rated pressure for 20 seconds. Thus, initially the pressure at the bottom of the tank is greater than the pump pressure and water will run out of the tank. At about 5 seconds, the pump begins to take effect and water is pumped back into the tank. The pressure continues to rise thereafter until the pressure at the bottom of the tank equals the pump pressure.

Example 6.6

The flywheel in the rotational mechanical circuit

shown in Fig. 6.19(a) has a moment of inertia $J = 50$ kg.m^2 and is initially turning clockwise at 10 rad/s. The spring with constant $K = 2$ N.m has an initial counterclockwise torque of 50 N.m. The damper, $B = 20$ N.m.s/rad. Find the velocity of the flywheel as a function of time, $[\omega(t)]$.

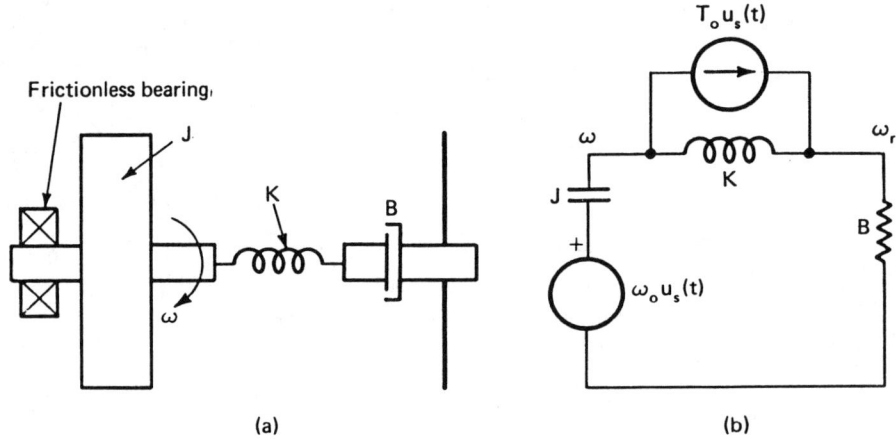

Figure 6.19 Rotational mechanical circuit (example 6.6)

Fig. 6.19(b) shows the analogous circuit. There are two sources: (i) the initial A-type storage of the flywheel, $10.U_s(t)$, and (ii) the initial T-type storage of the spring, $50.U_s(t)$. The angular velocity response will be the sum of the responses due to each source.

$$\omega(t) = \omega_1(t) + \omega_2(t) \qquad ..(6.133)$$

where $\omega_1(t)$ and $\omega_2(t)$ are the responses of the A-type and T-type sources respectively when acting alone. The homogeneous equation for this circuit has the form:

Response of Second Order Circuits 391

$$\frac{d^2T}{dt^2} + \frac{K}{B}\frac{dT}{dt} + \frac{K}{J}T = 0 \qquad \qquad ..(6.134)$$

For the components in this circuit, the characteristic equation is:

$$r^2 + 0.1\,r + 0.04 = 0 \qquad \qquad ..(6.135)$$

from which we find that the circuit is underdamped with

$$\delta = 0.25 \text{ and } \omega_N = 0.2 \text{ rad/s} \qquad ..(6.136)$$

Therefore,

$$\alpha = 0.05 \text{ s}^{-1} \text{ and } \omega_d = 0.194 \text{ rad/s} \qquad ..(6.137)$$

Flywheel Storage Contribution (Spring source suppressed)
(a) **Unit Step Response**
(i) Homogeneous Solution,

$$\omega_H = \varepsilon^{-\alpha t}\left[A_1 \sin \omega_d t + A_2 \cos \omega_d t\right] \qquad ..(6.138a)$$

(ii) Final Condition,

$$\omega(\infty) = 0 \text{ (always for storage)} \qquad ..(6.138b)$$

(iii) Initial Conditions,

$$\omega(0^+) = 1,\ \frac{d\omega}{dt}(0^+) = 0 \qquad ..(6.138c)$$

(iv) Unit Step Response,

$$\omega_{1s} = \varepsilon^{-0.05t}\left[0.258 \sin 0.194t + \cos 0.194t\right]$$

.. (6.138d)

(b) **Response to $10.U_s(t)$, (clockwise)**

$$\omega_1(t) = \varepsilon^{-0.05t}\left[2.58 \sin 0.194t + 10 \cos 0.194t\right]$$

.. (6.139)

Spring Storage Contribution (Flywheel source suppressed)

(a) **Unit Step Response**

(i) Homogeneous Solution,

$$\omega_H = \varepsilon^{-\alpha t}\left[A_3 \sin \omega_d t + A_4 \cos \omega_d t\right] \quad ..(6.140a)$$

(ii) Final Conditions,

$$\omega(\infty) = 0 \text{ (always for storage)} \quad ..(6.140b)$$

(iii) Initial Conditions.

$$\omega(\infty) = 0, \quad \frac{d\omega}{dt}(0^+) = \frac{1}{J} = 0.02 \quad ..(6.140c)$$

(iv) Unit Step Response,

$$\omega_{2s} = -\varepsilon^{-0.05t}\left[0.103 \sin 0.194t\right] \quad ..(6.140d)$$

(b) **Response to $50U_s(t)$, (counterclockwise)**

$$\omega_2(t) = -\left[5.15 \ \varepsilon^{-0.05t} \sin 0.194t\right] U_s(t) \quad ..(6.141)$$

Angular Velocity:

We sum the contributions. The polarity in the

circuit diagram already includes the proper directions. Thus:

$$\omega(t) = \varepsilon^{-0.05t}\left[-2.57 \sin 0.194t + 10 \cos 0.194t\right] U_s(t)$$

$$= 10.32 \, \varepsilon^{-0.05t} \cos(0.194t + \phi) \qquad ..(6.142a)$$

where $\phi = 14.4°$ ..(6.142b)

Fig. 6.20 shows the angular velocity response.

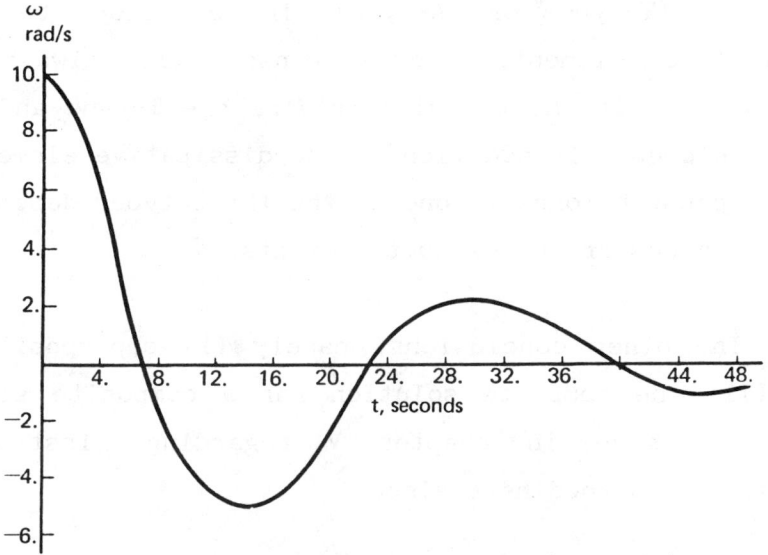

Figure 6.20 Angular velocity of flywheel (example 6.6)

6.6 SUMMARY AND DISCUSSIONS

In this chapter, we have considered second order circuits governed by second order differential equations. The particular solution depends on the type of input to the circuit. It is the natural response of such a circuit which has to be considered in detail. It can be one of three responses, namely:

(a) overdamped response, for which the roots of the characteristic polynomial are negative real and distinct,

(b) underdamped response, for which the roots of the characteristic polynomial are complex conjugates,

(c) critically damped response, for which the roots of the characteristic polynomial are negative real and coincident.

If the circuit contains two of the same type of storage elements (T-type or A-type) in addition to the dissipative element, the response will always be overdamped. If the circuit contains one T-type and one A-type element (in addition to the dissipative element), the response belongs to one of the three types depending on the values of the circuit elements.

The other conclusions, namely (i) superposition, and (ii) the complete solution for a composite signal function, made in chapter V regarding first-order circuits hold good here also.

6.7 PROBLEMS

6.1 The electrical circuit shown in Fig. P6.1 has L = 0.01 H, C = 4 μF, R = 100 Ω, and i(t) = 0.01 $U_s(t)$, A. Find $i_R(t)$ and $e_L(t)$.

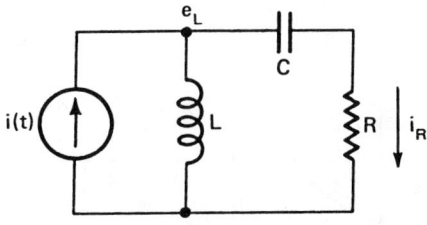

Figure P 6.1

6.2 In the electric circuit shown in Fig. P6.2, $e(t) = 100\ U_s(t)$, V, $L = 0.3125$ H, $C = 0.2\ \mu F$ and $R = 781.25\ \Omega$. Find $e_1(t)$ and $i_R(t)$.

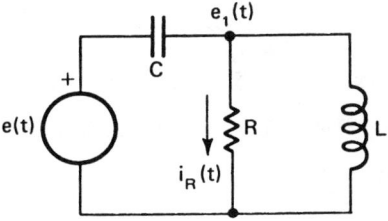

Figure P 6.2

6.3 In a parallel RLC electric circuit, the inductor has an initial current of 0.1 A, and the capacitor has an initial charge of $15(10^{-6})$ coulombs. If $L = 0.5$ H, $C = 0.25\ \mu F$ and $R = 667\ \Omega$, find the charge on the capacitor as a function of time.

6.4 An electric circuit consists of the parallel combination of $L = 0.133$ H and $C = 5\mu F$ connected in series with a resistance $R = 80\ \Omega$ to ground. The input is a voltage source, $e(t)$, shown in Fig. P6.4.

Find the voltage across the resistance as a function of time.

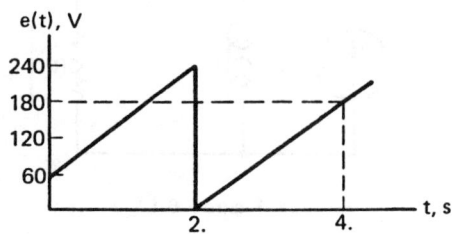

Figure P 6.4

6.5 A constant pressure pump is suddenly started to supply water to the fluid circuit shown in Fig. P6.5. The circuit consists of two resistive lines and two tanks.

(a) Find the characteristic equation of the circuit.
(b) Prove that the damping factor, δ, must always be greater than unity.

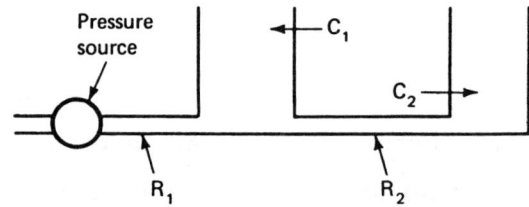

Figure P 6.5

6.6 A water tank has two outlet lines and one inlet line. One of the outlet lines is modeled as a pure resistance, $R = 12.5 \text{ N.s/m}^5$ and the other as a pure inertance, $L = 16.67 \text{ N.s}^2/\text{m}^5$. The inlet line has negligible resistance and inertance. It contains a flow pump which delivers $q\, U_r(t)$, m^3/s. The capacitance of the tank $C = 0.02\, m^5/N$. Find q, if the flow through the resistive line $q_R(2) = 31.29\, m^3/s$.

6.7 A pneumatic system, shown in Fig. P6.7, consists of a resistive line, an inertive line and a bellows capacitance. All pressures are initially at atmospheric pressure. The pump is suddenly turned on and produces a pressure input of $(10^{-4})\, U_s(t)$, Pa. If $L = 0.32(10^{10})\, N.s^2/m^5$, $C = 50(10^{-10})\, m^5/N$ and $R = 0.096(10^{10})\, N.s/m^5$, find the rate of flow through the inertive line.

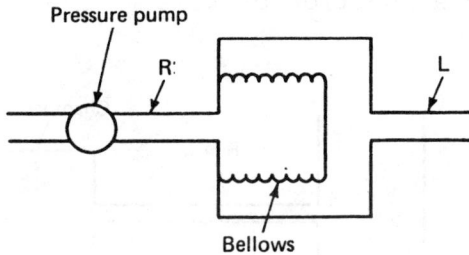

Figure P 6.7

6.8 A water tank is initially filled with water, when two outlet lines are suddenly opened. The capacitance of the tank is $C = 0.25\, m^5/N$. One outlet line is resistive and the other is inertive. Measurements show that the flow through the inertive line may be expressed as:

$$q_L(t) = 4000\,[\varepsilon^{-0.5t} - \varepsilon^{-2t}]\,U_s(t),\ m^3/s$$

Find the initial pressure at the bottom of the tank.

6.9 A crude pressure transducer can be constructed by attaching a strain gage to the flexible member covering a cylindrical chamber. A test is then performed to determine the response of the transducer. The test set-up consists of a long line [$L = 2(10^5)$ N.s^2/m^5, $R = 7(10^6)$ N.s/m^5] leading to the cylindrical chamber [$C = 167(10^{-10})$ m^5/N]. The line is connected to a shock tube [Fig. P6.9] which generates a step in pressure of 200 $U_s(t)$, Pa. Find an analytical expression for the pressure in the chamber as a function of time.

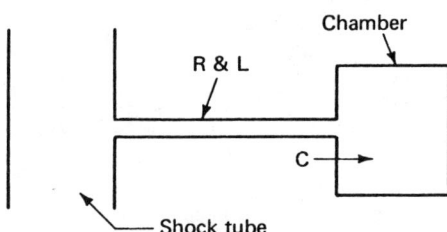

Figure P 6.9

6.10 The mass-spring system shown in Fig. P6.10 is subjected to a horizontal impulsive force of 100 $U_i(t)$, N. The mass is 200 kg and the spring constant is 3,200 N/m. If the mass is initially at rest, find the velocity of the mass as a function of time.

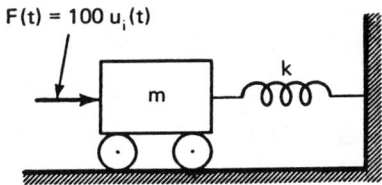

Figure P 6.10

6.11 The mechanical system shown in Fig. P6.11 is initially at rest. The mass, m = 10 kg, the spring, k = 20 N/m and the damper b = 20/3 N.s/m. Find the velocity of the mass, v(t), if the applied input force f(t) is:
(a) $f(t) = 10\ U_s(t)$, N
(b) $f(t) = 10\ U_r(t)$, N
(c) $f(t) = 10\ U_i(t)$, N
(d) $f(t) = 20\ U_s(t) - 10\ U_s(t-2)$, N

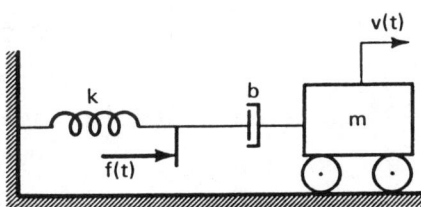

Figure P 6.11

6.12 In the mechanical system for Problem 6.11, the spring constant is changed to k = 10 N/m and the damper is changed to b = 25/3 N.s/m. Find v(t), if:

(a) $f(t) = 10 \, U_r(t) - 10 \, U_r(t-1)$, N

(b) $f(t) = 10 \, U_s(t-2) + 10 \, U_s(t-3)$, N

6.13 For the mechanical system shown in Fig. P6.13, m = 25 kg, k = 200 N/m and b = 100/3 N.s/m. The velocity input $v(t) = 5 \, U_i(t) - 5 \, u_i(t-2)$, m/s. Find:

(a) the velocity of the mass as a function of time, $v_m(t)$,

(b) the position of the mass as a function of time, $x_m(t)$.

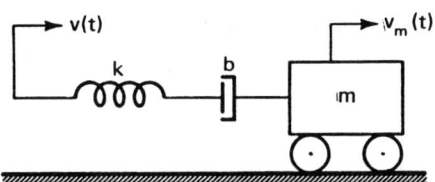

Figure P 6.13

6.14 A car is travelling along a straight road with constant forward velocity when it hits a small bump. The transverse velocity imparted to the car by the bump is shown in Fig. P6.14. The car suspension is modeled as a mass sitting atop the parallel combination of a damper and spring. If m = 1,500 kg, k = 60,000 N/m and b = 6,000 N.s/m, find $v_m(t)$, the transverse velocity of the mass.

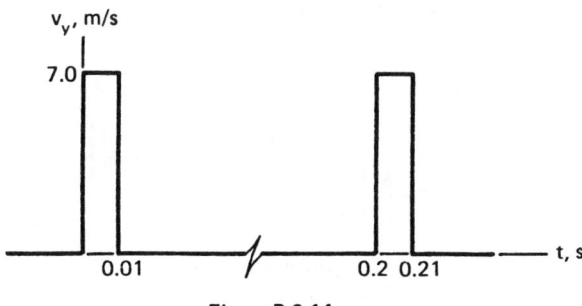

Figure P 6.14

6.15 A mass-spring-damper arrangement is shown in Fig. P6.15. The mass, m = 50 kg, the spring k = 200 N/m and the damper b = 200 N.s/m. The spring is initially stretched 2.025 m and then the mass is released from this position. The angle of the incline is 30 degrees. Find the velocity of the mass along the incline as a function of time.

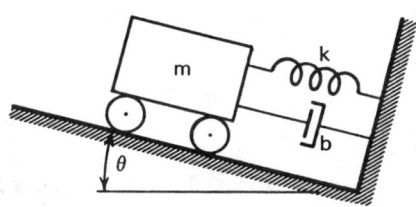

Figure P 6.15

6.16 The rotary mechanical system shown in Fig. 6.16 is subjected to an input torque of 1,350 $U_s(t)$, N.m. The moment of inertia of the flywheel J = 30 kg.m^2, the spring constant K = 30 N.m and the rotary damper B = 15 N.m.s. Express the angular velocity of the flywheel as a function of time.

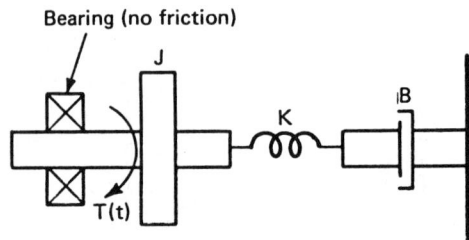

Figure P 6.16

6.17 For the rotational system shown in Fig. P6.17, $J = 4.5$ kg.m^2, $K = 27$ N.m/rad and $B = 22.5$ N.m.s. The angular velocity input $\omega(t) = 50\, U_s(t)$. rad/s. Find the angular velocity of the flywheel and the torque in the damper as functions of time.

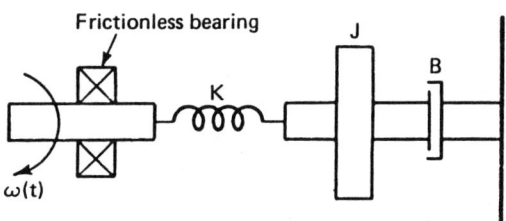

Figure P 6.17

6.18 In the rotary mechanical system shown in Fig. P6.18, the spring ($K = 82$ N.m/rad) has been initially turned through a clockwise angle of 3.5 radians. The flywheel ($J = 9$ kg.m^2) is subjected to a clockwise external torque of 315 $U_p(t)$, N.m. If the damper $B = 81$ N.m.s, find the angle $\theta_J(t)$ through which the flywheel turns.

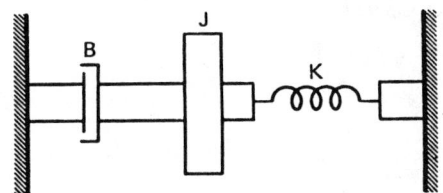

Figure P 6.18

6.19 For the rotational system shown in Fig. P6.19, J = 16 kg.m^2, K = 676 N.m/rad and B = 135.2 N.m.s. Initially with the clutch disengaged, the spring is twisted through an angle of 1 radian clockwise and the flywheel is rotating with an angular velocity of 6 rad/s counterclockwise. Find the angular velocity of the flywheel as a function of time, after the clutch has been engaged.

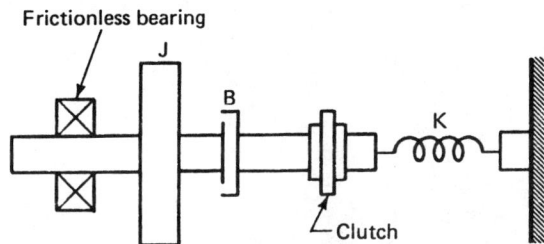

Figure P 6.19

6.20 Wind acting on the tail of a model rocket in flight acts as a torque input of $2(10^{-3})$ $U_s(t)$, N.m. This torque causes the rocket to rotate about an axis perpendicular to its length (pitch). The rocket can be modeled as a second-order system with J = $4(10^{-4})$ kg.m^2, K = 0.4 N.m and B = $2.53(10^{-3})$ N.m.s. The damping is caused by air resistance and the spring effect is due to a restoring tendency. The damping and spring are in parallel. Find the angular pitch velocity, $\omega(t)$, and $\omega(0.25)$.

6.21 The compartments of a container are perfectly insulated from the outside room. There is a barrier between the compartments and its has a thermal resistance of $0.02 \,^\circ$C/W. Both compartments are filled with water that is initially at $20\,^\circ$C. The smaller compartment has a thermal capacitance of (10^5) J/$^\circ$C and the larger compartment has a thermal capacitance of $2(10^5)$ J/$^\circ$C. A heating element in the smaller compartment is suddenly turned on and supplies 3,000 $U_s(t)$, W. Find the temperature in each compartment when $t = 1,000$ s.

CHAPTER 7

GENERALIZED IMPEDANCES AND SYSTEM FUNCTIONS

The model of any physical system with energy storage results in a differential equation. Thus the determination of system response requires a solution of a differential equation. This holds for both the mathematical and physical approaches given in Chapters V and VI. We may ask if it is possible to model a system by an algebraic equation from which the differential equation can be derived. To answer this question, we must consider the effect of the general exponential input function. This type of driving function in its most general mathematical form is not normally encountered by a system. However, this function has two special cases (step function and sine wave) which represent the two most common driving functions used in practice.

7.1 THE GENERAL EXPONENTIAL INPUT FUNCTION

The most general form for the exponential function is given in (2.31). If the time domain is restricted to $t > 0$, we may express the exponential function as:

$$g(t) = G\varepsilon^{st} \cdot U_s(t) \qquad ..(7.1)$$

where, in general, s is a complex number given by

$$s = \sigma + j\omega \qquad ..(7.2)$$

In complex form, (7.1) may be written as:

$$g(t) = G\varepsilon^{\sigma t}\left[\cos \omega t + j \sin \omega t\right] \qquad ..(7.3)$$

The real Re[ ] and imaginary Im[ ] parts of the general exponential function are therefore:

$$\text{Re}\{g(t)\} = G\varepsilon^{\sigma t} \cos \omega t \, U_s(t) \qquad ..(7.4a)$$

$$\text{Im}[g(t)] = G\varepsilon^{\sigma t} \sin \omega t \, U_s(t) \qquad ..(7.4b)$$

The shapes of the input signals given in (7.4) depend on the values of σ and ω. Fig. 7.1 shows the function for four possible combinations of σ and ω. The most common special cases are the step input [Fig. 7.1(a)] and the sine wave [Fig. 7.1(c)]. Less well known are the decaying exponential [Fig. 7.1(b)] and the decaying sine wave [Fig. 7.1(d)]. As a result, if we can find the response of a given system to the general exponential function of (7.1), we may, by specification of s, obtain the response for all the special cases. Although we do not specify the form of s in the following responses, the examples given in this chapter will consider the effect of a purely real value. We will reserve the specification of s as a pure imaginary value until Chapter VIII, when we discuss the frequency response.

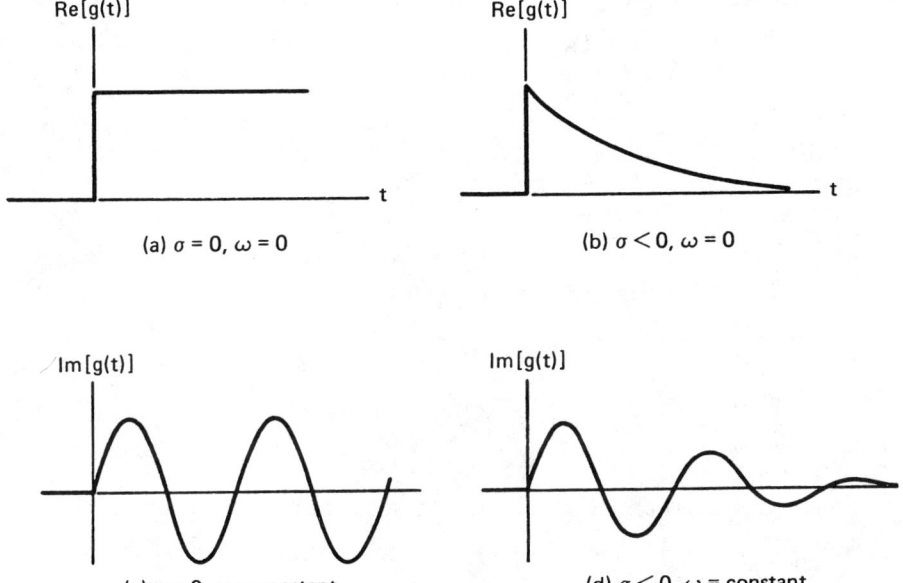

Figure 7.1 Some exponential functions

7.2 RESPONSE TO EXPONENTIAL FUNCTION

The differential equation for the general first-order circuit has been given in (5.1) and is:

$$\tau \frac{dy}{dt} + y = a_1 g(t) + a_2 \frac{dg(t)}{dt} \qquad ..(7.5a)$$

where, in this case

$$g(t) = G \varepsilon^{st} U_s(t) \qquad ..(7.5b)$$

We may obtain a complete solution for (7.5a) by decomposing the equation in the same manner as in (5.44). The result is:

$$\tau \frac{dy_1}{dt} + y_1 = a_1 G \varepsilon^{st} U_s(t) \qquad ..(7.6a)$$

$$\tau \frac{dy_2}{dt} + y_2 = a_2 G \frac{d}{dt}\left[\varepsilon^{st} U_s(t)\right] \qquad ..(7.6b)$$

$$y = y_1 + y_2 \qquad ..(7.6c)$$

We deal with (7.6a) first and recognize that

$$y_1 = y_{1H} + y_{1P} \qquad ..(7.7)$$

The homogeneous solution is:

$$A_1 \varepsilon^{-t/\tau} \qquad ..(7.8)$$

Generalized Impedances and System Functions 409

We assume a form for the particular solution, $A_2 \, \varepsilon^{st}$ and then evaluate A_2 by substituting this form into (7.5a). This procedure yields:

$$\tau s A_1 \varepsilon^{st} + A_2 \varepsilon^{st} = G a_1 \varepsilon^{st} \qquad ..(7.9)$$

from which we find that

$$A_2 = \frac{a_1 G}{(\tau s + 1)} \qquad ..(7.10)$$

and therefore:

$$y_{1P} = \frac{a_1 G}{\tau s + 1} \varepsilon^{st} \qquad ..(7.11)$$

The solution for $y_1(t)$ can now be formulated through the application of the initial condition $y(0^+) = 0$ and is:

$$y_1(t) = \frac{a_1 G}{\tau s + 1} \left[\varepsilon^{st} - \varepsilon^{-t/\tau} \right] U_s(t) \qquad ..(7.12)$$

Now,
$$y_2(t) = \frac{a_2}{a_1} \frac{dy_1}{dt} \qquad ..(7.13)$$

since the forcing function in (7.6b) is a linear function of the derivative of the forcing function in (7.6a). Thus

$$y_2(t) = \frac{a_2 G}{\tau s + 1}\left[s\, \varepsilon^{st} + \frac{1}{\tau}\varepsilon^{-t/\tau}\right] U_s(t) \quad\quad ..(7.14)$$

and the complete solution for (7.5) subject to an exponential input is

$$y(t) = G\left[\frac{a_1 + a_2 s}{\tau s + 1}\varepsilon^{st} + \frac{\frac{a_2}{\tau} - a_1}{\tau s + 1}\varepsilon^{-t/\tau}\right] U_s(t)$$
$$..(7.15)$$

When $s = 0$, the condition where the general exponential function becomes a step function, (7.15) reduces to the form given for the step response [Eq. (5.59)].

Example 7.1

A force $f(t) = F\,\varepsilon^{st}\, U_s(t)$ is applied to the mass-damper system shown in Fig. 7.2(a). Obtain an expression for the resulting velocity.

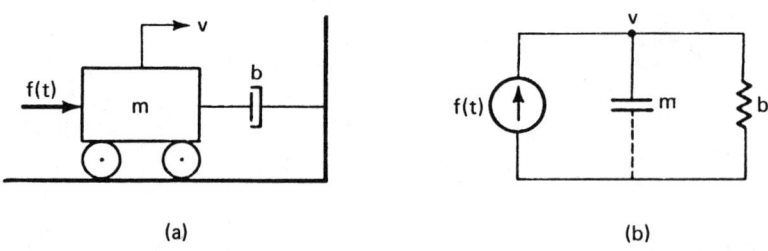

(a)   (b)

Figure 7.2 Mechanical circuit (example 7.1)

Generalized Impedances and System Functions

The analogous circuit diagram for this mechanical arrangement is shown in Fig. 7.2(b). The equilibrium equation is found as before in Example 5.6 as:

$$\frac{m}{b}\frac{dv}{dt} + v = \frac{1}{b} f(t) \qquad ..(7.16a)$$

$$= \frac{F}{b} \varepsilon^{st} U_s(t), \text{ for the forced response}$$
$$..(7.16b)$$

$$= 0, \text{ for the natural response} \qquad ..(7.16c)$$

The solution can be obtained directly from (7.15) by making the substitutions $a_1 = 1/b$, $a_2 = 0$, $G = F$ and $\tau = m/b$. However, let us go through the solution process for this case. As always the natural response (homogeneous solution) is an exponential of the form

$$v_H(t) = A_1 \varepsilon^{-t/\tau} \qquad ..(7.17)$$

In this system, the time constant

$$\tau = \frac{m}{b} \qquad ..(7.18)$$

The forced response (particular solution) of any linear system has the same shape as the driving function or its derivatives. Thus, for the mass damper-system under consideration, the forced response has the form:

$$v_p(t) = V \varepsilon^{st} \qquad ..(7.19)$$

where V is a constant that must be determined by substitution of the forced response form back into the differential equation. If we perform this operation, the result is:

$$(ms + b) V = F \quad \text{or} \quad V = \frac{F}{ms + b} \qquad ..(7.20)$$

Thus the complete solution for the velocity is:

$$v(t) = \frac{F}{ms + b} \varepsilon^{st} + A_1 \varepsilon^{-bt/m} \qquad ..(7.21)$$

where A_1 must be evaluated by application of the initial conditions. Since forces are constrained to finite values, the initial velocity of mass is the same as observed with step driving functions. That is:

$$v(0^+) = v(0^-) = 0 \qquad ..(7.22)$$

if the mass is at rest before the force is applied. From this condition, we find that:

$$A_1 = -\frac{F}{ms + b} \qquad ..(7.23)$$

and thus the complete solution for velocity is:

$$v(t) = \frac{F}{ms + b} \left[\varepsilon^{st} - \varepsilon^{-bt/m} \right] U_s(t) \qquad ..(7.24)$$

This result is consistent with the step response

of the system as may be observed by substituting the value s = 0.

For the case where the forcing function is the exponent,

$$s = -\frac{1}{\tau} = -\frac{b}{m} \qquad ..(7.25)$$

the above solution appears to be indeterminate (0/0). However, through the use of L'Hopitals' rule, we may show that:

$$v(t) = \lim_{s \longrightarrow -b/m} \left[\frac{\frac{d}{ds}(F\varepsilon^{st} - F\varepsilon^{-bt/m})}{\frac{d}{ds}(ms+b)} \right] \qquad ..(7.26)$$

and the solution is, therefore:

$$v(t) = \frac{Ft}{m}\varepsilon^{-bt/m} \qquad ..(7.27)$$

This form may be recognized as a special case of a critically damped second-order circuit [Eq. (6.32)]. The response of a first-order circuit driven by an exponential function with $s = -1/\tau$ may be, therefore, indistinguishable from the response of a critically damped second-order circuit, if there is no forcing function or the forced response to a step input is zero.

As a corollary of the discussion in Example 7.1, suppose we are given a response of the type:

$$y(t) = A_1 \varepsilon^{\alpha t} + A_2 \varepsilon^{\beta t} \qquad ..(7.28)$$

where α and β are negative. We are asked to determine what kind of a circuit produced such a response. There are actually three possibilities and we are unable to distinguish among them. The possibilities are:

(a) First-order circuit with time constant equal to $-1/\alpha$ and input exponential $\varepsilon^{\beta t}$.
(b) First-order circuit with time constant equal to $-1/\beta$ and input exponential $\varepsilon^{\beta t}$.
(c) Second-order circuit that is overdamped and has no forcing input function or the forced response to a step input is zero.

Second-Order Circuits

In general, second-order circuits produce the differential equation given in (6.1). This is:

$$\frac{d^2 y}{dt^2} + 2\delta\omega_N \frac{dy}{dt} + \omega_N^2 y = a_1 g(t) + a_2 \frac{dg(t)}{dt} + a_3 \frac{d^2 g(t)}{dt^2}$$

$$..(7.29)$$

If we apply an exponential input for $g(t)$ and decompose the resulting equation so that

$$y = y_1 + y_2 + y_3 \qquad ..(7.30a)$$

we obtain:

Generalized Impedances and System Functions

$$\frac{d^2y_1}{dt^2} + 2\delta\omega_N \frac{dy_1}{dt} + \omega_N^2 y_1 = a_1 G \varepsilon^{st} U_s(t) \qquad ..(7.30b)$$

$$\frac{d^2y_2}{dt^2} + 2\delta\omega_N \frac{dy_2}{dt} + \omega_N^2 y_2 = a_2 G \frac{d}{dt}\left[\varepsilon^{st} U_s(t)\right]$$
$$..(7.30c)$$

$$\frac{d^2y_3}{dt^2} + 2\delta\omega_N \frac{dy_3}{dt} + \omega_N^2 y_3 = a_3 G \frac{d^2}{dt^2}\left[\varepsilon^{st} U_s(t)\right]$$
$$..(7.30d)$$

The forcing function in (7.30c) is the derivative of the forcing function in (7.30b). Similarly, the forcing function in (7.30d) is the second derivative of the one in (7.30b). As a result, the complete solution can be expressed in terms of y_1 as:

$$y = a_1 y_1 + a_2 \frac{dy_1}{dt} + a_3 \frac{d^2 y_1}{dt^2} \qquad ..(7.31)$$

where y_1 is the response to $G \varepsilon^{st} U_s(t)$.

We now proceed to find the solution for y_1 as the sum of the homogeneous solution y_{1H} and the particular solution y_{1P}. As we have observed in Chapter VI, the homogeneous solution may take one of three forms depending on whether the system is overdamped, underdamped, or critically damped. To determine the particular solution, we must substitute $y_{1P} = A\,\varepsilon^{st}$ $U_s(t)$ into into (7.30b) and then evaluate A. This procedure leads to:

$$\left[s^2 + 2\delta\omega_N + \omega_N^2\right] A = G \qquad \qquad ..(7.32)$$

The solution for y is, therefore:

$$y_1(t) = \frac{G\,\varepsilon^{st}}{s^2 + 2\delta\omega_N + \omega_N^2} + y_H \qquad ..(7.33)$$

where $y_{1H} = y_H$, since there is only one homogeneous solution for each specific circuit. That is, $y_{1H} = y_{2H} = y_{3H} = y_H$. These constants may be evaluated from (7.33) by applying the initial conditions

$$y_1(0^+) = 0 \quad \text{and} \quad \frac{dy_1}{dt}(0^+) = 0 \qquad ..(7.34)$$

After the constants have been determined, the complete solution is obtained by substituting (7.33) into (7.31).

$$y(t) = \frac{G(a_1 + a_2 s + a_3 s^2)\,\varepsilon^{st}}{s^2 + 2\delta\omega_N s + \omega_N^2} +$$

$$a_1 y_H + a_2 \frac{dy_H}{dt} + a_3 \frac{d^2 y_H}{dt^2} \qquad ..(7.35)$$

Eq. (7.35) is merely meant to indicate the

formulation of a solution. In specific cases, each system should be solved without recourse to this general formulation.

7.3 IMPEDANCE OF BASIC COMPONENTS

It is important to notice that in (7.15) and (7.35), the use of the exponential forcing function resulted in the polynomials $(\tau s + 1)$ and $(s^2 + 2\delta\omega_N s + \omega_N^2)$ appearing in the forced response. These polynomials had previously been encountered as $(\tau r + 1)$ and $(r^2 + 2\delta\omega_N r + \omega_N^2)$ in the characteristic equations of first-order systems [Eq. (5..7)] and second-order systems [Eq. (6.18)]. The characteristic equations were used to obtain the natural responses of the systems. In each case, the factor r appeared as a result of a differentiation and the factor 1/r appeared as a result of an integration. In the case of exponential forcing functions, it should be evident that the factors s and 1/s appear for exactly the same reason. Thus it should not be surprising that the operations of differentiation and integration are expressible algebraically through the use of an operator. The choice of a symbol for the operator is quite arbitrary. We select the symbol s, since it is the most common in system and circuit analysis.

It should be possible, therefore, to find a method of writing equilibrium equations which are equally suitable for obtaining the natural response, the forced response or the complete response. We may accomplish this by representation of the basic components in algebraic terms rather than as time derivatives or integrals. The algebraic replacement used to describe the basic circuit components is called the **generalized impedance**.

The generalized impedance of an element is defined as the ratio of the amplitude of the exponential across variable to the amplitude of the exponential through variable. The generalized impedance, Z(s), is a function of the complex exponent, s, and is expressed mathematically as:

$$Z(s) = \frac{\text{Amplitude of an exponential across variable}}{\text{Amplitude of the corresponding through variable}}$$

..(7.36)

To derive the impedances of the three basic components, we begin with the elemental relations given in (3.1). The relations are:

$$av = K_1 \, (tv) \qquad \text{dissipative} \qquad ..(7.37a)$$

$$av = K_2 \frac{d}{dt} (tv) \qquad \text{T-type storage} \qquad ..(7.37b)$$

$$av = K_3 \int (tv) \, dt \qquad \text{A-type storage} \qquad ..(7.37b)$$

Now we assume that both the across variable difference and through variable can be represented by exponential functions. That is:

Generalized Impedances and System Functions 419

$$av = A\varepsilon^{st} \quad \quad \ldots(7.38a)$$

$$tv = T\varepsilon^{st} \quad \quad \ldots(7.38b)$$

When (7.38a) and (7.38b) are substituted into (7.37), the exponential terms cancel and we are left with a relationship between the amplitude A and T. If we then use the definition of generalized impedance as the ratio of A/T, we find that:

$$Z(s) = K_1, \quad \text{dissipative} \quad \ldots(7.39a)$$

$$Z(s) = K_2 s, \quad \text{T-type storage} \quad \ldots(7.39b)$$

$$Z(s) = K_3/s, \quad \text{A-type storage} \quad \ldots(7.39c)$$

Eqs. (7.39) are therefore the algebraic representation of the basic components that we had previously described in (7.37) with time differentials and integrals.

Table 7.1 lists the generalized impedances in the various physical media.

The reciprocal of the generalized impedance is sometimes a more convenient way to represent the basic elements. This is called the **generalized admittance**, $Y(s)$, and is defined as:

$$Y(s) = \frac{\text{Amplitude of an exponential through variable}}{\text{Amplitude of the corresponding across variable}}$$

$$\ldots(7.40)$$

TABLE 7.1 GENERALIZED IMPEDANCES OF ELEMENTS

	Dissipative	A-type	T-type
Mechanical Translation	1/b	1/ms	s/k
Mechanical Rotation	1/B	1/Js	s/k
Electric Circuit	R	1/Cs	Ls
Fluid System	R	1/Cs	Ls
Thermal System	R	1/Cs	--

It is obvious that

$$Y(s) = \frac{1}{Z(s)} \quad \quad ..(7.41)$$

Before proceeding further, it is important to acknowledge that traditionally the mechanical engineer has used the term **mechanical impedance** from a viewpoint of force; that is, larger the force required to produce movement, the larger the mechanical impedance. With this premise, mechanical impedance is given by F/V. Having chosen the concept of the through/across variable as the basic premise, this text used the electric circuit concept of resistance to the flow of current and generalizes it by taking **generalized impedance** as the ratio V/F. In some of the mechanical engineering literature, the ratio V/F is called the **mobility**. However, it will be found much simpler, if the term **generalized admittance** is used exclusively for the ratio

Generalized Impedances and System Functions   421

of through to across variable. This practice is followed throughout the text.

7.3.1 Impedances in Combination

Generalized impedances and admittances can be combined in much the same way that equivalent resistances are obtained for series and parallel connections of resistors. The essential difference is that the energy storage elements, being terms in s or 1/s, must have their impedance or admittance added algebraically so that in general, an equivalent impedance or admittance is a polynomial in s. The determination of a series or parallel connection must be made only by inspection of the analogous circuit. That is, it is only when the analogous circuit shows unequivocally that the flow of the through variable is identically the same for two or more elements that they may be considered as a series circuit. Similarly, it is only when the analogous circuit shows unequivocally that the across variable is identically the same for two or more elements that they may be considered as a parallel circuit.

Fig. 7.3(a) shows three impedances (Z_1, Z_2 and Z_3) interconnected so that there is a common through variable. This is, therefore, a series circuit. Suppose we wish to represent the circuit by the circuit with equivalent single impedance (Z_e) shown in Fig. 7.3(b). This requires that the applied potential (AV_{TOTAL}) produces the same through variable in both

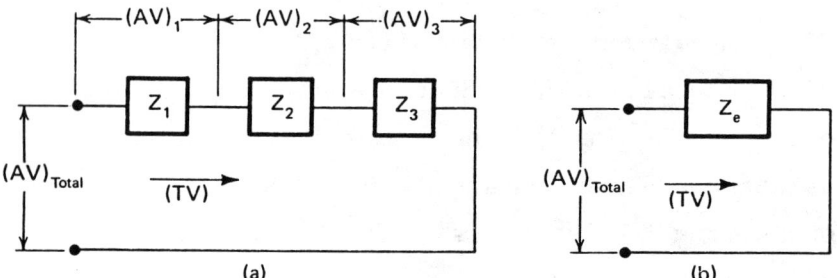

Figure 7.3 Impedances in series

circuits. The applied potential in Fig. 7.3(a) may be expressed as:

$$(AV)_{TOTAL} = (AV)_1 + (AV)_2 + (AV)_3 \qquad ..(7.42)$$

or in terms of impedances, (7.42) becomes

$$(AV)_{TOTAL} = Z_1(TV) + Z_2(TV) + Z_3(TV)$$
$$= (Z_1 + Z_2 + Z_3)(TV) \qquad ..(7.43)$$

The relation between the applied potential and through variable in the equivalent circuit of Fig. 7.3(b) is:

$$(AV)_{TOTAL} = Z_e (TV) \qquad ..(7.44)$$

A comparison of (7.43) and (7.44) shows that the equivalent impedance for a series circuit is:

Generalized Impedances and System Functions

$$Z_e = Z_1 + Z_2 + Z_3 \qquad ..(7.45)$$

In fact, this can be generalized as follows: If the impedances $Z_1, Z_2, Z_3, \ldots, Z_n$ are connected in series, the effective impedance will be

$$Z_e = Z_1 + Z_2 + Z_3 + \ldots + Z_n \qquad ..(7.46)$$

Let us now consider the equivalent impedance for a parallel circuit. Fig. 7.4(a) shows three impedances (Z_1, Z_2, Z_3) that have a common across variable and are therefore in parallel. To replace this circuit by a circuit with an equivalent single impedance (Z_e), Fig. 7.4(b), a through variable $(TV)_{TOTAL}$ must produce the same across variable in each circuit.

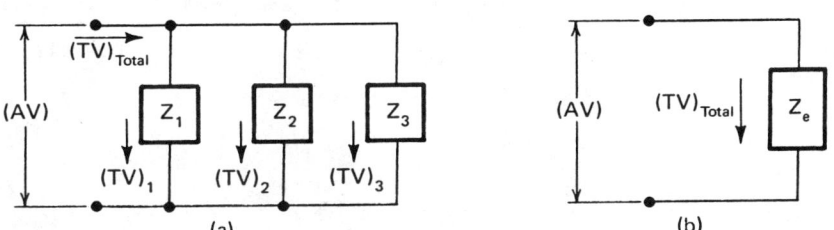

Figure 7.4 Impedances in parallel

For the parallel circuit in Fig. 7.4(a), we have

$$(TV)_{TOTAL} = (TV)_1 + (TV)_2 + (TV)_3 \qquad ..(7.47)$$

and in terms of impedances becomes:

$$(TV)_{TOTAL} = \frac{AV}{Z_1} + \frac{AV}{Z_2} + \frac{AV}{Z_3}$$

$$= \left\{ \frac{1}{Z_1} + \frac{1}{Z_2} + \frac{1}{Z_3} \right\} (AV) \quad \quad ..(7.48)$$

For the equivalent circuit shown in Fig. 7.4(b):

$$(TV)_{TOTAL} = \left(\frac{1}{Z_e}\right)(AV) \quad \quad ..(7.49)$$

From (7.48) and (7.49), we may observe that the equivalent impedance for the parallel circuit is:

$$\frac{1}{Z_e} = \frac{1}{Z_1} + \frac{1}{Z_2} + \frac{1}{Z_3} \quad \quad ..(7.50a)$$

or $\quad\quad Y_e = Y_1 + Y_2 + Y_3 \quad \quad ..(7.50b)$

This can be generalized as follows: If p admittances Y_1, Y_2, Y_3,......,Y_p are connected in parallel, the effective admittance will be:

$$Y_e = Y_1 + Y_2 + Y_3 + \ldots\ldots\ldots + Y_p \quad \quad ..(7.51)$$

Admittances in parallel are therefore additive just as impedances in series.

Fig. 7.5(a) shows admittances (Y_1, Y_2 and Y_3) in series and Fig. 7.5(b) shows an equivalent admittance (Y_e) representation. We may determine the relation between the equivalent admittance and the individual

Generalized Impedances and System Functions

admittances by proceeding directly from the series circuit summation of across variables given in (7.42). However, since we know that admittance is the reciprocal of impedance, we may use the series circuit given in

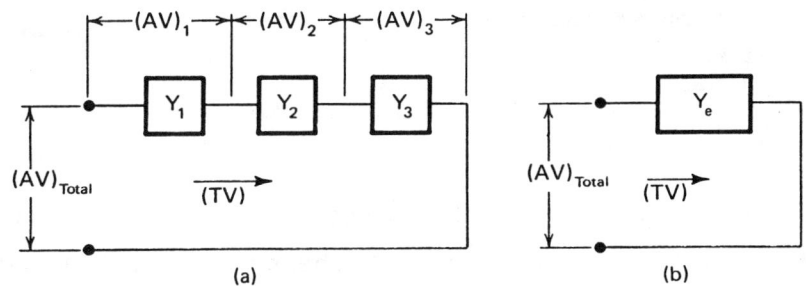

Figure 7.5 Admittances in series

(7.45) to obtain:

$$\frac{1}{Y_e} = \frac{1}{Y_1} + \frac{1}{Y_2} + \frac{1}{Y_3} \qquad ..(7.52)$$

Circuit components may be represented by either impedance or admittance, whichever is more convenient for a particular circuit. Care must be exercised, however, to designate clearly whether impedance or admittance is being used.

Example 7.2

A mass-damper system [Fig. 7.6(a)] is driven by a velocity input $v(t) = V \varepsilon^{st}$. The resulting force through the damper is $f(t) = F \varepsilon^{st}$. Find the equivalent impedance, Z_e = V/F for the system.

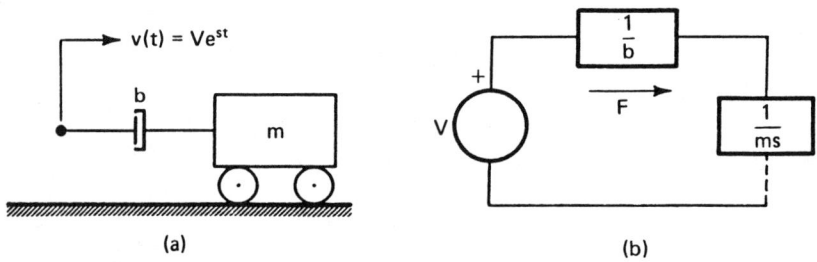

(a) (b)

Figure 7.6 Mechanical circuit (example 7.2)

Fig. 7.6(b) shows the impedance circuit diagram for the system. The impedance of the damper is 1/b and that of the mass is 1/ms. The impedances are connected in a simple series combination. Thus, according to (7.45), the equivalent impedance is the sum of the individual impedances. Hence,

$$Z_e = \frac{1}{b} + \frac{1}{ms} \qquad \qquad ..(7.53)$$

Example 7.3

The positions of the mass and the damper in Example 7.2 are reversed [Fig. 7.7(a)]. In this arrangement, the system is driven by a force input $f(t) = F\,\varepsilon^{st}$. The resulting velocity of the mass is expressible as $v(t) = V\,\varepsilon^{st}$. Find the equivalent impedance Z_e = V/F and the equivalent admittance Y_e = F/V for the system.

The impedance circuit for this arrangement is

Generalized Impedances and System Functions   427

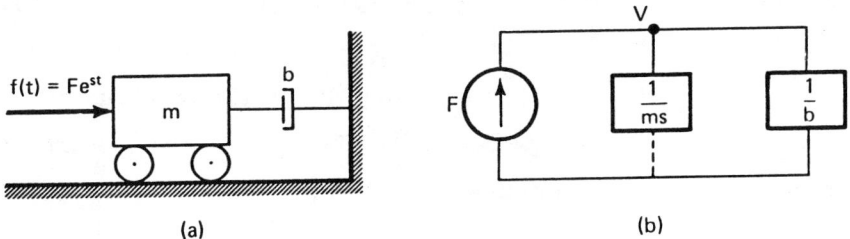

Figure 7.7 Mechanical circuit (example 7.3)

shown in Fig. 7.7(b). The components are in parallel. From (7.50), the equivalent impedance of a parallel circuit is:

$$\frac{1}{Z_e} = \frac{1}{Z_b} + \frac{1}{Z_m} = b + ms \qquad \qquad ..(7.54a)$$

or
$$Z_e = \frac{1}{b + ms} \qquad \qquad ..(7.54b)$$

The equivalent admittance can be found by inverting the equivalent impedance given above or from (7.50b). Thus:

$$Y_e = Y_b + Y_m = b + ms \qquad \qquad ..(7.55)$$

7.3.2 Driving Point Impedance and Admittance

The ratio of across-variable amplitude through variable amplitude as observed across the input terminals of a circuit not containing independent sources is called the driving point impedance Z_{dp}. We may obtain the driving point impedance by combining impedances in series and parallel until the entire circuit is represented by a single equivalent impedance. The driving point admittance is merely the reciprocal of the driving point impedance. The following examples will clarify the procedure.

Example 7.4

Fig. 7.8(a) shows an impedance circuit with impedances Z_1, Z_2, Z_3 and Z_4. Find the driving point impedance between the terminals 1-2.

The procedure is to apply the rules of series and parallel circuits [(7.45) and (7.50)] in sequence until only one equivalent impedance remains. Accordingly, we begin by combining the series connection of Z_3 and Z_4 into an equivalent impedance, $Z_3 + Z_4$. The circuit then appears as shown in Fig. 7.8(b). Now the parallel combination of Z_2 and $(Z_3 + Z_4)$ produces an equivalent impedance [according to (7.50)] of:

$$\frac{1}{Z_e} = \frac{1}{Z_2} + \frac{1}{Z_3 + Z_4} \qquad ..(7.56a)$$

or
$$Z_e = \frac{Z_2(Z_3 + Z_4)}{Z_2 + Z_3 + Z_4} \qquad ..(7.56b)$$

With this combination, the circuit reduces to the simple series circuit shown in Fig. 7.8(c). Finally, the addition of the two impedances in series permits us

Generalized Impedances and System Functions   429

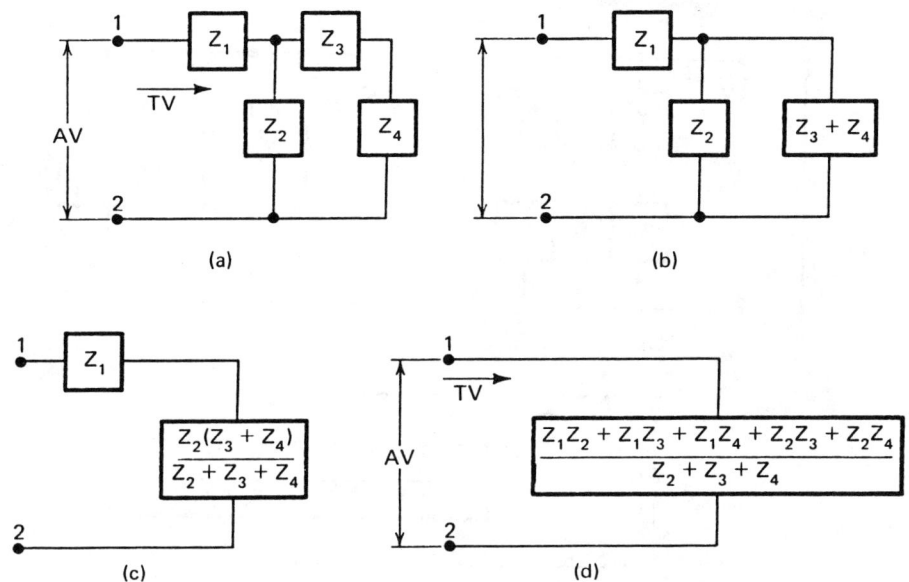

Figure 7.8 Determination of driving point impedance (example 7.4)

to represent the circuit by a single impedance [ig. 7.8(d)]. This is the driving point impedance

$$Z_{dp} = Z_1 + \frac{Z_2(Z_3 + Z_4)}{Z_2 + Z_3 + Z_4} \qquad \text{..(7.57a)}$$

$$= \frac{Z_1 Z_2 + Z_1 Z_3 + Z_1 Z_4 + Z_2 Z_3 + Z_2 Z_4}{Z_2 + Z_3 + Z_4} \qquad \text{..(7.57b)}$$

Example 7.5

Fig. 7.9(a) shows a circuit in which the components have been represented by their admittances Y_1, Y_2, Y_3 and Y_4. Find the driving point admittance

430   Modeling and Analysis of Linear Physical Systems

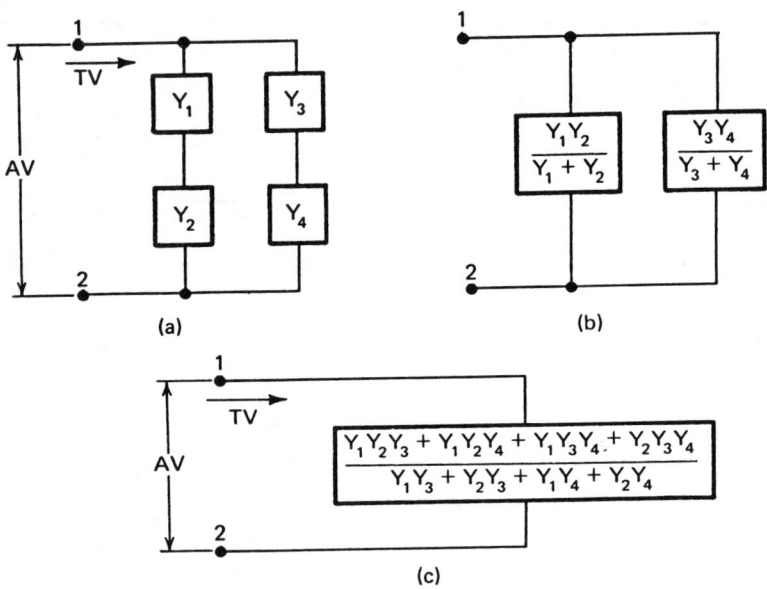

Figure 7.9 Determination of driving point admittance (example 7.5)

between the terminals 1-2.

We must apply the rules for combining admittances, namely (7.50) and (7.52). The series combinations of Y_1, Y_2 and Y_3, Y_4 are shown in Fig. 7.9(b). The two resulting admittances are in parallel and may be treated by direct addition. The result is:

$$Y_e = \frac{Y_1 Y_2}{Y_1 + Y_2} + \frac{Y_3 Y_4}{Y_3 + Y_4} \quad ..(7.58a)$$

$$= \frac{Y_1 Y_2 Y_3 + Y_1 Y_2 Y_4 + Y_1 Y_3 Y_4 + Y_2 Y_3 Y_4}{Y_1 Y_3 + Y_2 Y_3 + Y_1 Y_4 + Y_2 Y_4} \quad ..(7.58b)$$

The general method of determining the driving

Generalized Impedances and System Functions

point impedance or admittance of a circuit will be discussed in Section 7.6.

7.4 SYSTEM FUNCTIONS

In the previous examples, the ratio of the response to the excitation has been either a generalized admittance or a generalized impedance, since a suitable through variable was chosen as the response to an across variable excitation, or a suitable across variable was chosen as the response to a through variable excitation. In previous chapters, there have been examples where this has not been the case (e.g. excitation and response are both across variables or through variables). It is usual to define a System Function, $G(s)$, for the more general situation

$$G(s) = \frac{\text{Amplitude of exponential response}}{\text{Amplitude of exponential excitation}} = \frac{\text{Output}}{\text{Input}}$$
..(7.59)

The system function may therefore be a generalized admittance, a generalized impedance, the ratio of two across variables, or the ratio of two through variables. When the function $G(s)$ is defined as the ratio of two across variables or the ratio of two through variables, it is known as the **transfer function**. The general form is the ratio of two polynomials in s:

$$G(s) = \frac{b_m s^m + b_{m-1} s^{m-1} + \ldots\ldots\ldots + b_0}{a_n s^n + a_{n-1} s^{n-1} + \ldots\ldots\ldots + a_0}$$
..(7.60a)

$$= \frac{(s-s_a)(s-s_b)(s-s_c)\ldots\ldots\ldots\ldots\ldots(s-s_m)}{(s-s_1)(s-s_2)(s-s_3)\ldots\ldots\ldots\ldots\ldots\ldots(s-s_n)}$$

..(7.60b)

where a_n, a_{n-1}, ..., b_m, b_{m-1},...... are the coefficients of the polynomials and s_a, s_b,...., s_1, s_2,..... refer to the roots of the polynomials.

Although, strictly speaking, a system function is obtained for the forced response only, it is a simple matter to **recover** the original differential equation by recalling the association between s and d/dt, and between 1/s and $\int dt$.

It should be noted that a system function implies a single exponential source. As a result the system function concept is valid only when the storage elements have no initial energy. The reason for this is that initial storage is modeled with an additional source (Chapter IV).

There are two situations, the potential divider and the flow divider, which often appear in circuits and are readily treated with system functions. Fig. 7.10 shows the potential and flow divider circuits. The potential divider circuit of Fig. 7.10(a) is basically a series circuit and has only a single through variable associated with it. The system function that relates

Generalized Impedances and System Functions

the across variable output, $(AV)_o$, and the across variable input, $(AV)_i$, is

$$G(s) = \frac{(AV)_o}{(AV)_i} = \frac{Z_2}{Z_1 + Z_2} \qquad ..(7.61a)$$

$$= \frac{Y_1}{Y_1 + Y_2} \qquad ..(7.61b)$$

where (7.61b) represents the system function for the potential divider in terms of admittances.

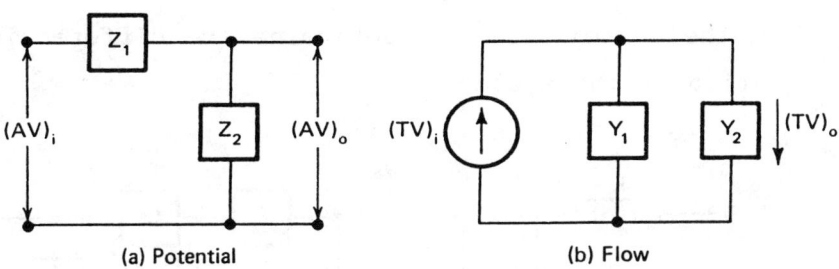

(a) Potential   (b) Flow

Figure 7.10 Potential and flow divider circuits

The flow divider circuit of Fig. 7.10(b) is a parallel circuit. There is only one across variable difference across both impedances. The system function for the ratio of the through variable output, $(TV)_o$, to through variable input, $(TV)_i$, is:

$$G(s) = \frac{(TV)_o}{(TV)_i} = \frac{Y_2}{Y_1 + Y_2} \qquad ..(7.62a)$$

$$= \frac{Z_1}{Z_1 + Z_2} \qquad \qquad ..(7.62b)$$

where (7.62b) is the system function for the flow divider circuit in terms of impedances.

Example 7.6

Fig. 7.11(a) shows an RLC series circuit. The output $e_o(t)$, is the voltage across the capacitance. The input is supplied by an ideal voltage source, $e_i(t)$.
(a) Find the system function that relates the output signal to the input signal.
(b) From the system function determine the differential equation of the system.

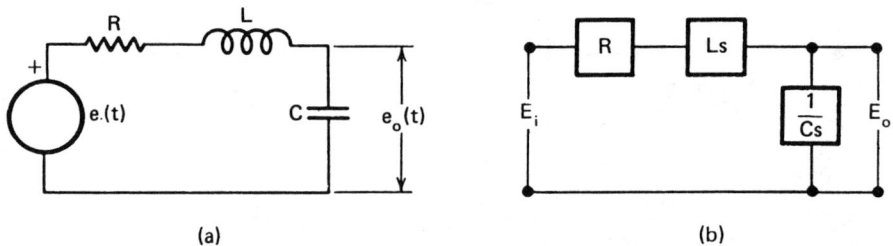

Figure 7.11 Electric circuit (example 7.6)

(a) It is convenient to redraw the circuit diagram with impedance blocks rather than component symbols. This is illustrated in Fig. 7.11(b). We may now place this circuit in potential divider form by noting that the resistance and inductance in series produce an impedance, R + Ls. The system function now follows directly from the application of the potential divider

Generalized Impedances and System Functions

equation (7.61a). Thus:

$$\frac{E_o}{E_i} = \frac{\frac{1}{Cs}}{(R + Ls) + \frac{1}{Cs}} = \frac{1}{LCs^2 + RCs + 1} \qquad \ldots (7.63)$$

(b) Cross multiplication of the system function yields:

$$LCs^2 E_o + RCs E_o + E_o = E_i \qquad \ldots (7.64)$$

Now we may return the time domain by recalling that s represents differentiation with respect to time. Therefore, we may write directly that:

$$LC \frac{d^2 e_o}{dt^2} + RC \frac{de_o}{dt} + e_o = e_i \qquad \ldots (7.65)$$

Example 7.7

In a fluid system, the system function that relates the output pressure, P_o, to the input pressure, P_i, is:

$$g(s) = \frac{P_o}{P_i} = \frac{s + 2}{s + 4} \qquad \ldots (7.66)$$

If $p_i(t) = U_s(t)$, find $p_o(t)$ for this system.

The first step is to transfer the system function into the time domain and recover the governing differential equation. We accomplish this by cross-multiplying the system function so that:

$$s P_o + 4 P_o = s P_i + 2 P_i \qquad \ldots (7.67)$$

This algebraic equation may be transferred now into the time domain with the result that:

$$\frac{dp_o}{dt} + 4 p_o = \frac{dp_i}{dt} + 2 p_i \qquad ..(7.68)$$

Now we may solve the differential equation by decomposing as in Chapter V [Eq.(5.44)] and Chapter VI. Thus:

$$\frac{dp_{o1}}{dt} + 4 p_{o1} = 2 p_i = 2 U_s(t) \qquad ..(7.69a)$$

$$\frac{dp_{o2}}{dt} + 4 p_{o2} = \frac{dp_i}{dt} = U_i(t) \qquad ..(7.69b)$$

$$p_o = p_{o1} + p_{o2} \qquad ..(7.69c)$$

When decomposition is used, recall that the initial conditions on the separated equations are always zero, that is, $p_{o1}(0^+) = 0$. The solution for p_{o1} is therefore:

$$p_{o1} = \frac{1}{2} (1 - \varepsilon^{-4t}) U_s(t) \qquad ..(7.70)$$

We may determine the solution for p_{o2} by noting that:

$$p_{o2} = \frac{1}{2} \frac{d}{dt} (p_{o1}) \qquad ..(7.71)$$

Thus:

Generalized Impedances and System Functions 437

$$p_{o2} = \frac{1}{2}\left[\frac{1}{2} 4 \varepsilon^{-4t}\right] U_s(t) = \varepsilon^{-4t} U_s(t) \quad ..(7.72)$$

The required $p_o(t)$ is the sum of p_{o1} and p_{o2} so that:

$$p_o(t) = \frac{1}{2}\left[1 + \varepsilon^{-4t}\right] U_s(t) \quad ..(7.73)$$

Alternately, a solution can be obtained, using the Laplace transforms, described in chapter 9.

When a system function is expressed as an output over an input, the characteristic equation of the system can be obtained by equating the denominator of the system function to zero. Thus, from (7.60a), the general characteristic equation of a system is:

$$a_n s^n + a_{n-1} s^{n-1} + \ldots\ldots\ldots + a_o = 0 \quad ..(7.74)$$

The roots of this equation, s_1, s_2, ..., s_n are called the **poles** of the system function. Although excitation by an exponential function, ε^{st}, where s is exactly equal to one of the poles (s_1, s_2, ..., s_n) would appear to produce an infinite response, this is not necessarily the case. In Example 7.1, we observed that when the forcing exponential was equal to a root of the characteristic equation, the first-order system responded as a critically damped second-order system. In general, excitation of any system with a value of s equal to one of the system poles produces a dynamic response which has the same form as that of the next higher order system with two (or more) coincident poles. It may also be observed from (7.60b) that excitation at any root of the numerator of the system function (s_a, s_b, s_c) produces zero forced response. These roots are (s_a, s_b, s_c, ...) are therefore called the **zeros** of the system function.

It is common to think of **poles** and **zeros** in graphical terms by plotting their position on the complex plane. The result is that the general form of the natural response of a system may be associated directly with the position of the poles on the complex s-plane. Fig. 7.12 shows the correspondence between pole location and time response.

In each case, the figure on the left gives a pole position on the s-plane ($s = \sigma + j\omega$) and the figure on the right gives the resulting contribution of that pole arrangement to the transient response. A simple pole ($s = -\sigma_1$) near the $j\omega$-axis as shown in Fig. 7.12(a) leads to a slowly decaying exponential in the time domain. When the pole moves further away from the $j\omega$-axis ($s = -\sigma_1$) as in Fig. 7.12(b), the resulting time function decays more rapidly. Fig. 7.12(c) shows the complex poles ($s = -\sigma_1 \pm j\omega_1$) and the corresponding damped sine wave response. When the poles move further away from the real axis as in Fig. 7.12(d) ($s = -\sigma_1 \pm j\omega_2$), the frequency of the response increases but the decay rate is the same.

Generalized Impedances and System Functions

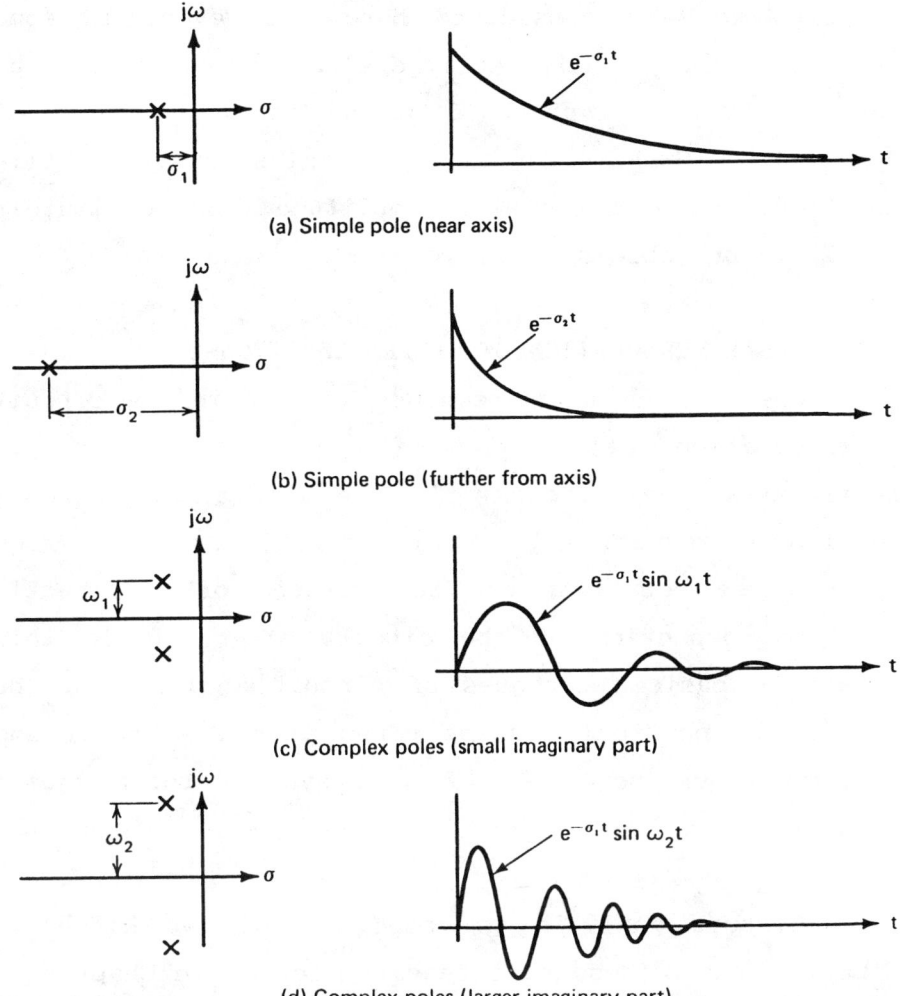

(a) Simple pole (near axis)

(b) Simple pole (further from axis)

(c) Complex poles (small imaginary part)

(d) Complex poles (larger imaginary part)

Figure 7.12 Correspondence of time response to pole positions

For the physical circuits which emanate from the combination of elemental components (resistance, inductance, capacitance), the system poles always lie on or to the left of the $j\omega$-axis. Thus the corresponding time response either remains at a fixed amplitude or decays. This is the condition that occurs in all the circuits we have considered here. However, in some systems, (e.g., with amplifiers), the poles may be located in the right-half plane (to the right of $j\omega$-axis). Such pole locations give rise to time functions with increasing amplitude and ultimately result in unstable systems.

7.5 GENERAL FORMULATION OF SYSTEM EQUATIONS

One of the main reasons for seeking a circuit representation of a physical system is that it facilitates the process of determining the circuit equilibrium equations. Simple algorithms exist from which these equations can be written, often directly from an inspection of the circuit diagram. In this case, the basic techniques of circuit analysis can be applied to physical systems other than electrical and the analogous properties of such systems become quite evident.

Circuit analysis is based on the two Kirchhoff circuit laws. These have been considered in Chapter IV [Eqs. (4.1) and (4.2)] but are restated here in the form which is most convenient for the analysis of more complicated circuits than had previously been considered. For Kirchhoff's Current Law (KCL):

$$\sum (TV)_P = \sum (TV)_S \qquad ..(7.75a)$$

Eq.(7.75a) means that the sum of all through variables leaving a junction through the passive elements connected directly to the junction, $(TV)_P$, is equal to the sum of all through variables entering the junction from through variable sources, $(TV)_S$.

For Kirchhoff's Voltage Law (KVL):

$$\sum (AV)_P = \sum (AV)_S \qquad ..(7.75b)$$

which states that the sum of the across variable drops across passive elements, $(AV)_P$, around closed circuit path, equals the sum of across variable increases due to across variable sources, $(AV)_S$.

Up to this point, the most complicated system which has been described in detail, by means of one or two equilibrium equations, has had only three elements. We are now in a position to develop a systematic procedure to obtain sets of equilibrium equations for any system, no matter how complicated, using its analogous circuit diagram as the common intermediate step. The circuit diagram is at first simply a graphical representation of the interactions which take place among all the elements of a system in qualitative terms. However, each interaction can be modeled by means of a simple elemental mathematical relationship and thus the circuit effectively becomes a graphical representation of the equilibrium equations. The problem is therefore to find a method whereby the information contained in the circuit diagram may be quantified quickly and reliably in the form of equilibrium differential equations. If we use the ideas developed in this chapter, we will be able to derive the equations more confidently and with less effort. We, therefore, imply that all the driving functions are exponential, and use the generalized impedance notation. It should be clearly understood that this does not in any way limit the validity or generality of the resulting equations.

7.5.1 Determination of Analogous Impedance Circuits

Since the circuit diagram is the basis on which our unified approach rests, it is appropriate to review the process by which the circuit was obtained in previous chapters. Let us consider the mass-spring-damper system in Fig. 7.13(a).

The first step is to identify all parts which are constrained to move at the same velocity and show the direction chosen to be positive. Number these points since they will form the junctions or nodes of the circuit diagram. In this case, there are three independent velocities, v_1, v_2, and v_3 and a reference velocity, v_g. The equations are usually easier to write and the results easier to interpret, if positive velocity is taken in the same direction at each part of the system. Now draw the **skeleton** of the circuit, i.e., the three nodes corresponding to the three parts of the system at which the velocities are different and the

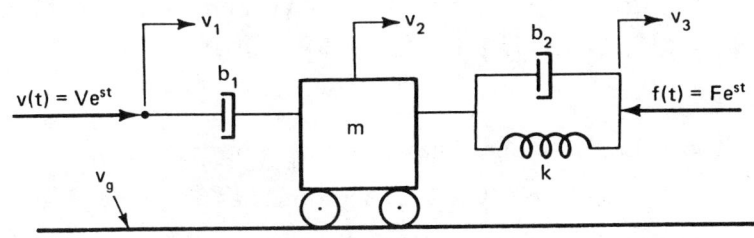

(a) Mechanical arrangement

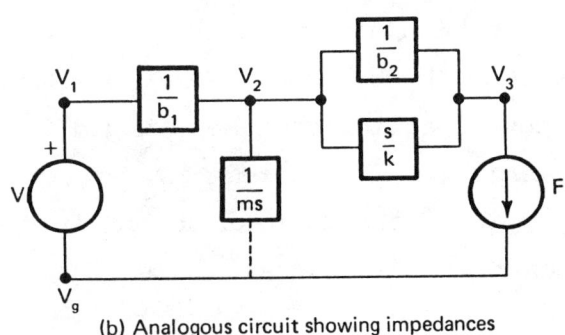

(b) Analogous circuit showing impedances

Figure 7.13 Determination of analogous circuit

datum or the reference node. In Fig. 7.13(b), these are indicated as V_1, V_2, V_3 and V_g. The impedance blocks corresponding to the passive elements are added by noting the velocity difference which is required in the force-velocity relation of the element. The damper, b_1, models an effect dependent on the relative velocity between nodes 1 and 2. Therefore, the impedance block ($1/b_1$) is inserted between these nodes. The damper, b_2 and the spring, k, both depend on the relative velocity between nodes 2 and 3. The mass, on the other hand, models an effect dependent on the absolute velocity of node 2. To make this a two-terminal component, we use the ground as a fictitious terminal and connect the mass impedance block (1/ms) between V_2 and V_g.

Now, the branches having the sources can be added. The velocity source at node 1 constrains node 1 to move to the right at a velocity V relative to the datum. Therefore, on the circuit diagram, this is represented by a velocity source connected between node 1 and the datum. It is positive at node 1, because the source constrains node 1 to move in the positive direction. The force source acting at node 3 would have moved node 3 in the negative direction, if it had been acting separately. Thus the **flow of force** into node 3 must also be negative and the arrow is directed away from node 3.

7.5.2 Node-to-datum Equation Formulation

The node-to-datum equations are usually referred to as node equations. The method is a very convenient way of formulating equations when the desired output is of the across variable type. When applying the node-to-datum method, the coefficients turn out simplest, if admittance parameters are used. Accordingly, consider the circuit shown in Fig. 7.14. Suppose it is the analogous circuit for a particular fluid system that has exponential flow sources with amplitudes Q_1 and Q_2. We seek to find the equilibrium equations that describe the exponential pressure

Generalized Impedances and System Functions 445

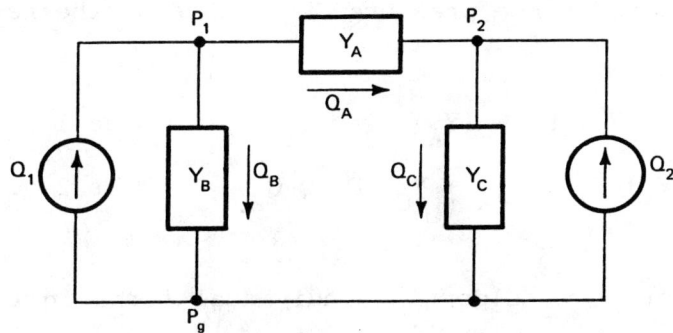

Figure 7.14 Analogous circuit involving admittances

amplitudes P_1 and P_2.

We begin by applying (7.75a) (KCL) to nodes 1 and 2.

$$\text{Node 1} \longrightarrow Q_A + Q_B = Q_1 \qquad ..(7.76a)$$
$$\text{Node 2} \longrightarrow Q_C - Q_A = Q_2 \qquad ..(7.76b)$$

where flows away from the node are positive on the left-hand side of the equation and negative on the right-hand side. Once all the node equations are written (in this case the two equations (7.76a) and (7.76b)), we formulate the component relations. The three component relations are:

$$(P_1 - P_2) Y_A = Q_A \qquad ..(7.77a)$$
$$(P_1 - P_g) Y_B = Q_B \qquad ..(7.77b)$$
$$(P_2 - P_g) Y_C = Q_C \qquad ..(7.77c)$$

Now, if (7.77) are substituted into (7.76) and we note that the reference pressure, P_g, is zero, the result is:

$$(Y_A + Y_B) P_1 - (Y_A) P_2 = Q_1 \quad \longleftarrow \text{Node 1} \quad ..(7.78a)$$
$$-(Y_A) P_1 + (Y_A + Y_C) P_2 = Q_2 \quad \longleftarrow \text{Node 2} \quad ..(7.78b)$$

Let us consider the individual terms in these two equations and try to interpret them.

Node 1: $(Y_A + Y_B) P_1$ = total flow away from node 1 due to P_1 only, all other pressures set to zero.

$-(Y_A) P_2$ = total flow away from node 1 due to P_2 only, all other pressures set to zero.

Q_1 = total flow into node 1 from all flow sources connected to node 1.

Node 2: $(Y_A + Y_C) P_2$ = total flow away from node 2 due to P_2 with all other variables set to zero.

$-(Y_A) P_1$ = total flow away from node 2 due to P_1 with all other variables set to zero.

Q_2 = total flow into node 2 from all flow sources connected to node 2.

At this stage, it is valuable to note the

coefficients in the equations, and the matrix form of the equations is useful for this purpose.

$$\begin{array}{c} \text{Node 1} \longrightarrow \\ \text{Node 2} \longrightarrow \end{array} \begin{bmatrix} (Y_A + Y_B) & -Y_A \\ -Y_A & (Y_A + Y_C) \end{bmatrix} \begin{bmatrix} P_1 \\ P_2 \end{bmatrix} = \begin{bmatrix} Q_1 \\ Q_2 \end{bmatrix}$$

..(7.79)

In the array of admittances, the terms on the main diagonal are positive and all others are negative. This is true in general for node-to-datum equilibrium equations.

The general form is:

$$\begin{bmatrix} Y_{11} & Y_{12} \\ Y_{21} & Y_{22} \end{bmatrix} \begin{bmatrix} P_1 \\ P_2 \end{bmatrix} = \begin{bmatrix} Q_1 \\ Q_2 \end{bmatrix} \quad ..(7.80)$$

Note that Y_{11} is the sum of the admittances of the passive branches connected to node 1 and Y_{22} is the sum of the admittances at node 2. Y_{12} is the negative of the admittance of the branch connecting nodes 1 and 2. If every element connecting two nodes is **bilateral** (that is, the pressure-flow relationship is the same for flow in both directions), $Y_{12} = Y_{21}$ or $Y_{jk} = Y_{kj}$ and the matrix is symmetrical.

With some practice, it is a simple matter to write the equilibrium equations of many systems directly in this form (that is, **by inspection** of the system circuit) thus reducing the likelihood of error. We will clarify the procedure in the following two examples.

Example 7.8

The mechanical arrangement shown in Fig. 7.15(a) consists of two masses, three dampers and one spring. There are two known force inputs (f_1 and f_2). Present the node-to-datum equations in matrix form for this system.

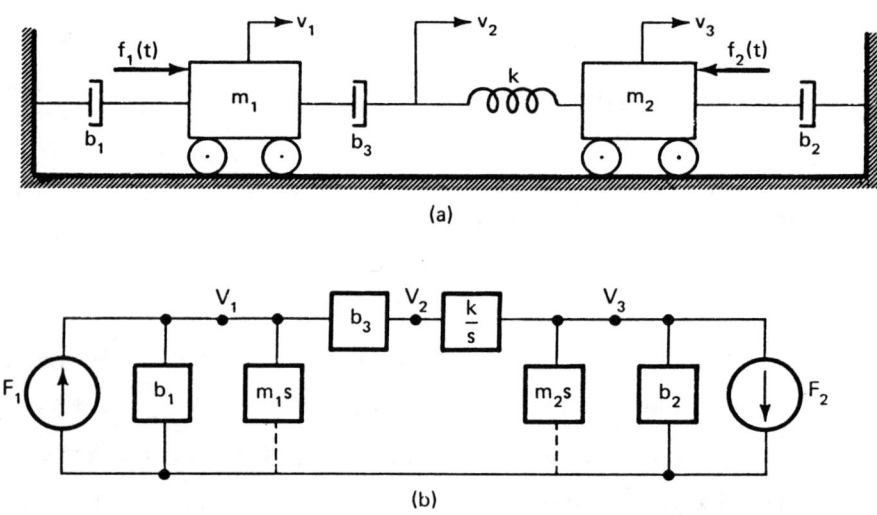

Figure 7.15 Mechanical circuit (example 7.8)

Fig. 7.15(b) shows the circuit representation of the mechanical system. Since there are three unknown velocities, the required matrix must be of the form:

Generalized Impedances and System Functions

$$\begin{array}{c} \text{Node 1} \rightarrow \\ \text{Node 2} \rightarrow \\ \text{Node 3} \rightarrow \end{array} \begin{bmatrix} Y_{11} & Y_{12} & Y_{13} \\ Y_{21} & Y_{22} & Y_{23} \\ Y_{31} & Y_{32} & Y_{33} \end{bmatrix} \begin{bmatrix} V_1 \\ V_2 \\ V_3 \end{bmatrix} = \begin{bmatrix} F_1 \\ 0 \\ -F_2 \end{bmatrix}$$

with the column indicators:
$V_2=0, V_3=0$ under column 1;
$V_1=0, V_3=0$ under column 2;
$V_1=0, V_2=0$ under column 3.

..(7.81)

The column vector of unknowns has the exponential amplitudes at the velocity nodes; these are designated as V_1, V_2 and V_3. The column vector of knowns consists of exponential amplitudes of the force inputs. The force amplitude entering node 1 due to the force inputs is merely F_1. There is no force input entering node 2 directly and thus the zero in the known column vector at node 2. For node 3, we may observe that force amplitude, F_2, acts away from the node. As a result, the force amplitude is $-F_2$.

We may now determine the admittance matrix. Element Y_{11} represents the admittance from terminal 1 (V_1) with all other terminals grounded ($V_2 = V_3 = 0$). This is the parallel combination of b_1, $m_1 s$, and b_3, so that:

$$Y_{11} = b_1 + b_3 + m_1 s \qquad ..(7.82)$$

In a similar way, Y_{22} represents the equivalent admittance from terminal 2 (V_2) with all other terminals grounded ($V_1 = V_3 = 0$). Thus:

$$Y_{22} = b_3 + \frac{k}{s} \qquad \qquad ..(7.83)$$

Similarly, Y_{33} applies to terminal 3 (V_3) with the other terminals grounded ($V_1 = V_2 = 0$). Thus,

$$Y_{33} = b_2 + m_2 s + \frac{k}{s} \qquad \qquad ..(7.84)$$

The other admittances are merely the negative of all common terms. For example, to find Y_{12} and Y_{21}, there is only one common term, b_3, in Y_{11} and Y_{22}. Thus,

$$Y_{12} = Y_{21} = -b_3 \qquad \qquad ..(7.85)$$

To find Y_{23} and Y_{32}, the only common term in Y_{22} and Y_{33} is k/s and thus

$$Y_{23} = Y_{32} = -\frac{k}{s} \qquad \qquad ..(7.86)$$

Since there are no common terms between Y_{11} and Y_{33}, the admittances Y_{13} and Y_{31} are zero. Thus:

$$Y_{13} = Y_{31} = 0 \qquad \qquad ..(7.87)$$

Generalized Impedances and System Functions

The complete matrix is therefore:

$$\begin{bmatrix} b_1 + b_3 + m_1 s & -b_3 & 0 \\ -b_3 & b_3 + \dfrac{k}{s} & -\dfrac{k}{s} \\ 0 & -\dfrac{k}{s} & b_2 + m_2 s + \dfrac{k}{s} \end{bmatrix} \begin{bmatrix} V_1 \\ V_2 \\ V_3 \end{bmatrix} = \begin{bmatrix} F_1 \\ 0 \\ -F_2 \end{bmatrix}$$

..(7.88)

To express the amplitude at any terminal in terms of the basic components and the known input amplitudes, we must use determinants. Thus,

$$V_1 = -\frac{\begin{vmatrix} F_1 & -b_3 & 0 \\ 0 & b_3 + \dfrac{k}{s} & \dfrac{k}{s} \\ & -\dfrac{k}{s} & b_3 + m_2 s + \dfrac{k}{s} \end{vmatrix}}{\begin{vmatrix} b_1 + b_3 + m_1 s & -b_3 & 0 \\ -b_3 & b_3 + \dfrac{k}{s} & -\dfrac{k}{s} \\ 0 & -\dfrac{k}{s} & b_2 + m_2 s + \dfrac{k}{s} \end{vmatrix}}$$

..(7.89)

If we evaluate the determinants, we get

$$V_1 = \frac{1}{\Delta} \left\{ F_1 \left[(b_2 b_3 + m_2 k) + b_3 m_2 s + \frac{k}{s} (b_2 + b_3) \right] \right.$$

$$\left. - F_2 \frac{b_3 k}{s} \right\} \qquad ..(7.90a)$$

where

$$\Delta = b_3 m_1 m_2 s^2 + (b_1 b_3 m_2 + b_2 b_3 m_1 + m_1 m_2 k) s$$

$$+ \frac{k}{s} (b_1 b_2 + b_2 b_3 + b_1 b_3)$$

$$+ (b_1 b_2 b_3 + b_1 m_2 k + b_2 m_1 k + b_3 m_1 k + b_3 m_2 k)$$

$$\qquad ..(7.90b)$$

Similarly,

$$V_2 = \frac{1}{\Delta} \left\{ F_1 (b_2 b_3 + m_2 b_3 s + \frac{b_3 k}{s}) - \right.$$

$$\left. F_2 \frac{k}{s} (b_1 + b_3 + m_1 s) \right\} \qquad ..(7.90c)$$

and

$$V_3 = \frac{1}{\Delta} \left\{ F_1 \frac{b_3 k}{s} - F_2 \left[(b_1 b_3 + m_1 k) + b_3 m_1 s \right. \right.$$

$$\left. \left. + \frac{k}{s} (b_1 + b_3) \right] \right\} \qquad ..(7.90d)$$

Generalized Impedances and System Functions

Example 7.9

Find the node-to-datum equations in matrix form for the electrical circuit shown in Fig. 7.16(a). Also express the amplitude E_2 in terms of the components and input sources.

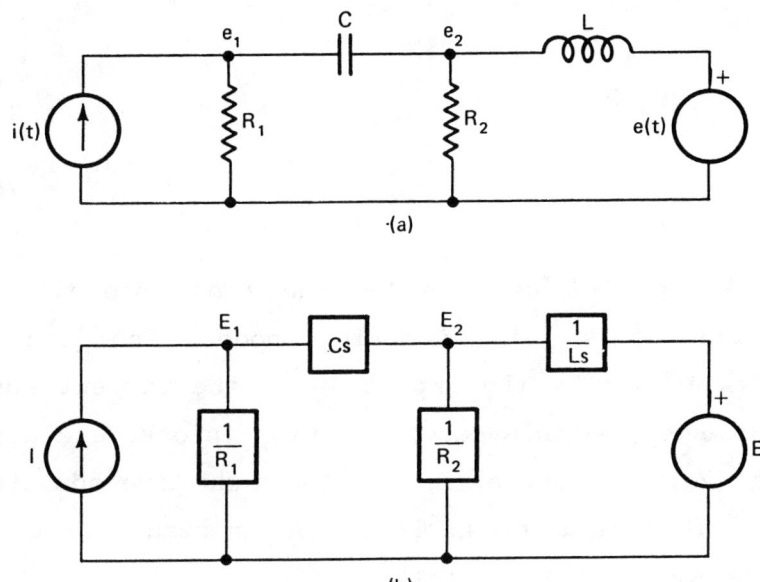

Figure 7.16 Electrical circuit (example 7.9)

Fig. 7.16(b) shows the admittance diagram representation of the electrical circuit. There is a total of four nodes or terminals. However, two of them (the ground terminal and the voltage source terminal, E) are already defined. Thus, there are only two independent nodes and the admittance matrix will be of the form:

$$\begin{bmatrix} Y_{11} & Y_{12} \\ Y_{21} & Y_{22} \end{bmatrix} \begin{bmatrix} E_1 \\ E_2 \end{bmatrix} = \begin{bmatrix} I \\ E/Ls \end{bmatrix} \begin{matrix} \leftarrow \text{node 1} \\ \leftarrow \text{node 2} \end{matrix}$$

with unknowns $\begin{bmatrix} E_1 \\ E_2 \end{bmatrix}$ (where $E_2=0$, $E_1=0$ indicated) and knowns $\begin{bmatrix} I \\ E/Ls \end{bmatrix}$

..(7.91)

First, let us consider the flow into the nodes from the inputs. The flow into node 1 from the input sources is merely the amplitude of the current source, I. However, in this case, the flow into node 2 depends on the voltage source, E, and the inductive admittance, 1/Ls. This is a flow, E/Ls. As a result, the known column vector is as indicated above.

Matrix admittance element, Y_{11} is the parallel combination of $1/R_1$ and Cs. This is the condition that exists, when all the terminals except E_1 are grounded. Thus:

$$Y_{11} = \frac{1}{R_1} + Cs \qquad ..(7.92)$$

To find the matrix element Y_{22}, we reverse the procedure. Now all terminals except E_2 are grounded. The admittance from node 2 in this case is the parallel

Generalized Impedances and System Functions

combination of C, $1/R_2$, and $1/Ls$, or

$$Y_{22} = \frac{1}{R_2} + \frac{1}{Ls} + Cs \qquad \ldots (7.93)$$

In this case, there is only one common term between Y_{11} and Y_{12}, and it is Cs, so that:

$$Y_{12} = Y_{21} = -Cs \qquad \ldots (7.94)$$

and the complete matrix representation is:

$$\begin{bmatrix} \dfrac{1}{R_1} + Cs & -Cs \\ -Cs & \dfrac{1}{R_2} + \dfrac{1}{Ls} + Cs \end{bmatrix} \begin{bmatrix} E_1 \\ E_2 \end{bmatrix} = \begin{bmatrix} I \\ E/Ls \end{bmatrix}$$

$$\ldots (7.95)$$

The amplitude of E_2 in terms of determinants is:

$$E_2 = \frac{\begin{vmatrix} \dfrac{1}{R_1} + Cs & I \\ -Cs & E/Ls \end{vmatrix}}{\begin{vmatrix} \dfrac{1}{R_1} + Cs & -Cs \\ -Cs & \dfrac{1}{R_2} + \dfrac{1}{Ls} + Cs \end{vmatrix}} \qquad \ldots (7.96)$$

and evaluating the determinants, we get

$$E_2 = \frac{(\frac{1}{R_1} + Cs)(\frac{E}{Ls}) - (-Cs)(I)}{\frac{1}{R_1}(\frac{1}{R_2} + \frac{1}{Ls} + Cs) + Cs(\frac{1}{R_2} + \frac{1}{Ls})} \qquad ..(7.97)$$

7.5.3 Loop and Mesh Equations

The loop and mesh equations method is a convenient means of formulating equations, when the unknowns are of the through variable type. This method favors the use of impedance parameters. Let us consider, therefore, the circuit shown in Fig. 7.17. The circuit represents a particular fluid circuit that has input pressure sources of exponential type with amplitudes P_1 and P_2. We want to find equations that describe the exponential flows through all the impedance elements.

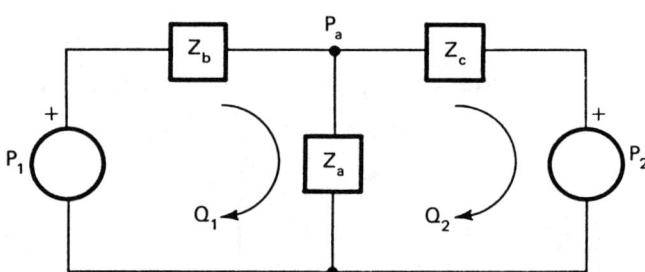

Figure 7.17 Analogous circuit involving impedances

In this case, there are three distinct impedances and three distinct flows. However, only two of them are independent. For example, if Q_1 is the flow through Z_b

Generalized Impedances and System Functions 457

and Q_2 is the flow through Z_c, then the flow through Z_a is $(Q_1 - Q_2)$. The flows Q_1 and Q_2 are the mesh flows and are the most convenient choice for the unknowns, since we may apply (7.75b) around the closed mesh. If we apply (7.75b) (KVL) around meshes 1 and 2, we get:

Mesh 1 → $(P_1 - P_a) + (P_a - P_g) = (P_1 - P_g)$ ..(7.98a)

Mesh 2 → $(P_g - P_a) + (P_a - P_2) = (P_g - P_2)$ ..(7.98b)

The meshes are the open spaces in the impedance circuit. We always take the flow in each mesh in a clockwise direction. We may also substitute the component relations into the mesh equations (7.98a) and (7.98b) to obtain:

Mesh 1 → $Z_b Q_1 + Z_a (Q_1 - Q_2) = P_1$ ..(7.99a)

Mesh 2 → $Z_a (Q_2 - Q_1) + Z_c Q_2 = -P_2$ ..(7.99b)

or if we group the flows together in (7.99), the result is:

Mesh 1 → $(Z_a + Z_b) Q_1 - Z_a Q_2 = P_1$ ..(7.100a)

Mesh 2 → $-Z_a Q_1 + (Z_a + Z_c) Q_2 = -P_2$ ..(7.100b)

We may interpret the terms in (7.100) as follows:

In Mesh 1: $(Z_a + Z_b) Q_1$ = sum of the pressure drops across each impedance in the direction of flow and due to Q_1 only. All other flows are set to zero.

$-Z_a Q_2$ = sum of the pressure drops across impedances, in the direction of Q_1, but due to flow Q_2. All other flows are set to zero.

In Mesh 2: $-Z_a Q_1$ = sum of the pressure drops in mesh 2 flow direction due to flow Q_1. All other flows are set to zero.

$(Z_a + Z_c) Q_2$ = sum of the pressure drops due to flow Q_2. All other flows are set to zero.

In matrix form, (7.100) may be written as:

$$\begin{matrix} \text{Mesh 1} \rightarrow \\ \\ \text{Mesh 2} \rightarrow \end{matrix} \begin{bmatrix} Z_a + Z_b & -Z_a \\ \\ -Z_a & Z_a + Z_c \end{bmatrix} \begin{bmatrix} Q_1 \\ \\ Q_2 \end{bmatrix} = \begin{bmatrix} P_1 \\ \\ P_2 \end{bmatrix}$$

with unknowns Q_1, Q_2 and knowns P_1, P_2; $Q_2 = 0$ and $Q_1 = 0$ indicated under the respective columns.

..(7.111)

In the impedance array, the terms on the main

Generalized Impedances and System Functions

diagonal are positive and all others are negative. This is true for mesh equations, when the assumed flow directions are all clockwise or all counterclockwise. The general form for the mesh equations is:

$$\begin{bmatrix} Z_{11} & Z_{12} \\ Z_{21} & Z_{22} \end{bmatrix} \begin{bmatrix} Q_1 \\ Q_2 \end{bmatrix} = \begin{bmatrix} P_1 \\ P_2 \end{bmatrix} \quad \ldots (7.112)$$

The element Z_{11} is the sum of all impedances in mesh 1 and Z_{22} is the sum of all impedances in mesh 2. The impedances Z_{12} and Z_{21} are equal and are the negative of all the common terms in Z_{11} and Z_{22} provided all the circulating flows are taken in the same direction (e.g., all clockwise).

We may also now formulate the equilibrium equations for flow variables by the mesh method and by inspection of the impedance circuit. We will demonstrate the procedure in the following examples.

Example 7.10

Fig. 7.18 shows an electrical circuit that consists of two resistances, a capacitance, an inductance and two sources. In Fig. 7.18(a), the current source is on the left and the voltage source is on the right. The same circuit with sources reversed in position is shown in Fig. 7.18(b). In each case, find the relationship between the current through the capacitance and the sources.

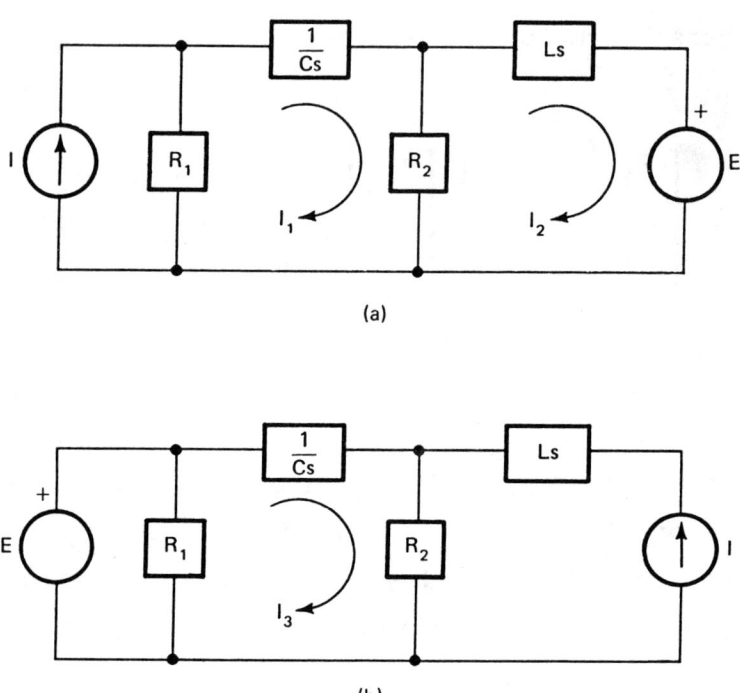

Figure 7.18 Circuits with impedances shown (example 7.10)

There are three meshes in the circuit shown in Fig. 7.18(a). However, the current is prescribed in the left-most mesh. Thus, there are only two unknown currents and the mesh matrix has the form:

$$\begin{matrix} \text{mesh } 1 \rightarrow \\ \\ \text{mesh } 2 \rightarrow \end{matrix} \begin{bmatrix} Z_{11} & Z_{12} \\ Z_{21} & Z_{22} \end{bmatrix} \begin{bmatrix} I_1 \\ I_2 \end{bmatrix} = \begin{bmatrix} R_1 I \\ -E \end{bmatrix} \quad ..(7.113)$$

where the right-hand side shows the effect of the sources. In mesh 1 (I_1), the current source (I) contributes a voltage rise, $R_1 I$. In mesh 2 (I_2), the voltage source (E) causes a voltage drop and thus is described as $-E$.

Now let us consider the impedance matrix. The term Z_{11} represents the impedance around mesh 1 and is:

$$Z_{11} = R_1 + R_2 + \frac{1}{Cs} \qquad \ldots (7.114)$$

Similarly, the term Z_{22} is the total impedance around mesh 2. This is:

$$Z_{22} = R_2 + Ls \qquad \ldots (7.115)$$

There is only one common factor between Z_{11} and Z_{22}, and that is the resistance, R_2. Thus,

$$Z_{12} = Z_{21} = -R_2 \qquad \ldots (7.116)$$

The complete matrix representation is:

$$\begin{bmatrix} R_1 + R_2 + \frac{1}{Cs} & -R_2 \\ -R_2 & R_2 + Ls \end{bmatrix} \begin{bmatrix} I_1 \\ I_2 \end{bmatrix} = \begin{bmatrix} R_1 I \\ -E \end{bmatrix} \qquad \ldots (7.117)$$

The current through the capacitor is I_1. We,

therefore, solve the equations for I_1. The solution has the form of the ratio of two determinants.

$$I_1 = \frac{\begin{vmatrix} R_1 I & -R_2 \\ -E & R_2 + Ls \end{vmatrix}}{\begin{vmatrix} R_1 + R_2 + \frac{1}{Cs} & -R_2 \\ -R_2 & R_2 + Ls \end{vmatrix}} \quad \quad ..(7.118a)$$

$$= \frac{R_1 I (R_2 + Ls) - (-E)(-R_2)}{(R_1 + R_2 + \frac{1}{Cs})(Ls) + R_2(R_2 + \frac{1}{Cs})} \quad \quad ..(7.118b)$$

Now let us turn our attention to Fig. 7.18(b). Once again, there are three meshes. In this case, two of them are special cases. The only unknown current is through the capacitance (I_3). Furthermore, the components R_1 and L have no effect on the unknown current. That is, the contribution to mesh (I_3) from the leftmost mesh is always a voltage rise, E, regardless of the magnitude of R_1. The contribution to mesh (I_3) from the rightmost mesh is always a voltage drop, $R_2 I$, and it is independent of L. The matrix thus degenerates to:

$$Z_{33} I_3 = E - R_2 I \quad \quad ..(7.119a)$$

Generalized Impedances and System Functions   463

and
$$Z_{33} = R_2 + \frac{1}{Cs} \quad \quad \quad ..(7.119b)$$

so that
$$I_3 = \frac{E - R_2 I}{R_2 + \frac{1}{Cs}} \quad \quad \quad ..(7.120)$$

We may observe that the current through the capacitor is vastly different, when the sources are reversed.

Example 7.11

Fig. 7.19(a) shows a mechanical circuit with two dampers and a mass. This mechanical circuit may be

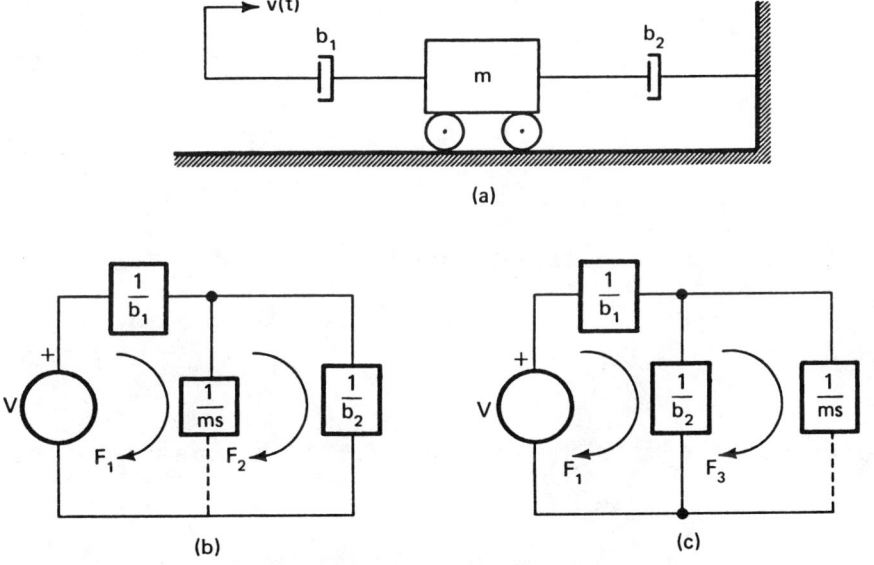

Figure 7.19 Mechanical circuit (example 7.11)

represented by either of the two impedance circuits shown in Figs. 7.19(b) and (c). Show that the force acting to accelerate the mass is the same in both circuits.

In circuit of Fig. 7.19(b), the mesh equations matrix has the form:

$$\begin{bmatrix} Z_{11} & Z_{12} \\ Z_{21} & Z_{22} \end{bmatrix} \begin{bmatrix} F_1 \\ F_2 \end{bmatrix} = \begin{bmatrix} V \\ 0 \end{bmatrix} \quad ..(7.121)$$

where the source has no contribution in mesh (F_2) and contributes a velocity rise (V) to mesh (F_1). The sum of the impedances around the meshes yields:

$$Z_{11} = \frac{1}{b_1} + \frac{1}{ms} \quad ..(7.122a)$$

and

$$Z_{22} = \frac{1}{b_2} + \frac{1}{ms} \quad ..(7.122b)$$

and the common term is $1/ms$ so that

$$Z_{12} = Z_{21} = -\frac{1}{ms} \quad ..(7.122c)$$

Therefore, the matrix equation (7.121) becomes:

Generalized Impedances and System Functions

$$\begin{bmatrix} \dfrac{1}{b_1} + \dfrac{1}{ms} & -\dfrac{1}{ms} \\ -\dfrac{1}{ms} & \dfrac{1}{b_2} + \dfrac{1}{ms} \end{bmatrix} \begin{bmatrix} F_1 \\ F_2 \end{bmatrix} = \begin{bmatrix} V \\ 0 \end{bmatrix} \qquad \text{..(7.123)}$$

In this circuit of Fig. 7.19(b), the force applied to the mass is $F_M = F_1 - F_2$. We must, therefore, solve the matrix for F_1 and F_2 and then subtract the results.

$$F_1 = \frac{\begin{vmatrix} V & -\dfrac{1}{ms} \\ 0 & \dfrac{1}{b_2} + \dfrac{1}{ms} \end{vmatrix}}{\begin{vmatrix} \dfrac{1}{b_1} + \dfrac{1}{ms} & -\dfrac{1}{ms} \\ -\dfrac{1}{ms} & \dfrac{1}{b_2} + \dfrac{1}{ms} \end{vmatrix}} \qquad \text{..(7.124a)}$$

$$F_1 = \frac{V\left(\dfrac{1}{b_2} + \dfrac{1}{ms}\right)}{\dfrac{1}{b_1 b_2} + \dfrac{1}{ms}\left(\dfrac{1}{b_1} + \dfrac{1}{b_2}\right)} \qquad \text{..(7.124b)}$$

Similarly,

$$F_2 = \frac{\begin{vmatrix} \frac{1}{b_1} + \frac{1}{ms} & V \\ -\frac{1}{ms} & 0 \end{vmatrix}}{\begin{vmatrix} \frac{1}{b_1} + \frac{1}{ms} & -\frac{1}{ms} \\ -\frac{1}{ms} & \frac{1}{b_2} + \frac{1}{ms} \end{vmatrix}} \quad \ldots (7.125a)$$

$$= \frac{\frac{V}{ms}}{\frac{1}{b_1 b_2} + \frac{1}{ms}\left(\frac{1}{b_1} + \frac{1}{b_2}\right)} \quad \ldots (7.125b)$$

Thus:

$$F_M = F_1 - F_2 = \frac{\frac{V}{b_2}}{\frac{1}{b_1 b_2} + \frac{1}{ms}\left(\frac{1}{b_1} + \frac{1}{b_2}\right)} \quad \ldots (7.126)$$

Let us solve the same problem by using the circuit shown in Fig. 7.19(c). The mesh equations by inspection are:

$$\begin{bmatrix} \frac{1}{b_1} + \frac{1}{b_2} & -\frac{1}{b_2} \\ -\frac{1}{b_2} & \frac{1}{b_2} + \frac{1}{ms} \end{bmatrix} \begin{bmatrix} F_1 \\ F_2 \end{bmatrix} = \begin{bmatrix} V \\ 0 \end{bmatrix} \quad \ldots (7.127)$$

Generalized Impedances and System Functions   467

Note that this matrix is different from the matrix determined for Fig. 7.19(b). In this circuit, the force applied to the mass is F_3 alone. Thus, if we solve for F_3, we obtain:

$$F_M = F_3 = \frac{\begin{vmatrix} \frac{1}{b_1} + \frac{1}{b_2} & V \\ -\frac{1}{b_2} & 0 \end{vmatrix}}{\begin{vmatrix} \frac{1}{b_1} + \frac{1}{b_2} & -\frac{1}{b_2} \\ -\frac{1}{b_2} & \frac{1}{b_2} + \frac{1}{ms} \end{vmatrix}} \quad ..(7.128a)$$

$$= \frac{\frac{V}{b_2}}{\frac{1}{b_1 b_2} + \frac{1}{ms}(\frac{1}{b_1} + \frac{1}{b_2})} \quad ..(7.128b)$$

and the results are the same in both the circuits.

7.6 SOME PROPERTIES OF LINEAR SYSTEMS

Much of the simplification of the process of writing equilibrium equations in Section 7.5 is due to the basic property of the elements of a linear system: the responses are directly proportional to the driving function (or its derivative or integral). Indeed, this is why the equivalent impedance of a series connection is simply the sum of the individual impedances. It should, therefore, be expected that these basic attributes of the elements which make up a system should result in characteristic properties of a linear system. Some of these properties are expressed as Circuit Theorems which greatly simplify the task of eliminating unwanted variables. Others are useful in that they give guidance as to the kinds of measurements which must be taken to obtain a valid model of a system.

Consider first a set of node equations, the basic form of which we have seen is:

$$[Y][V] = [F] \qquad ..(7.129)$$

In the case of a circuit having three independent nodes, the details are:

$$\begin{bmatrix} Y_{11} & Y_{12} & Y_{13} \\ Y_{21} & Y_{22} & Y_{23} \\ Y_{31} & Y_{32} & Y_{33} \end{bmatrix} \begin{bmatrix} V_1 \\ V_2 \\ V_3 \end{bmatrix} = \begin{bmatrix} F_1 \\ F_2 \\ F_3 \end{bmatrix} \qquad ..(7.130)$$

The solution has the basic form:

$$[V] = [Y]^{-1}[F] = [z][F] \qquad ..(7.131)$$

where $$[z] = \begin{bmatrix} z_{11} & z_{12} & z_{13} \\ z_{21} & z_{22} & z_{23} \\ z_{31} & z_{32} & z_{33} \end{bmatrix} \quad ..(7.132a)$$

$$= \frac{1}{\Delta} \begin{bmatrix} \Delta_{11} & -\Delta_{21} & \Delta_{31} \\ -\Delta_{12} & \Delta_{22} & -\Delta_{32} \\ \Delta_{13} & -\Delta_{23} & \Delta_{33} \end{bmatrix} \quad ..(7.132b)$$

$$\Delta = \text{the determinant of } [Y] \quad ..(7.132c)$$

and $\quad \Delta_{jk} = $ the co-factor of $Y_{jk} \quad ..(7.132d)$

The terms in the inverse matrix have some physical significance, in that they are driving point and transfer impedances. That is, the term z_{jj} is the impedance between node j and ground, when the sources connected to all other nodes are suppressed (that is, set to zero), which can be recognized as the driving point impedance discussed in Section 7.3. The term z_{jk} is the ratio of the across variable between node j and ground to the through variable entering node k, when all other sources are suppressed, usually called the transfer impedance.

It is, therefore, theoretically possible to take appropriate measurements on a system to yield values for the entire array of deriving point and transfer impedances so that the matrix [z] may be measured. By inverting it, the matrix [Y] of the corresponding node equations can then be obtained. Reversal of the procedure to obtain the node equations by inspection of the circuit will then produce a circuit model of the system.

In practice, for the systems we are considering, it may not be necessary to carry out this inversion process, because the form of the circuit model can be determined, knowing the number of variables required. However, this does indicate the form of the measurements which must be taken on a system, including the necessary constraints to be applied.

It should be evident that if we consider a set of mesh equations having the form:

$$[Z][F] = [V] \qquad ..(7.133)$$

the solution is:

$$[F] = [Z]^{-1}[V] = [y][V] \qquad ..(7.134)$$

where [y] is the array of driving point and transfer admittances.

Generalized Impedances and System Functions

Note that there is no simple relation between [y] (that is, the driving point and transfer admittances) and [Y] (that is, the admittance array of the node equations). Nor is there any simple relation between the driving point and transfer impedance matrix [z] and the mesh equation impedance matrix [Z].

7.6.1 Superposition Theorem

Returning the (7.131), we can write out the solution for V_1 as:

$$V_1 = z_{11} F_1 + z_{12} F_2 + z_{13} F_3 \qquad \ldots (7.135)$$

We note that for this system having three sources, the solution has three components each of which is the value of V_1, while the other sources are set to zero. **The Superposition Theorem uses this property of a linear system to obtain the total response as the sum of the responses to each driving function or source taken one at a time, while all other sources are supressed.**

Example 7.12

A circuit consisting of two resistors is driven by a voltage source and a current source as shown in Fig. 7.20(a). Find the voltage across R_2.

The procedure is simply to obtain the values of E_2, when only one source is acting at a time, and to add these solutions. In this case, there are two sources.

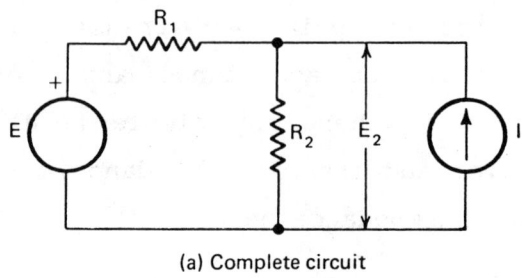

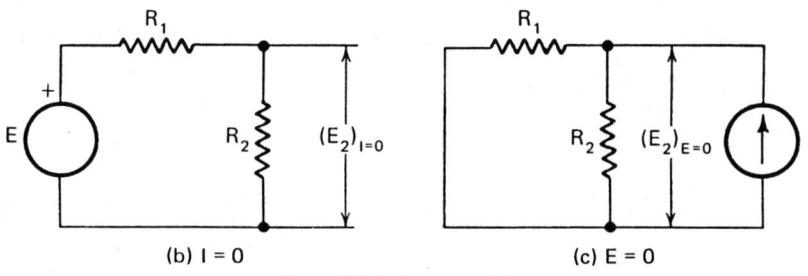

Figure 7.20 Superposition

So the problem is to find E_2 in each of Figs. 7.20(b) and (c) and add them.

From Fig. 7.20(b), the first component of E_2 can be found directly by noting that this is a voltage divider. Hence

$$E_{21} = \frac{R_2}{R_1 + R_2} E \qquad \qquad ..(7.136)$$

From Fig. 7.20(c), the second component of E_2 is found by noting that this is the voltage across the parallel combination of R_1 and R_2. Hence

$$E_{22} = \frac{R_1 R_2}{R_1 + R_2} I \qquad ..(7.137)$$

The complete response is the sum of those two expressions:

$$E_2 = E_{21} + E_{22}$$

$$= \frac{R_2}{R_1 + R_2}\left[E + R_1 I \right] \qquad ..(7.138)$$

7.6.2 Reciprocity Theorem

We have previously noted that when all of the elements in a system or circuit are bilateral, the terms Y_{jk} and Y_{kj} are equal in the node equations and Z_{jk} and Z_{kj} are equal in the mesh equations, and therefore the matrices [Y] and [Z] are symmetrical. As a result, their inverses, [z] and [y], are also symmetrical.

Although this is of little value as an aid to analyze a circuit, a corollary of the theorem is of some value in modeling. If a set of measurements on a system shows that any pair of transfer impedances or admittances is not reciprocal (that is, $z_{jk} \neq z_{kj}$ or $y_{jk} \neq y_{kj}$), the system cannot be modeled using only bilateral elements and independent sources such as those being considered in this text.

7.6.3 Thevenin's Theorem

Thevenin's Theorem is a very convenient method of obtaining a simple circuit which is equivalent to a section of a complete circuit. To some extent, it is an extension of the modeling of real sources as an ideal source associated with an impedance.

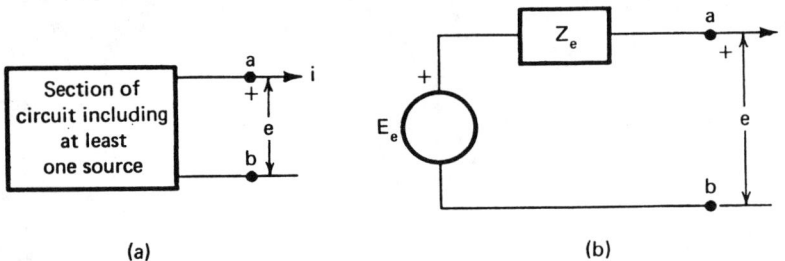

Figure 7.21 Equivalent circuit

The section of the circuit to be modeled or simplified will always have at least one source and a pair of terminals at which it is connected to the remainder of the complete circuit. This is shown symbolically in Fig. 7.21(a). The theorem states simply that the circuit shown in Fig. 7.21(b) will produce exactly the same current-voltage relationships at terminals a and b provided:

(a) E_e is the voltage between a and b, when $i = 0$, (that is, the open circuit voltage),

(b) Z_e is the impedance between the terminals a and b, when all the sources within the section of the circuit are suppressed (that is, the driving point impedance).

Example 7.13

Obtain the system function E_L/E for the circuit shown in Fig. 7.22(a).

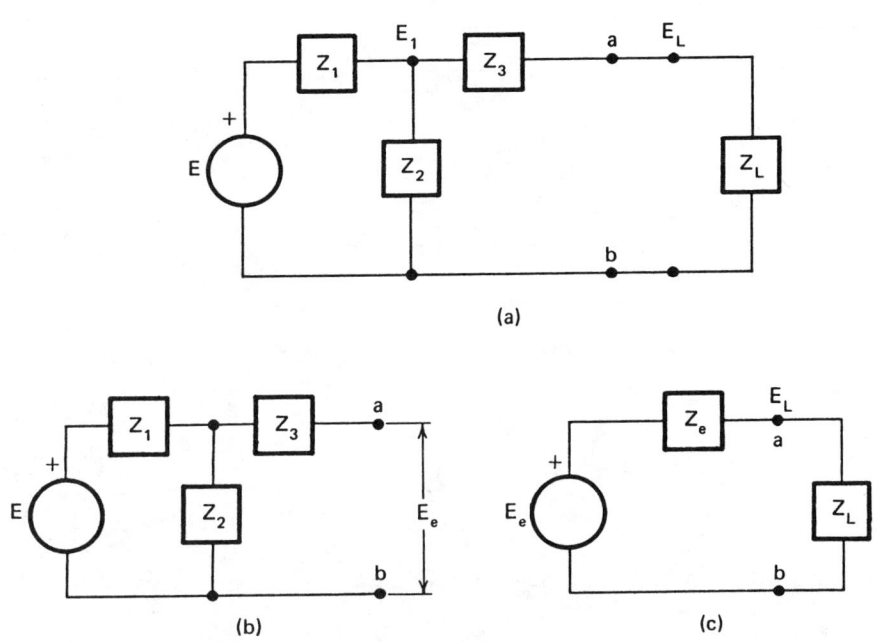

Figure 7.22 Thevenin equivalent circuit (example 7.13)

The procedure is to obtain the Thevenin equivalent of the section of the original circuit reproduced in Fig. 7.22(b). Since there is no current flow through Z_3, the open circuit voltage is obtained directly from the voltage divider relation:

$$E_e = \frac{Z_2}{Z_1 + Z_2} E \qquad \ldots (7.139)$$

The equivalent impedance is the driving point impedance

between the terminals a and b, when the source is suppressed and this is readily seen to be the equivalent of Z_3 in series with the parallel combination of Z_1 and Z_2. Hence,

$$Z_e = Z_3 + \frac{Z_1 Z_2}{Z_1 + Z_2}$$

$$= \frac{Z_1 Z_2 + Z_2 Z_3 + Z_3 Z_1}{Z_1 + Z_2} \qquad \ldots (7.140)$$

Using these values in Fig. 7.22(c), we obtain:

$$\frac{E_L}{E_e} = \frac{Z_L}{Z_e + Z_L} \qquad \ldots (7.141a)$$

and substituting for E_e and Z_e gives

$$\frac{E_L}{E} = \frac{Z_2}{Z_1 + Z_2} \cdot \frac{Z_L}{Z_e + Z_L}$$

$$= \frac{Z_2 Z_L}{Z_1 Z_2 + Z_2 Z_3 + Z_3 Z_1 + Z_L(Z_1 + Z_2)} \qquad \ldots (7.141b)$$

This circuit is sufficiently simple to check this answer by obtaining the solution to the node equations of the entire circuit. We have

$$\begin{bmatrix} \dfrac{1}{Z_1} + \dfrac{1}{Z_2} + \dfrac{1}{Z_3} & -\dfrac{1}{Z_3} \\ -\dfrac{1}{Z_3} & \dfrac{1}{Z_3} + \dfrac{1}{Z_L} \end{bmatrix} \begin{bmatrix} E_1 \\ E_L \end{bmatrix} = \begin{bmatrix} \dfrac{E}{Z_1} \\ 0 \end{bmatrix} \qquad ..(7.142)$$

The solution is

$$E_L = \dfrac{\dfrac{E}{Z_3 Z_1}}{\left(\dfrac{1}{Z_1} + \dfrac{1}{Z_2} + \dfrac{1}{Z_3}\right)\left(\dfrac{1}{Z_3} + \dfrac{1}{Z_L}\right) - \dfrac{1}{Z_3^2}} \qquad ..(7.143a)$$

$$= \dfrac{Z_2 Z_L E}{Z_1 Z_2 + Z_2 Z_3 + Z_1 Z_3 + Z_L(Z_1 + Z_2)} \qquad ..(7.143b)$$

In some cases, the process of obtaining the Thevenin equivalent by the simple procedures used in Example 7.13 becomes rather complicated and it is simpler to obtain the open circuit voltage and driving point impedance directly from a set of node (or mesh) equations. This process is shown in the following example.

Example 7.14

Find an expression for the current through the resistor R in the circuit shown in Fig. 7.23(a).

In this case, the section of the circuit for which the Thevenin equivalent is required consists of all

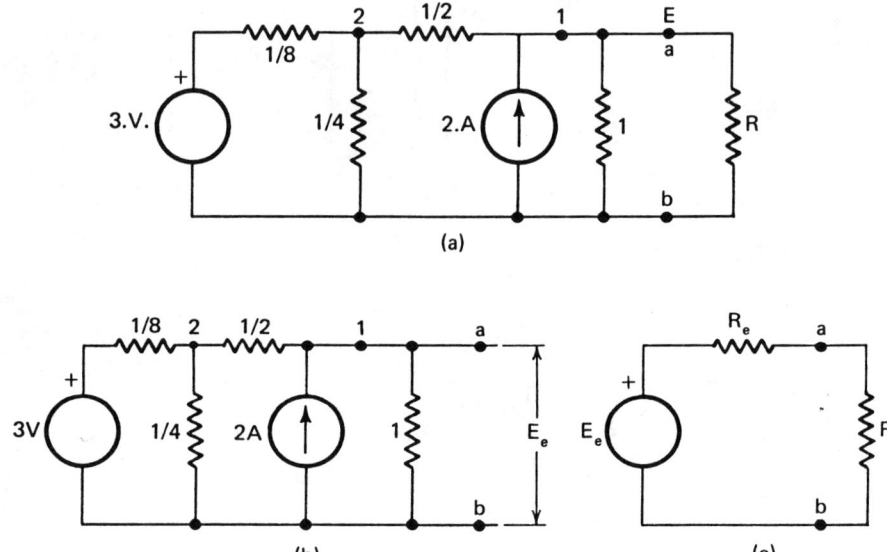

Figure 7.23 Electrical circuit (example 7.14)

elements and sources other than R. For clarity, this is shown in Fig. 7.23(b), but normally should not be necessary. The node equations are:

$$\begin{bmatrix} 3 & -2 \\ -2 & 14 \end{bmatrix} \begin{bmatrix} E_1 \\ E_2 \end{bmatrix} = \begin{bmatrix} 2 \\ 24 \end{bmatrix} \quad ..(7.144)$$

The open circuit voltage, E_e, for this circuit is:

$$E_1 = \frac{1}{\Delta} \begin{vmatrix} 2 & -2 \\ 24 & 14 \end{vmatrix} \quad ..(7.145a)$$

Generalized Impedances and System Functions 479

where

$$\Delta = \begin{vmatrix} 3 & -2 \\ -2 & 14 \end{vmatrix} = 38 \qquad ..(7.145b)$$

That is:

$$E_e = E_1 = 2 \text{ V} \qquad ..(7.146)$$

The equivalent impedance is the driving point impedance between node 1 and ground. The simplest way to obtain this is the replacement of the actual sources by a vector having only one source injecting the current I at node 1. That is, the equations are now:

$$\begin{bmatrix} 3 & -2 \\ -2 & 14 \end{bmatrix} \begin{bmatrix} E_1 \\ E_2 \end{bmatrix} = \begin{bmatrix} I \\ 0 \end{bmatrix} \qquad ..(7.147)$$

The required driving point impedance is the ratio:

$$\frac{E_1}{I} = Z_{11} = \frac{14}{38} = \frac{7}{19} \text{ ohm} \qquad .(7.148)$$

Thus the Thevenin equivalent shown in Fig. 7.23(c) has

$E_e = 2$ V and $R_e = 7/19$ ohm, and hence the current through resistor R is given by:

$$I = \frac{2}{R + \frac{7}{19}} \qquad \qquad ..(7.149)$$

A simple application of Thevenin's Theorem which is often very useful is that of transforming a through variable source and the impedance connected across its terminals [Fig. 7.24(a)] to an equivalent series connected across variable source and impedance [Fig. 7.24(b)]. The circuits are equivalent, provided $(AV)_e = Z$ (TV) and $Z_e = Z$.

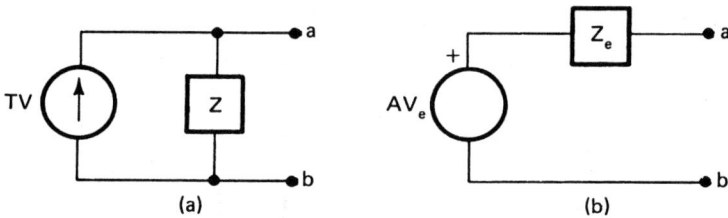

Figure 7.24 Application of Thevenin's Theorem. (a) Original circuit; (b) Thevenin equivalent circuit

7.6.4 Norton's Theorem

This is the dual of Thevenin's Theorem in that the equivalence for a section of a circuit is expressed in terms of a through variable source and shunting admittance. This is indicated in Fig. 7.25, where Fig. 7.25(b) is the equivalent of Fig. 7.25(a), provided:

Generalized Impedances and System Functions   481

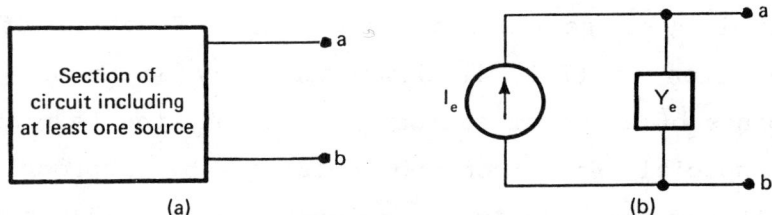

Figure 7.25 Norton equivalent circuit

(a) I_e is the current flow, when the voltage between a and b is zero (that is, the short-circuit current).

(b) Y_e is the admittance between the terminals a and b, when all the sources within the section of the circuit are suppressed (that is, the driving point admittance).

The Norton equivalent can be obtained by series-parallel combinations, or by solutions of mesh equations, or by a transformation of the Thevenin equivalent.

In all the equivalent circuits we have considered, it is important to note that some information is lost, when working with the equivalent. The equivalence is valid only at the terminals we have designated a and b. In effect, these equivalences are convenient methods to eliminate unwanted variables thereby simplifying some of the mathematical manipulations which may be required.

7.7 SUMMARY AND DISCUSSIONS

In this chapter, the case when the system is excited by a generalized exponential function of the form given in (7.1) is discussed. By determining the response of a system to such a function, the response to any special form can be obtained by appropriately specifying s, which is a complex variable. This permits the definition of generalized impedances for the basic components, namely, the dissipative element, the A-type and T-type storage elements. The system functions, namely the driving point impedance (or admittance) and the transfer functions are defined. It is shown how they can be obained by analyzing the given network. The methods discussed are the mesh and the nodal analysis. Some theorems which aid in the analysis of circuits are also discussed. Superposition theorem is useful when the circuit contains several sources and is the basis for linear systems. If a branch response is required, either Thevenin's or Norton's theorem can be used. These theorems are also useful in the determination of the conditions under which maximum power can be transferred between the source and the load.

7.8 PROBLEMS

7.1 Find the driving point impedance, $Z_{dp}(s)$, for the circuit diagram shown in Figs. P7.1(a) and P7.1(b).

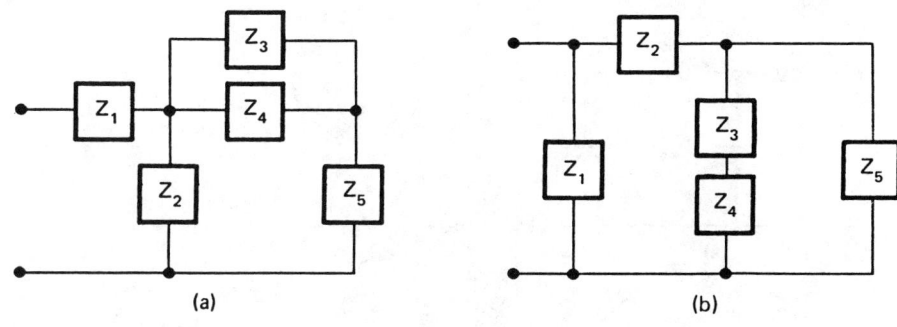

Figure P 7.1

7.2 Find the driving point admittance, $Y_{dp}(s)$, for the circuit diagrams shown in Figs. P7.2(a) and P7.2(b).

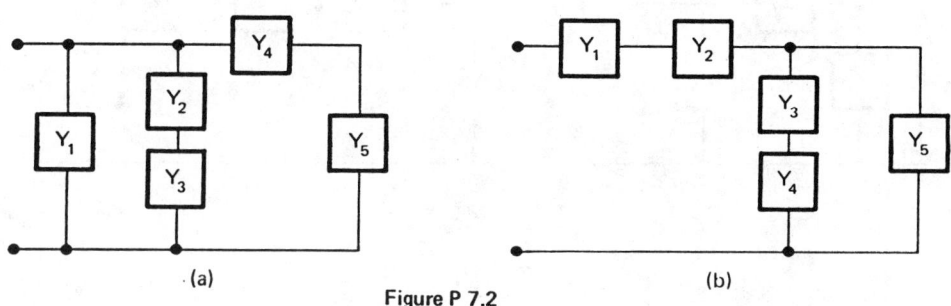

Figure P 7.2

7.3 Determine the driving point impedance, $Z_{dp}(s)$, and the driving point admittance, $Y_{dp}(s)$, for the electric circuits shown in Figs. P7.3(a) and P7.3(b).

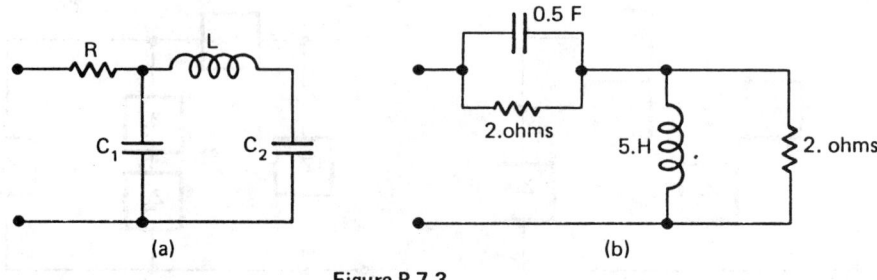

Figure P 7.3

7.4 Derive the system function (output/input) for the circuits shown in Figs. P7.4(a) and P7.4(b).

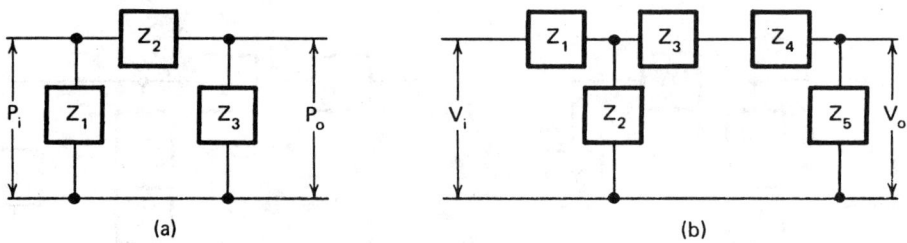

Figure P 7.4

7.5 Find the system function $[G(s) = E_o/E_i]$ for the electric circuits given in Figs. P7.5(a) and P7.5(b).

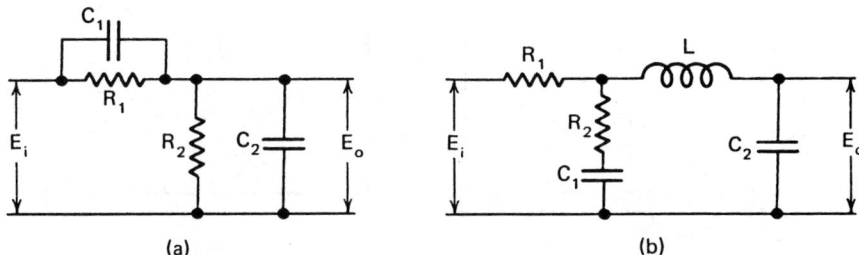

Figure P 7.5

7.6 The input and output voltages are indicated on the electric circuit in Fig. P7.6. Find the system function, $G(s) = E_o/E_i$.

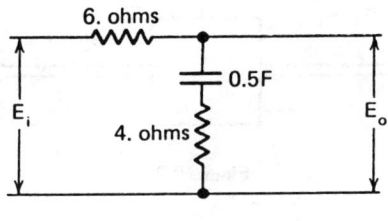

Figure P 7.6

7.7 A fluid system is shown schematically in Fig. P7.7. The section at the left end of R_1 is designated as terminal 1. Find the driving point admittance (from terminal 1 to g) as a function of s.

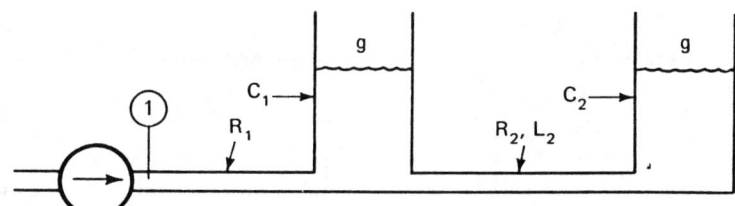

Figure P 7.7

7.8 In the fluid system [Fig. P7.8], an air pump pressurizes two tanks. Find the system function $G(s) = P_o/P_i$.

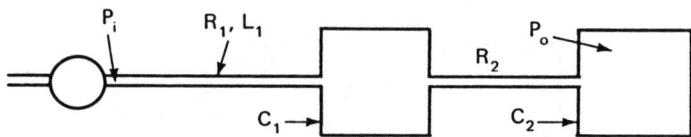

Figure P 7.8

7.9 A constant flow pump supplies water to a reservoir with a drain in it. The system can be modeled as indicated in Fig. P7.9. Find the system function that describes the ratio of output flow to input flow.

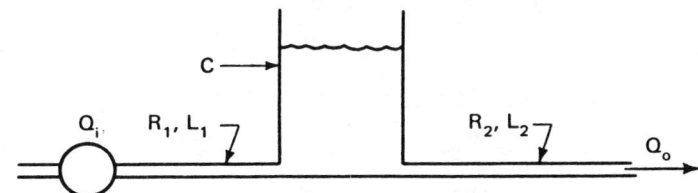

Figure P 7.9

7.10 Derive the driving point impedance, Z(s), (from terminal 1 to g) for the mechanical system given in Figs. P7.10(a) and P7.10(b).

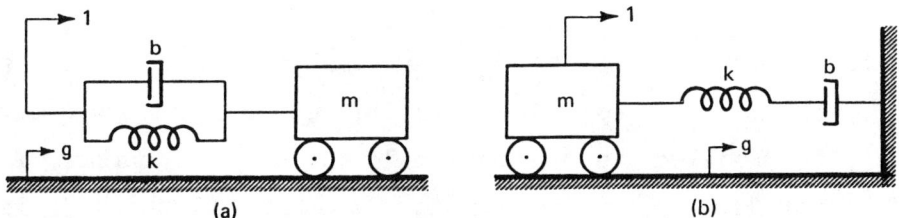

Figure P 7.10

7.11 A force $f(t) = F\varepsilon^{st}$ is applied to mass 1 of Fig. P7.11. The velocity of mass 3 is assumed to have the form $v_3(t) = V_3 \varepsilon^{st}$. Find the system function (V_3/F) in terms of m_1, m_2, m_3, b, k and s.

Figure P 7.11

7.12 The input to a mechanical system [Fig. P7.12] is a velocity of the form $v(t) = V \varepsilon^{st}$. The output is designated as the force transmitted through the damper b_2 and it is assumed to be $f_b(t) = F \varepsilon^{st}$. Determine the system function with these signals as input and output.

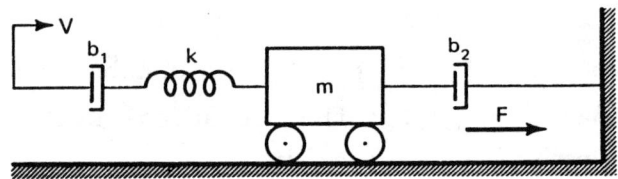

Figure P 7.12

7.13 A torque (T_i) is applied to flywheel 1 and is transmitted to flywheel 2 through a shaft that may be modeled as a torsional spring and damper in parallel. The torque developed in flywheel 2 is designated as T_o [Fig. P7.13]. Find the system function, $G(s) = T_o/T_i$.

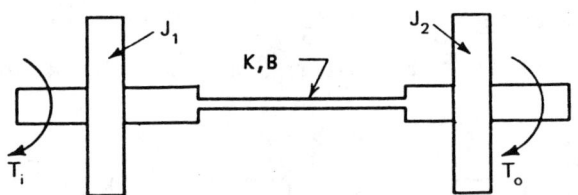

Figure P 7.13

7.14 The ratio of output force to input velocity for a mechanical system is given as:

Generalized Impedances and System Functions 489

$$\frac{F_o}{V_i} = \frac{1}{s^2 + 12s + 100}$$

The input velocity may be described by $v_i(t) = 2 U_i(t)$. What is the output force as a function of time?

7.15 The system function for a particular fluid system is:

$$\frac{P_o}{P_i} = \frac{s + 1}{s + 2}$$

If $p_i(t) = 3 U_r(t)$, find $p_o(t)$.

7.16 The output flow is related to the input pressure in a fluid circuit by the system function

$$\frac{Q_o}{P_i} = \frac{s + 4}{s^2 + 3s + 2}$$

Determine $q_o(t)$, if $p_i(t) = U_s(t)$.

7.17 An electric circuit has the system function:

$$\frac{E_o}{E_i} = \frac{s}{s^2 + 4s + 4}$$

An input $e_i(t) = 5 U_s(t)$ is applied. What is the output voltage, $e_o(t)$?.

7.18 A mechanical system has a system function that has been derived as:

$$\frac{V_o}{V_i} = \frac{s}{s + 10}$$

Find $v_o(t)$, if $v_i(t) = 10\, U_s(t) + 5\, U_s(t - 2)$.

7.19 Write a set of node equations in matrix form for the physical systems shown in Figs. P7.19(a), P7.19(b), P7.19(c) and P7.19(d).

7.20 For the network shown in Fig. P7.20, find the output voltage, $e_o(t)$, if the input voltage $e_i(t) = U_s(t)$.

Hint: Use node equations to find system function E_o/E_i and then derive the differential equation.

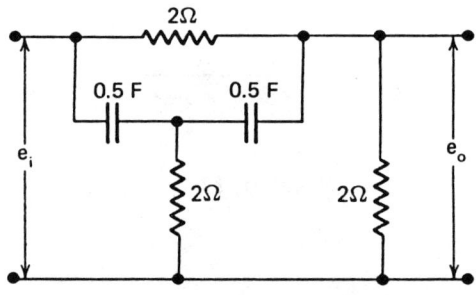

Figure P 7.20

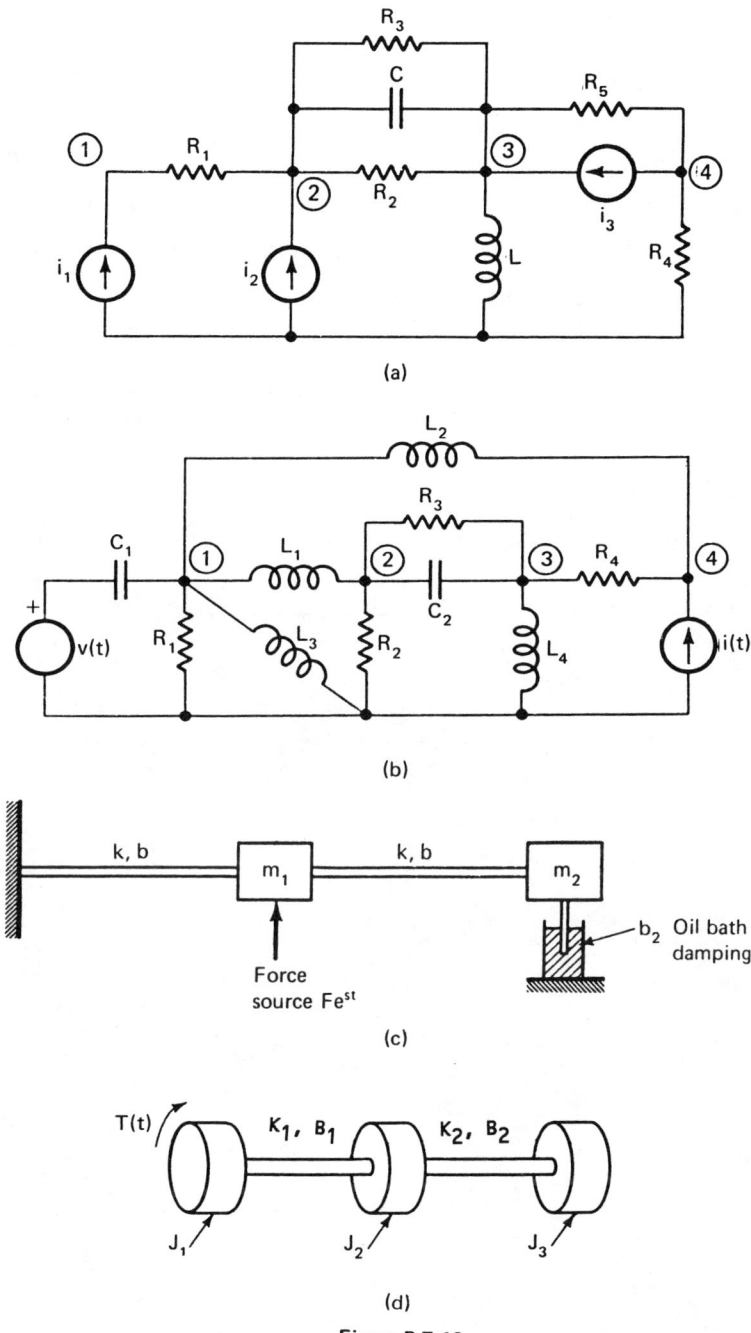

Figure P 7.19

7.21 Write a set of mesh equations in matrix form for the physical circuits indicated in Figs. P7.21(a), P7.21(b), P7.21(c) and P7.21(d).

7.22 For the resistive network shown in Fig. P7.22, the source voltage e_s = 100 V. Using a Thevenin equivalent circuit, determine the voltage, e_3, in terms of the unknown resistance, R.

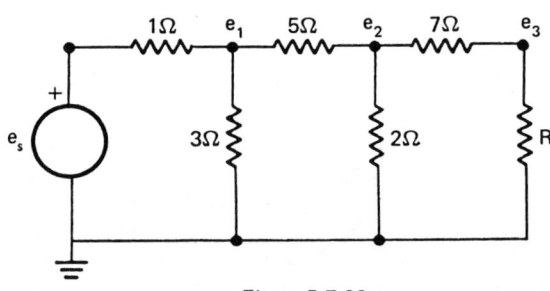

Figure P 7.22

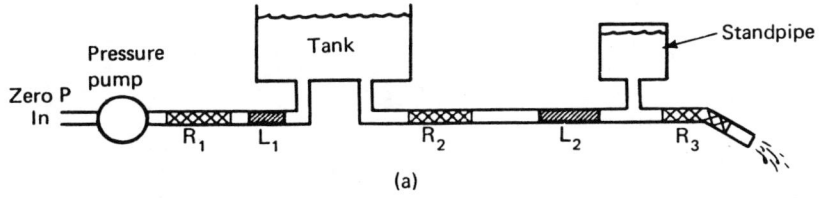

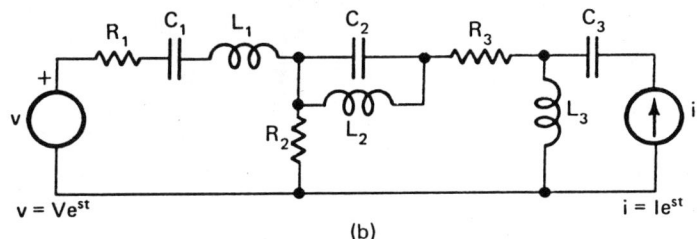

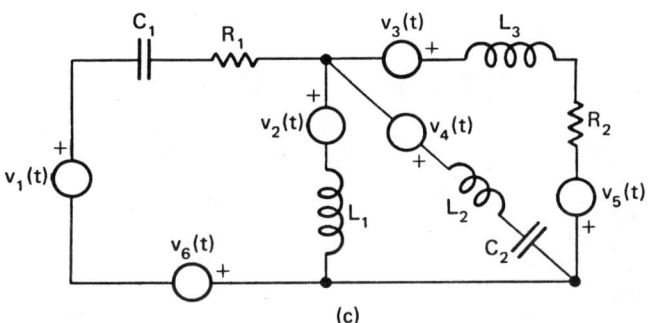

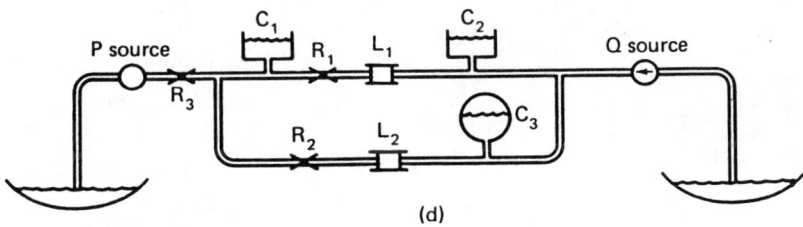

Figure P 7.21

CHAPTER 8

SINUSOIDAL RESPONSE

In Chapter V, we noted that the response of a circuit is the time variation that occurs in one signal variable as the result of a time variation in another signal variable. We may also express the idea of response as the output for a specific input. The common factor in any of these responses is that the inputs and the outputs are both functions of time.

When a periodic function (e.g., sine or cosine) is applied as an input to a circuit, we might term the output as the sinusoidal (or cosinusoidal) response. (This will be proved analytically in the next section). Here again, the input and the output would be functions of time. This would then be consistent with our previous treatment of other input functions. In this chapter, we will apply sine and cosine inputs to physical circuits. Upon examination of the responses to these inputs, we will notice a transient (non-periodic) portion of the response and a periodic portion of the response. In some situations, such as analyzing fault currents in a power system, the complete response is important and must be determined. However, for many applications, the transient portion is of no particular interest and therefore it is not evaluated. The periodic portion is usually more important and is designated as the **sinusoidal steady-state** response.

In the sinusoidal steady-state, when the input to a linear system has a sine (or cosine) waveform, the output also has a sine (or cosine) waveform at the same frequency. The sinusoidal steady state response is distinguished by the amplitude and the phase shift of the output waveform relative to the input waveform. This relative amplitude and phase shift property of the sinusoidal steady state response is a function of the waveform frequency. As a result, the amplitude-frequency and the phase shift-frequency relations completely specify the response. These relations are referred to as the **frequency response**.

We will begin the development of the sinusoidal response relations by applying direct mathematical methods. Then we will use impedance and system functions concepts to obtain the same results.

8.1 RESPONSE TO SINE AND COSINE FUNCTIONS
8.1.1 First Order Circuits

From (5.1), the most general differential equation for first order linear time invariant circuits is:

$$\tau \frac{dy}{dt} + y = a_1 g(t) + a_2 \frac{dg(t)}{dt} \qquad ..(8.1)$$

When the input function, $g(t)$, is a sine wave function of amplitude, G, and frequency, ω, then

$$g(t) = G \sin \omega t \qquad ..(8.2)$$

Eq. (8.1) may, therefore, be written as:

$$\tau \frac{dy}{dt} + y = a_1 G \sin \omega t + a_2 \omega G \cos \omega t \qquad ..(8.3a)$$

with $\qquad y(0^+) = 0 \qquad ..(8.3b)$

(The initial condition is zero for the sine input, if there is no initial storage). The complete solution for y consists of the summation of the homogeneous solution and the particular solution [Eq. (5.4)]. Thus:

$$y = y_H + y_P \qquad ..(8.4)$$

We know from Chapter V that there is only one homogeneous solution for first order circuits. That homogeneous solution has the form given in (5.9), which is $A_1 \varepsilon^{-t/\tau}$.

To find the particular solution, we follow the standard mathematical procedure of undetermined coefficients. This method may be used, when the input function has only a finite number of independent derivatives. The method consists of summing a term proportional to the input function with terms proportional to all possible derivatives of the input functions. In this case, since the sine and cosine functions are derivatives of each other, we have only two terms in the solution. Thus we assume a particular solution of the form:

$$y_P = A_2 \sin \omega t + A_3 \cos \omega t \qquad ..(8.5)$$

where the coefficients A_2 and A_3 are to be determined by substituting (8.4) into (8.3). That operation yields:

$$\left[A_2 - \omega\tau A_3 \right] \sin \omega t + \left[A_3 + \omega\tau A_2 \right] \cos \omega t$$
$$= a_1 G \sin\omega t + a_2 \omega G \cos\omega t \qquad ..(8.6)$$

If we equate the coefficients of sine and cosine terms on each side of (8.6), we obtain the simultaneous equations:

$$A_2 - \omega\tau A_3 = a_1 G \qquad ..(8.7a)$$
$$A_3 + \omega\tau A_2 = a_2 \omega G \qquad ..(8.7b)$$

Solution of (8.7) produces:

Sinusoidal Response

$$A_2 = \frac{G\left[a_1 + a_2\omega^2\tau\right]}{\left[1 + \omega^2\tau^2\right]} \qquad ..(8.8a)$$

$$A_3 = \frac{G\omega\left[a_2 - a_1\tau\right]}{\left[1 + \omega^2\tau^2\right]} \qquad ..(8.8b)$$

Thus, the particular solution of (8.6) becomes:

$$y_P = \frac{G}{1+\omega^2\tau^2}\left[(a_1+a_2\omega^2\tau)\sin\omega t + (a_2-a_1\tau)\omega\cos\omega t\right] \qquad ..(8.9)$$

Eq. (8.9) may also be expressed in terms of a phase shifted sine wave as indicated in Chapter II. Since the input $g(t) = G\sin\omega t$, we would like to represent the forced output (particular response) as $A\sin(\omega t+\phi)$, where A is an amplitude and ϕ is a phase shift. In those terms, (8.9) becomes

$$y_P = \frac{G(a_1^2 + a_2^2\omega^2)^{1/2}}{(1+\omega^2\tau^2)^{1/2}}\sin(\omega t + \phi) \qquad ..(8.10a)$$

where $\quad \phi = \tan^{-1}\dfrac{(a_2 - a_1\tau)\omega}{a_1 + a_2\omega^2\tau} \qquad ..(8.10b)$

and
$$\sin \phi = \frac{(a_2 - a_1\tau)\omega}{\sqrt{(a_1^2 + a_2^2\omega^2)(1+\omega^2\tau^2)}} \qquad ..(8.10c)$$

The complete solution for y(t) in (8.3) now has the form:

$$y(t) = A_1 \varepsilon^{-t/\tau} + \frac{G(a_1^2+a_2^2\omega^2)^{1/2}}{(1+\omega^2\tau^2)^{1/2}} \sin(\omega t+\phi) \qquad ..(8.11)$$

We may evaluate the constant A_1 in (8.11) by applying the initial condition (8.3b):

$$y(0^+) = A_1 + \frac{G(a_1^2+a_2^2\omega^2)^{1/2}}{(1+\omega^2\tau^2)^{1/2}} \sin \phi = 0 \qquad ..(8.12a)$$

from which we obtain:

$$A_1 = \frac{-G(a_1^2+a_2^2\omega^2)^{1/2}}{(1+\omega^2\tau^2)^{1/2}} \sin \phi = \frac{G\omega(a_1\tau - a_2)}{(1+\omega^2\tau^2)} \qquad ..(8.12b)$$

As a result, the complete output response for a sine wave input is:

$$y(t) = \frac{G\omega(a_1\tau-a_2)}{1+\omega^2\tau^2} \varepsilon^{-t/\tau} + \frac{G(a_1^2+a_2^2\omega^2)^{1/2}}{(1+\omega^2\tau^2)^{1/2}} \sin(\omega t+\phi)$$
$$..(8.13)$$

where ϕ is defined in (8.10b).

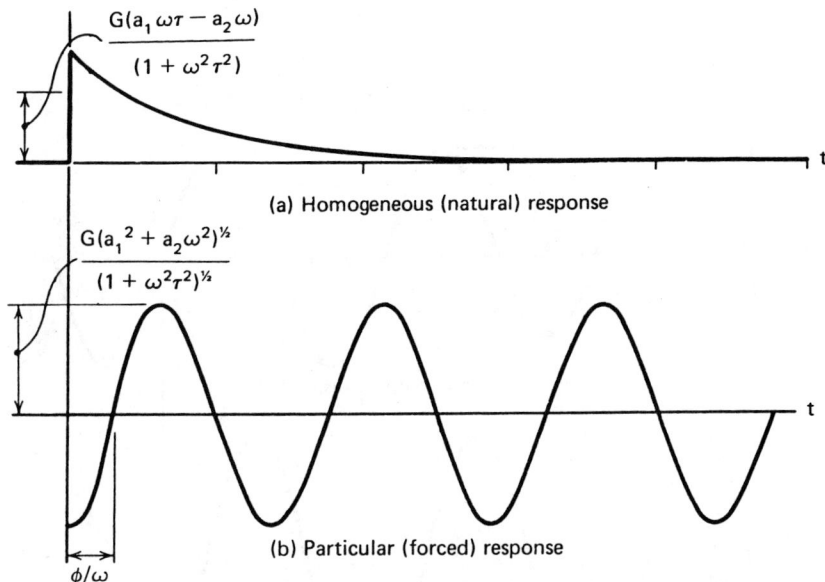

Figure 8.1 Typical natural and forced response of first order circuits

The first term on the right-hand side of (8.13) is the natural response of the system. The second term on the right-hand side is the forced response of the system. Fig. 8.1 is a graphical representation of the natural and forced responses. The natural response [Fig. 8.1(a)] is merely a decaying exponential function. As we have observed previously, this function is less than 5% of its initial value after three time constants ($t = 3\tau$). For times greater than this, therefore, the contribution of the natural response is negligible. The forced response [Fig. 8.1(b)] has the form of a pure sine wave. It is, however, displaced in time by ϕ/ω as indicated in (8.13).

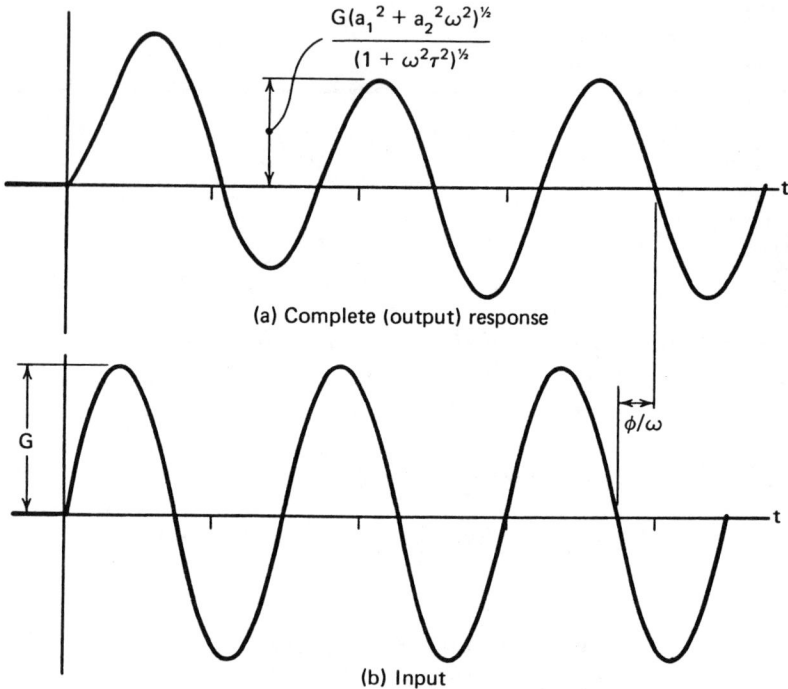

Figure 8.2 Output and input of first order circuit

Fig. 8.2 shows the complete response and the input waveform. Initially, the complete response is affected by the natural response. This initial period is sometimes referred to as the transient period. After the natural contribution becomes small, both the input and the output are sine waves of the same frequency. This region is known as the sinusoidal steady state. In most cases, when there is a sine wave input, it is a continuing function that is maintained for a long time. We are, therefore, generally concerned only with the forced portion of the response. Indeed, in most cases,

Sinusoidal Response

the system has settled into the sinusoidal steady state before measurements or observations are made.

In the sinusoidal steady state, the output sine wave differs from the input sine wave in phase, ϕ, and amplitude. The magnitude, M, expresses the ratio of output amplitude to input amplitude. The sinusoidal steady state is then completely described by M and ϕ. For the first order circuit, these parameters have the values:

$$M = \frac{(a_1^2 + a_2\omega^2)^{1/2}}{(1 + \omega^2\tau^2)^{1/2}} \qquad ..(8.14a)$$

$$\phi = \tan^{-1}\frac{\omega(a_2 - a_1\tau)}{a_1 + a_2\omega^2\tau} \qquad ..(8.14b)$$

These magnitude-frequency and phase-frequency relations are called the frequency response.

There is no difference in frequency response between a sine wave input and a cosine wave input. We may think of the cosine input as a sine wave displaced by $t = -\pi/2\omega$. In this case, the output will also be displaced by the same time period. The magnitude and phase of output relative to input, however, remain unchanged.

The following examples will demonstrate the frequency response of some first-order circuits.

Example 8.1

Fig. 8.3 shows a first-order electrical RC circuit. If the supply voltage has the form $e_s(t) = E \cos \omega t$, find:

(a) the complete response, $e(t)$,

(b) the sinusoidal steady state response,

(c) the frequency response.

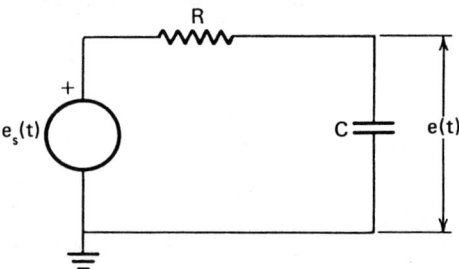

Figure 8.3 First order electrical circuit (example 8.1)

The differential equation that represents the RC series circuit is:

$$RC \frac{de}{dt} + e = e_s(t) = E \cos \omega t \qquad ..(8.15a)$$

with

$$e(0^+) = 0 \qquad ..(8.15b)$$

The complete solution for the output voltage, $e(t)$, consists of the summation of the homogeneous solution and the particular solution. The homogeneous solution for a first-order circuit always has the form:

$$e_H = A_1 \varepsilon^{-t/RC} \qquad ..(8.16)$$

Sinusoidal Response

To find the particular solution, we assume that:

$$e_P = A_2 \sin \omega t + A_3 \cos \omega t \qquad ..(8.17)$$

Now the substitution of the assumed particular solution into the differential equation yields:

$$\left[A_2 - \omega RC A_3\right] \sin \omega t + \left[\omega RC A_2 + A_3\right] \cos \omega t = E \cos \omega t \qquad ..(8.18)$$

If the coefficients of like terms are equated, the result is:

$$A_2 - \omega RC \, A_3 = 0 \qquad ..(8.19)$$

and
$$\omega RC \, A_2 + A_3 = E \qquad ..(8.20)$$

The solution of these simultaneous equations provides A_2 and A_3 for the particular solution. The result is:

$$e_P = \frac{E}{\left[1 + (\omega RC)^2\right]} \left[\omega RC \sin \omega t + \cos \omega t\right] \qquad ..(8.21a)$$

$$= \frac{E}{\left[1 + (\omega RC)^2\right]^{1/2}} \cos(\omega t - \phi) \qquad ..(8.21b)$$

where $\phi = + \tan^{-1} \omega RC \qquad ..(8.21c)$

and
$$\cos \phi = \frac{1}{\sqrt{1+(\omega RC)^2}} \quad \quad ..(8.21d)$$

The complete solution for the voltage across the capacitor then has the form:

$$e(t) = A_1 \varepsilon^{-t/RC} + \frac{E}{\left[1 + (\omega RC)^2\right]^{1/2}} \cos(\omega t - \phi)$$
$$..(8.22)$$

The constant A_1 may now be determined from the initial condition with the result that the complete solution (part a) is:

$$e(t) = \frac{-E\varepsilon^{-t/RC} \cos \phi + E \cos(\omega t - \phi)}{\sqrt{1+(\omega RC)^2}} \quad \quad ..(8.23a)$$

$$e(t) = \frac{-E\varepsilon^{-t/RC}}{\left[1 + (\omega RC)^2\right]^{1/2}} + \frac{E \cos(\omega t - \phi)}{\left[1 + (\omega RC)^2\right]^{1/2}} \quad \quad ..(8.23b)$$

To obtain the sinusoidal steady state response, we should recognize that the exponential term in the complete response approaches zero. Therefore, the sinusoidal steady state response is the same as the forced (particular) response and the answer to part (b) is:

$$e(t) = \frac{E}{\left[1 + (\omega RC)^2\right]^{1/2}} \cos(\omega t - \phi) \quad \quad ..(8.24)$$

The frequency response (part c) is:

$$M = \frac{E}{\left[1 + (\omega RC)^2\right]^{1/2}} \left(\frac{1}{E}\right) = \frac{1}{\left[1 + (\omega RC)^2\right]^{1/2}} \quad ..(8.25)$$

This same result could have been reached by using the general expression for the frequency response of a first order system [Eqs. (8.14)] with $a_1 = 1$, $a_2 = 0$, and $\tau = RC$. The parameter ωRC represents a normalized angular frequency. Fig. 8.4 shows the magnitude, M and the phase, ϕ, plotted against this normalized frequency. The scale on the left side represents the magnitude. At zero frequency, the capacitor has infinite impedance and the voltage of the output is equal to the input voltage. The magnitude is, therefore, equal to unity for dc signals. As the frequency increases, the output continually decreases. At very high frequencies, the magnitude approaches zero. Here the capacitor behaves as a low impedance and the output is near ground potential. The scale on the right represents the phase shift in degrees. At low frequencies, the input and the output are almost in phase. At the frequency $\omega = 1/RC$, the phase lags by 45 degrees. This series RC circuit is

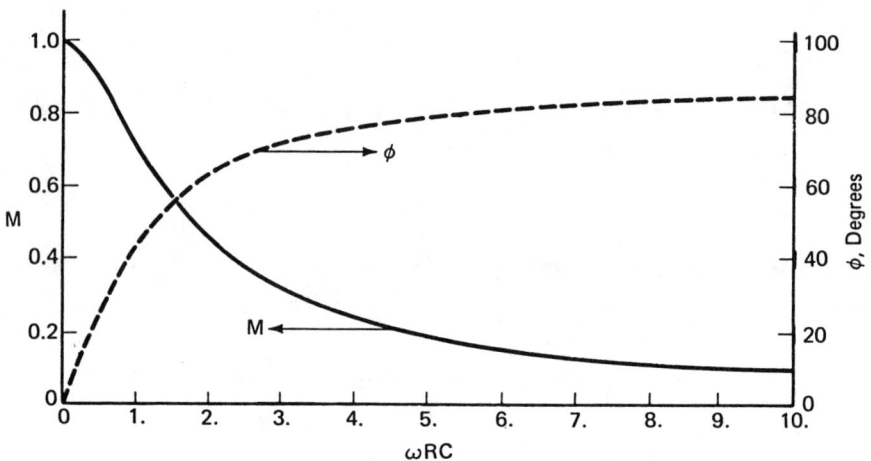

Figure 8.4 Magnitude and phase response for first order series circuit (example 8.1)

often referred to as a first order lag, because of its phase-frequency characteristics.

As mentioned previously, the frequency response is the same for sine or cosine inputs. However, the transient portion of the complete response (the constant A_1) is not equal for both input waveforms.

8.1.2 Second-Order Circuits

The most general differential equation for linear time invariant second-order circuits is given in (6.1) as:

$$\frac{d^2y}{dt^2} + 2\delta\omega_N \frac{dy}{dt} + \omega_N^2 y = a_1 g(t) + a_2 \frac{dg(t)}{dt} + a_3 \frac{d^2g(t)}{dt^2}$$

..(8.26)

Sinusoidal Response

When the input function is a sine wave $g(t) = G \sin \omega t$, (8.26) becomes:

$$\frac{d^2 y}{dt^2} + 2\delta \omega_N \frac{dy}{dt} + \omega_N^2 y = Gd_1 \sin \omega t + Gd_2 \cos \omega t$$

..(8.27a)

where
$$d_1 = (a_1 - a_2 \omega^2)$$
..(8.27b)

and
$$d_2 = a_2 \omega$$
..(8.27c)

The complete solution for (8.27a) is the sum of the homogeneous and particular solutions. In Chapter VI, we learned that there are three possible homogeneous solutions for the general second-order circuit. The solution that is appropriate depends on whether the circuit is underdamped, overdamped, or critically damped. For an underdamped or overdamped circuit, the homogeneous solution has the form given in (6.20) or (6.22):

$$Y_H = A_1 \varepsilon^{r_1 t} + A_2 \varepsilon^{r_2 t}$$
..(8.28)

where r_1 and r_2 are the roots of the characteristic equation. To determine the particular solution, we once again apply the method of undetermined coefficients. We, therefore, assume as in (8.5) a particular solution, y_p, that is:

$$y_P = A_3 \sin \omega t + A_4 \cos \omega t \qquad \ldots (8.29)$$

If (8.29) is substituted into (8.27a), the result is:

$$\left[c_1 A_3 - c_2 A_4 \right] \sin \omega t + \left[c_2 A_3 + c_1 A_4 \right] \cos \omega t =$$

$$Gd_1 \sin \omega t + Gd_2 \cos \omega t \qquad \ldots (8.30a)$$

where
$$c_1 = (\omega_N^2 - \omega^2) \qquad \ldots (8.30b)$$

and
$$c_2 = 2\delta\omega_N \qquad \ldots (8.30c)$$

Now, if the coefficients of like terms on each side of (8.30a) are equated, we obtain the simultaneous equations:

$$c_1 A_3 - c_2 A_4 = Gd_1 \qquad \ldots (8.31a)$$

$$c_2 A_3 + c_1 A_4 = Gd_2 \qquad \ldots (8.31b)$$

The coefficients A_3 and A_4 are determined from (8.31). The particular solution [Eq. (8.29)] thus becomes:

$$y_P = \frac{G}{c_1^2 + c_2^2} \left[(d_1 c_1 + d_2 c_2) \sin \omega t + (d_2 c_1 - d_1 c_2) \cos \omega t \right]$$

Sinusoidal Response

$$= \frac{G(d_1^2 + d_2^2)^{1/2}}{(c_1^2 + c_2^2)^{1/2}} \sin(\omega t + \phi) \qquad ..(8.32a)$$

where $\quad \phi = \tan^{-1} \dfrac{d_2 c_1 - d_1 c_2}{d_1 c_1 + d_2 c_2} \qquad ..(8.32b)$

The complete solution is the sum of (8.28) and (8.32a). Hence

$$y = A_1 \varepsilon^{r_1 t} + A_2 \varepsilon^{r_2 t} + \frac{G(d_1^2 + d_2^2)^{1/2}}{(c_1^2 + c_2^2)^{1/2}} \sin(\omega t + \phi)$$
$$..(8.33)$$

The constants A_1 and A_2 are now determined from the initial conditions. However, when we are concerned only with the sinusoidal steady state, we need not find A_1 and A_2. In the sinusoidal steady state (t greater than five times the larger time constant), the effects of the exponential terms in (8.33) fade away. As a result, the sinusoidal steady state and the particular solution [Eq. (8.32a)] are the same. Thus, the frequency response relations for a second order circuit are:

$$M = \left[\frac{d_1^2 + d_2^2}{c_1^2 + c_2^2} \right] = \left[\frac{(a_1 - a_3 \omega^2)^2 + (a_2 \omega)^2}{(\omega_N^2 - \omega^2)^2 + (2\delta\omega\omega_N)^2} \right] \qquad ..(8.34a)$$

$$\phi = \tan^{-1} \frac{d_2 c_1 - d_1 c_2}{d_1 c_1 + d_2 c_2}$$

$$= \tan^{-1} \frac{(a_2 \omega)(\omega_N^2 - \omega^2) - (a_1 - a_2 \omega^2)(2\delta \omega \omega_N)}{(a_1 - a_3 \omega^2)(\omega_N^2 - \omega^2) + (a_2 \omega)(2\delta \omega \omega_N)}$$

..(8.34b)

Eq. (8.34a) expresses the frequency response for all second order circuits. The magnitude response given in (8.34a) gets large without bound, when the frequency of the input sine wave is the same as the natural frequency of the circuit ($\omega = \omega_N$) and the circuit has no damping ($\delta = 0$). Both conditions are required for this result. From a mathematical viewpoint, the particular and homogeneous solutions under these conditions contain identical terms. Thus to derive the complete solution, we must multiply the assumed particular solution by time. From a physical viewpoint, these conditions are equivalent to exciting an undamped circuit with its own natural frequency. In theory, the output amplitude of oscillation would grow without bound. As a practical matter, however, the component models would no longer be linear and the amplitude of the oscillations would settle at some large value (limit cycle) provided that the system is not damaged in the process.

The following example demonstrates the frequency response of a specific second order circuit. Other second order circuits may be treated in the same way.

Sinusoidal Response   513

Example 8.2

A pneumatic signal generator supplies air to a closed chamber through a long line [Fig. 8.5(a)]. The pressure waveform of the generator is $p(t) = P\cos\omega t$. If the chamber has a capacitance, C, and the line has resistance, R, and Inertance, L, determine the frequency response of the flow, $q(t)$.

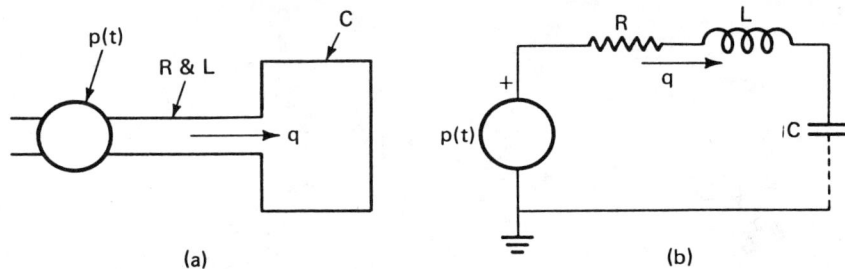

Figure 8.5 Fluid circuit (example 8.2)

Fig. 8.5(b) shows the analogous circuit diagram for the fluid circuit. The differential equation representing the flow is:

$$\frac{d^2q}{dt^2} + \frac{R}{L}\frac{dq}{dt} + \frac{1}{LC}q = \frac{1}{L}\frac{dp(t)}{dt} = -\frac{\omega P}{L}\sin\omega t \quad ..(8.35)$$

Since we want to find the frequency response, we need only consider the particular solution, q_P. If we assume that:

$$q_P = A_1 \sin\omega t + A_2 \cos\omega t \quad ..(8.36)$$

then substitution into the differential equation yields

$$\left[(\frac{1}{LC} - \omega^2) A_1 - \frac{\omega R}{L} A_2\right] \sin \omega t + \left[\frac{\omega R}{L} A_1 + \right.$$

$$\left.(\frac{1}{LC} - \omega^2) A_2\right] \cos \omega t = -\frac{\omega P}{L} \sin \omega t \quad ..(8.37)$$

If we equate the coefficients of the sine and the cosine terms on each side of the equation, we obtain that:

$$q_P = \frac{\omega P/L}{\left[(\frac{1}{LC} - \omega^2)^2 + (\frac{\omega R}{L})^2\right]} \left[-(\frac{1}{LC} - \omega^2) \sin \omega t + \right.$$

$$\left.(\frac{\omega R}{L}) \cos \omega t \right] \quad ..(8.38a)$$

$$= \frac{\omega P/L}{\left[(\frac{1}{LC} - \omega^2)^2 + (\frac{\omega R}{L})^2\right]^{1/2}} \cos(\omega t + \phi) \quad ..(8.38b)$$

where $\phi = \tan^{-1} \frac{(\frac{1}{LC} - \omega^2)}{\omega R/L}$ \quad ..(8.38c)

Now, by definition:

$$\omega_N = (\frac{1}{LC})^{1/2} \quad ..(8.39a)$$

and $\quad \delta_N = \frac{R}{2L\omega} \quad ..(8.39b)$

so that:

Sinusoidal Response

$$q_P = \frac{\omega P/L}{\left[(\omega_N^2 - \omega^2)^2 + (2\delta\omega\omega_N)^2\right]^{1/2}} \cos(\omega t + \phi) \quad ..(8.40)$$

Thus:

$$M = \frac{\omega/L}{\left[(\omega_N^2 - \omega^2)^2 + (2\delta\omega\omega_n)^2\right]^{1/2}}$$

$$= \frac{(C/L)^{1/2} (\omega/\omega_N)}{\left[1 + (4\delta^2 - 2)(\omega/\omega_N)^2 + (\omega/\omega_N)^4\right]^{1/2}} \quad ..(8.41a)$$

$$\phi = \tan^{-1} \frac{\left[1 - (\omega/\omega_N)^2\right]}{2\delta(\omega/\omega_N)} \quad ..(8.41b)$$

In this case, the magnitude function is not dimensionless, since input is a pressure and the output is a flow. However, the parameter $(L/C)^{1/2}M$ is dimensionless. Fig. 8.6 shows this parameter plotted against the normalized frequency ω/ω_N for various cases of the damping factor, δ. This is the magnitude-frequency response for this circuit. Note that the response has a peak at the natural frequency ($\omega/\omega_N = 1$), when the damping factor is small.

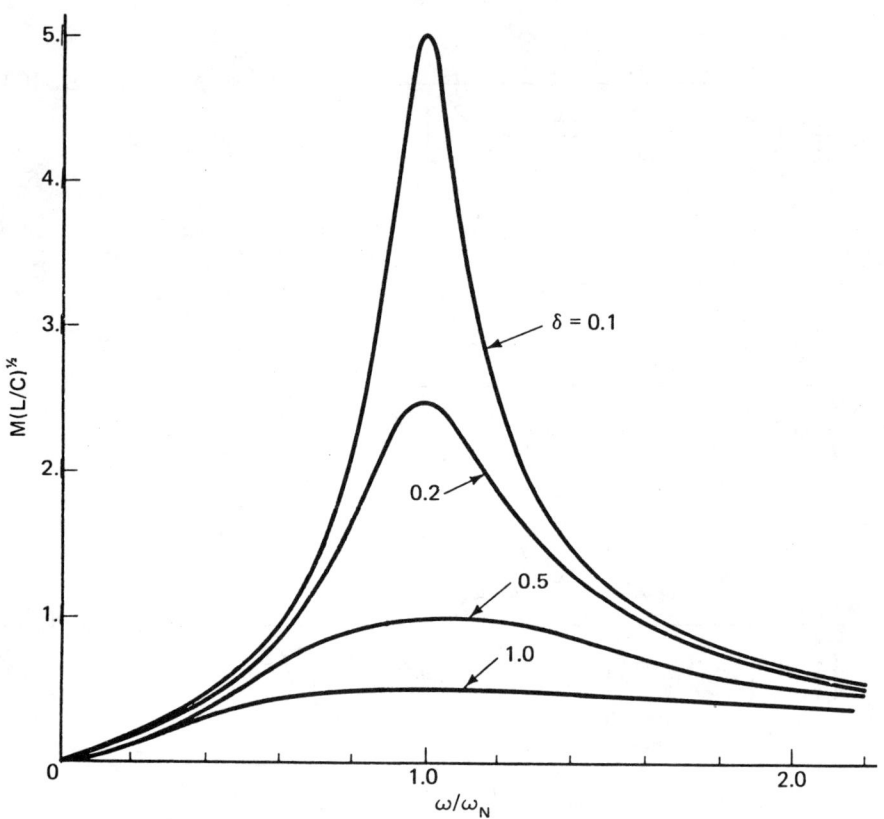

Figure 8.6 Magnitude-frequency response (example 8.2)

Fig. 8.7 shows the phase-frequency response. As the frequency increases, the phase goes from a 90 degree lead to a 90 degree lag. When the input frequency is the same as the natural frequency (ω/ω_N), the input and the output are in phase. Although we present magnitude and phase [Figs. 8.6 and 8.7] to indicate frequency response, one relation or the other is usually sufficient. In the circuits we deal with in this text (minimum phase circuits), the shape of one plot is

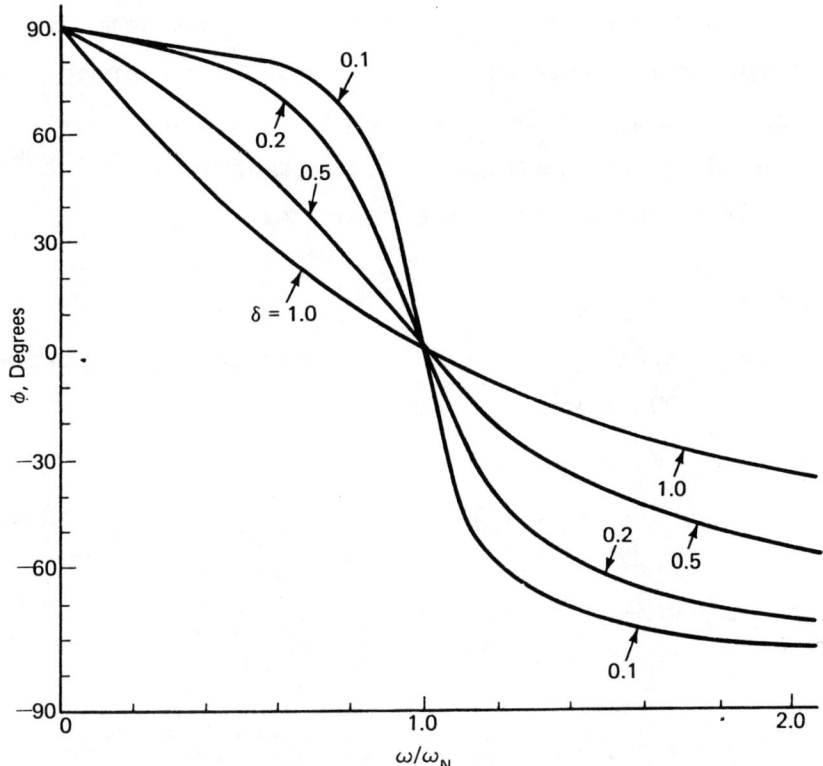
Figure 8.7 Phase-frequency response (example 8.2)

evidence of the shape of the other. Thus, for example, when phase changes rapidly in a narrow frequency band, the accompanying amplitude has a large peak. Conversely, a large peak on the magnitude plot means a rapid phase change on the phase plot.

The results presented in this example could have been determined directly from (8.34a) with $a_2 = 1/L$ and $a_1 = a_3 = 0$.

8.2 SINUSOIDAL STEADY STATE

The procedure just described to obtain the total response to a sinusoidal input is very cumbersome as far as the forced or steady-state response is concerned. If we recall from Chapter VII that a sinusoid may be represented by the real or the imaginary part of $\varepsilon^{j\omega t}$, the solution is obtained more rapidly.

Example 8.3

Fig. 8.8(a) shows a mechanical mass-damper circuit. If the applied force is

$$f(t) = F \cos \omega t \qquad \qquad ..(8.42)$$

find the steady-state velocity of the mass.

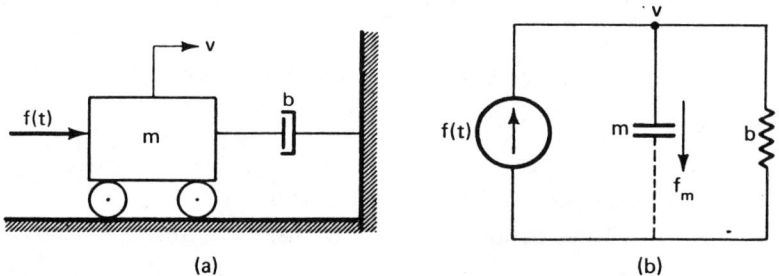

Figure 8.8 Mechanical circuit (example 8.3)

The analogous circuit is shown in Fig. 8.8(b). The equilibrium equation is:

$$m \frac{dv}{dt} + bv = \text{Re}\left[F \, \varepsilon^{j\omega t}\right] \qquad ..(8.43a)$$

Sinusoidal Response

and the steady-state velocity must have the form:

$$v(t) = \text{Re}\left[V \, \varepsilon^{j\omega t}\right] \qquad ..(8.43b)$$

Using the complete driving function $F \, \varepsilon^{j\omega t}$ and substituting gives:

$$j\omega m V + bV = F \qquad ..(8.44)$$

from which:

$$V = \frac{F}{b + j\omega m} = \frac{F}{Y \, \varepsilon^{j\phi}} \qquad ..(8.45a)$$

where

$$Y = \sqrt{b^2 + (\omega m)^2} \qquad ..(8.45b)$$

and

$$\tan \phi = \frac{\omega m}{b} \qquad ..(8.45c)$$

That is:

$$v(t) = \text{Re}\left[\frac{F}{Y \, \varepsilon^{j\phi}} \, \varepsilon^{j\omega t}\right]$$

$$= \frac{F}{Y} \text{Re}\left[\varepsilon^{j(\omega t - \phi)}\right] = \frac{F}{Y} \cos(\omega t - \phi) \qquad ..(8.46)$$

Note that if only the amplitude ($M = F/Y$) and phase (ϕ) of the steady-state response are required, the final part of this solution is quite unnecessary. Fig. 8.9 shows a plot of the magnitude and phase response for this mechanical circuit.

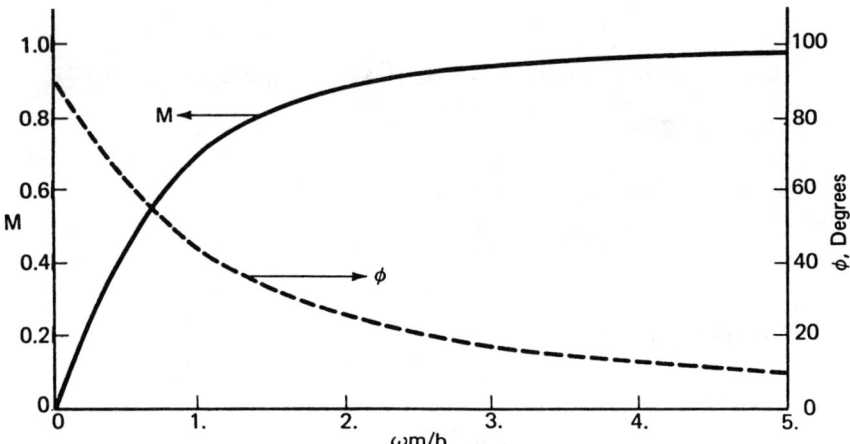

Figure 8.9 Magnitude and phase response for mechanical circuit (example 8.3)

Although the procedure used in this example is more convenient than that in Section 8.1, it is possible to improve it still further by noting the nature of a system function. Suppose, for example, that the input to a circuit has the form

$$x(t) = X \, \varepsilon^{st} \qquad \qquad ..(8.47a)$$

and the output has the form

$$y(t) = Y \, \varepsilon^{st} \qquad \qquad ..(8.47b)$$

The system function is, by definition:

Sinusoidal Response

$$g(s) = \frac{Y}{X} \qquad \qquad ..(8.48)$$

For the sinusoidal steady-state, we have already noted that we are dealing with the particular case of an exponential function where $s = j\omega$. This is an application of Euler's equation [Eq. (2.34)]:

$$\varepsilon^{j\omega t} = \cos \omega t + j \sin \omega t \qquad ..(8.49)$$

Thus, when input and output are cosine waves, we may write:

$$x(t) = \mathrm{Re}\left[X \, \varepsilon^{j\omega t}\right] \qquad ..(8.50a)$$

$$y(t) = \mathrm{Re}\left[Y \, \varepsilon^{j\omega t}\right] \qquad ..(8.50b)$$

If (8.48) is substituted into (8.50), the result is:

$$y(t) = \mathrm{Re}\left[X \, G(j\omega) \, \varepsilon^{j\omega t}\right] \qquad ..(8.51)$$

In terms of the cosine functions, (8.51) is equivalent to:

$$y(t) = X \, |G(j\omega)| \, \cos(\omega t + \phi) \qquad ..(8.52)$$

where $|G(j\omega)|$ is the magnitude of the transfer function and ϕ is the angle of $G(j\omega)$. The magnitude ratio of output y(t) to input x(t) is therefore:

$$M = \frac{X|G(j\omega)|}{X} = |G(j\omega)| \quad \quad ..(8.53)$$

and the phase may be expressed as:

$$\phi = \angle G(j\omega) \quad \quad ..(8.54)$$

To find the frequency response magnitude and phase, we merely substitute $j\omega$ for s in the transfer function and then perform the operations required in (8.53) and (8.54).

Example 8.4

As an illustration, let us reconsider the fluid circuit of Example 8.2 The transfer function $G(s)$ is:

$$G(s) = \frac{Q}{P} = \frac{Cs}{LCs^2 + RCs + 1} \quad \quad ..(8.55)$$

If we replace s by $j\omega$, the result is:

$$G(j\omega) = \frac{j\omega C}{(1 - \omega^2 LC) + j\omega RC} \quad \quad ..(8.56)$$

From the rules of complex algebra, we may express $G(j\omega)$ as:

Sinusoidal Response

$$G(j\omega) = \frac{\omega C}{\left[(1 - \omega^2 LC)^2 + (\omega RC)^2\right]^{1/2}} \angle 90° - \tan^{-1}\frac{\omega RC}{1-\omega^2 LC}$$

..(8.57)

and by algebraic manipulation with

$$\omega_N = (1/LC)^{1/2} \qquad ..(8.58a)$$

and

$$2\delta\omega_N = R/L \qquad ..(8.58b)$$

the above equation becomes:

$$G(j\omega) = \frac{(C/L)^{1/2}(\omega/\omega_N)}{\left[1+(4\delta^2-2)(\omega/\omega_N)^2 + (\omega/\omega_N)^4\right]^{1/2}} \angle \theta \qquad ..(8.59a)$$

with

$$\phi = \tan^{-1}\frac{[1-(\omega/\omega_N)^2]}{2\delta(\omega/\omega_N)} \qquad ..(8.59b)$$

and we can recognize the amplitude and phase of $G(j\omega)$ as M and ϕ from Example 8.2.

Example 8.5

The mass-spring damper system shown in Fig. 8.10(a) is subjected to an oscillatory force

$$f(t) = F\cos\omega t, \text{ N} \qquad ..(8.60)$$

Find the amplitude and phase of the forces in each element.

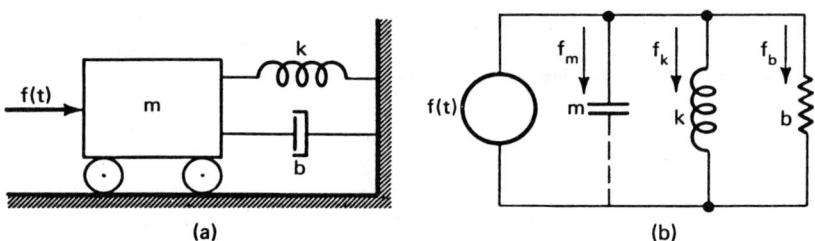

Figure 8.10 Mechanical circuit (example 8.5)

The analogous circuit is shown in Fig. 8.10(b) from which the equilibrium equation is obtained:

$$(ms + b + \frac{k}{s}) V = F \qquad ..(8.61)$$

The solution is:

$$V = \frac{Fs}{ms^2 + bs + k} \qquad ..(8.62)$$

and the required forces are:

$$F_m = msV = \frac{mFs^2}{ms^2 + bs + k} \qquad ..(8.63a)$$

Sinusoidal Response

$$F_b = bV = \frac{bFs}{ms^2 + bs + k} \quad \quad ..(8.63b)$$

$$F_k = \frac{k}{s} V = \frac{kF}{ms^2 + bs + k} \quad \quad ..(8.63c)$$

Replacing s by jω gives:

$$F_m(j\omega) = \frac{-\omega^2 mF}{(k - \omega^2 m) + j\omega b} \quad \quad ..(8.64a)$$

$$F_b(j\omega) = \frac{j\omega bF}{(k - \omega^2 m) + j\omega b} \quad \quad ..(8.64b)$$

$$F_k(j\omega) = \frac{kF}{(k - \omega^2 m) + j\omega b} \quad \quad ..(8.64c)$$

Thus:

$$F_m(j\omega) = \frac{\omega^2 mF}{\sqrt{(k-\omega^2 m)^2 + (\omega b)^2}} \underline{/180° - \phi} \quad ..(8.65a)$$

$$F_b(j\omega) = \frac{\omega bF}{\sqrt{(k-\omega^2 m)^2 + (\omega b)^2}} \underline{/90° - \phi} \quad ..(8.65b)$$

$$F_k(j\omega) = \frac{kF}{\sqrt{(k-\omega^2 m)^2 + (\omega b)^2}} \underline{/-\phi} \quad ..(8.65c)$$

where
$$\tan \phi = \frac{\omega b}{k - \omega^2 m} \qquad ..(8.65d)$$

Normally, this form is adequate and with experience does convey the complete information on the response. However, if the time functions are required, it is simply a matter of noting that since the driving functions was $F \cos\omega t$, the responses must be:

$$F_b(j\omega) = \frac{\omega^2 mF}{\sqrt{(k-\omega^2 m)^2 + (\omega b)^2}} \cos(\omega t + 180° - \phi)$$

$$..(8.66)$$

The magnitude and phase responses for this mechanical circuit are shown in Fig. 8.11 for the case where $b = \sqrt{km}$.

8.3 POWER IN THE SINUSOIDAL STATE

For all the physical systems we have considered except thermal systems, the through and across variables were chosen so that not only would the equilibrium equations have the same mathematical form but also their product would be power. In the case of the dissipative elements, this power normally appears as heat. In the case of the storage elements, the power is the rate at which they are storing energy. If we use the polarity convention consistently, the power flow into any of the passive elements is given by:

Sinusoidal Response

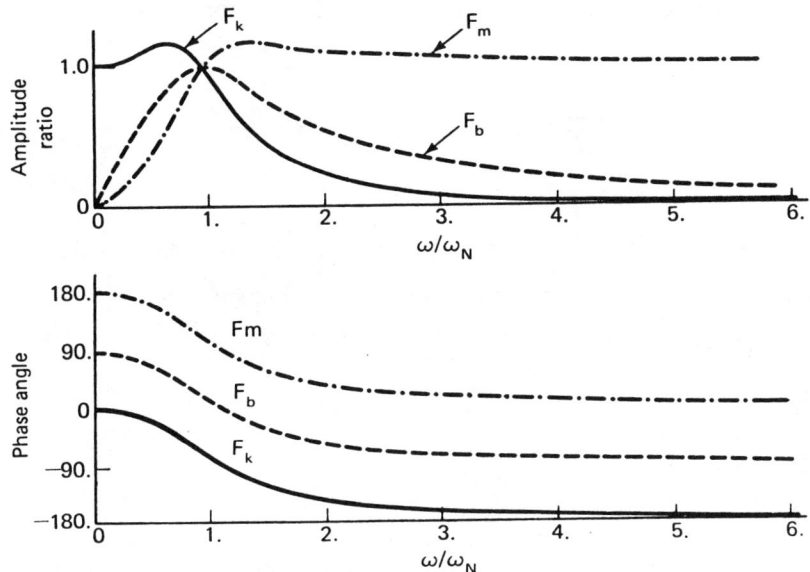

Figure 8.11 Magnitude and phase frequency response (example 8.5)
$(b = \sqrt{km})$

$$P(t) = [AV(t)][TV(t)] \qquad ..(8.67)$$

where $AV(t)$ and $TV(t)$ are the instantaneous values of the across and through variables respectively.

In the sinusoidal steady-state, both AV and TV are sinusoids and thus the power flow is not constant. For example, the power dissipated in a resistance is:

$$P(t) = EI \cos^2 \omega t \qquad ..(8.68)$$

if $\qquad e(t) = E \cos \omega t \qquad ..(8.69a)$

and $\qquad i(t) = I \cos \omega t \qquad ..(8.69b)$

This expression must be modified in order to interpret it. By double angle relation, this becomes:

$$P(t) = \frac{EI}{2} (1 + \cos 2\omega t) \quad \quad ..(8.70)$$

Thus, the power flow is oscillating at twice the frequency of the current and the voltage, but there is a time average value given by:

$$P_{av} = \frac{EI}{2} \quad \quad ..(8.71)$$

The factor of 1/2 has usually been considered troublesome and may be eliminated by working with effective values of current and voltage so that:

$$P_{av} = E_{eff} I_{eff} \quad \quad ..(8.72)$$

where
$$E_{eff} = E/\sqrt{2} \quad \quad ..(8.73a)$$
and
$$I_{eff} = I/\sqrt{2} \quad \quad ..(8.73b)$$

The effective values are also known as the rms values since more generally they can be shown to be the square root of the average of the square of the voltage or current. In much of the literature, (8.72) does not have any subscripts and appears simply as P = EI. Care is therefore required when writing and interpreting expressions involving power in the sinusoidal steady

Sinusoidal Response

state. In addition, it may be noted that:

$$Z(j\omega) = \frac{E}{I} = \frac{E_{eff}}{I_{eff}} \qquad ..(8.74)$$

Therefore, all equations involving impedances or admittances may be used with either effective values or amplitudes.

When one or more storage elements is present, the expression for power must be modified still further. If we consider a simple series circuit consisting of a resistor and an inductor, we have:

$$Z(j\omega) = R + j\omega L \qquad ..(8.75)$$

and hence

$$I(j\omega) = \frac{E}{R + j\omega L} = \frac{E}{|Z|} \angle{-\phi} \qquad ..(8.76)$$

where $\quad |Z| = \sqrt{R^2 + (\omega L)^2} \qquad ..(8.77a)$

and $\quad \tan \phi = \frac{\omega L}{R} \qquad ..(8.77b)$

Thus, if $e(t) = E \cos\omega t \qquad ..(8.78a)$

the current must be given by:

$$i(t) = \frac{E}{|Z|} \cos(\omega t - \phi) \quad \quad ..(8.78b)$$

and the power is:

$$P(t) = EI \cos \omega t \cos(\omega t - \phi)$$

$$= \frac{EI}{2} \left[\cos \phi + \cos(2\omega t - \phi) \right] \quad ..(8.79)$$

Again the time variation in power is at twice the excitation frequency, but the average value is now:

$$P_{av} = \frac{EI}{2} \cos \phi = E_{eff} I_{eff} \cos \phi \quad ..(8.80)$$

The ratio of the average power to the product of current and voltage is usually called the **power factor** which in this case is the cosine of the phase angle ϕ.

Example 8.6

If we return to Example 8.3 where F is the amplitude of the applied force, determine the power dissipated.

The velocity amplitude and phase were found to be:

$$V = \frac{F}{\sqrt{b^2 + (\omega m)^2}} \quad ..(8.81a)$$

and

$$\phi = \tan^{-1} \frac{\omega m}{b} \quad ..(8.81b)$$

Sinusoidal Response

Thus, the average power entering the system is:

$$P = V_{eff} F_{eff} \cos \phi$$

$$= \left[\frac{F_{eff}^2}{\sqrt{b^2 + (\omega m)^2}}\right]\left[\frac{b}{\sqrt{b^2 + (\omega m)^2}}\right]$$

$$= \frac{F_{eff}^2 b}{b^2 + (\omega m)^2} = V_{eff}^2 b \qquad \ldots (8.82)$$

Since an ideal mass may only store energy, the average power associated with it must be zero and hence it is essential that any expression for the average power entering a system should equal the average power dissipated in the dissipative elements. Also, it can be noted that the expressions $I^2 R$ and E^2/R and their analogs may be used as correct power relations provided I is the effective value of the current flowing through the resistance R or E is the effective value of the voltage across it.

8.3.1 Maximum Power Transfer

There are occasions when it is necessary to choose a dissipative element so that the power dissipated in it is the maximum possible with a given source. In Fig. 8.12, $E_{e(eff)}$ and Z_e may be considered as the Thevenin equivalent of any circuit. For the sinusoidal steady

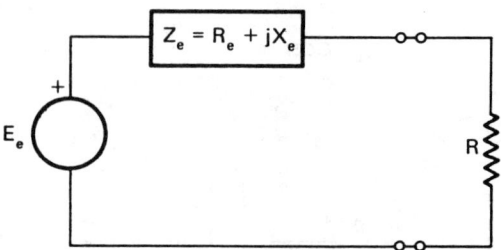

Figure 8.12 Circuit reduction for maximum power transfer

state, $Z_e(s)$ becomes $Z_e(j\omega)$ and at any one frequency this can be simplified into a real part (equivalent resistance) and an imaginary part (equivalent reactance), $R_e + jX_e$. Note that the values of R_e and X_e usually depend on the frequency.

Hence the current can be expressed, using effective values as:

$$I_{eff}(j\omega) = \frac{E_{e(eff)}}{R_e + jX_e + R} \quad \quad ..(8.83)$$

The power dissipated in R is:

$$P = \frac{E_{e(eff)}^2 \, R}{(R + R_e)^2 + X_e^2} \quad \quad ..(8.84)$$

The maximum value of P is found in the usual manner by differentiating with respect to R, equating to zero and solving for R. The result is that power is maximum when:

Sinusoidal Response   533

$$R = \sqrt{R_e^2 + X_e^2} \qquad \qquad ..(8.85)$$

That is, maximum power is dissipated in R when it is numerically the same as the magnitude of $Z_e(j\omega)$. For the particular case where $X_e = 0$, the condition $R = R_e$ gives maximum power transfer.

If, in addition to adjusting the value of R, it is possible to have either a capacitor or an inductor connected in series with the resistor, the condition for maximum power is modified. That is, the current becomes:

$$I_{eff}(j\omega) = \frac{E_{e(eff)}}{Z_e(j\omega) + Z(j\omega)}$$

$$= \frac{E_{e(eff)}}{(R_e + R) + j(X_e + X)} \qquad ..(8.86)$$

and maximum power is obtained when Z is the complex conjugate of Z_e. That is:

$$R = R_e, \quad X = -X_e \qquad \qquad ..(8.87)$$

Example 8.7

A mass-damper system is shown in Fig. 8.13(a). It is subjected to an oscillatory force of frequency 1.592 Hz and amplitude 12.5 N. If $m_1 = 3$ kg, $m_2 = 0.96$ kg and

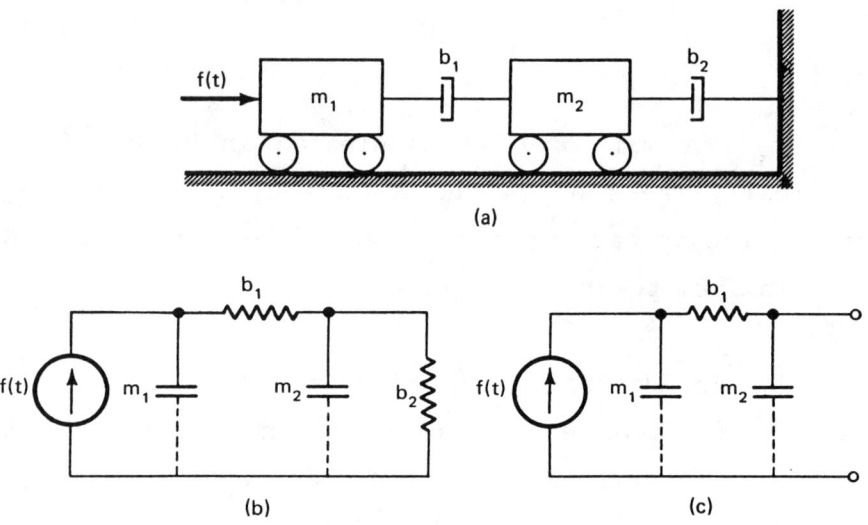

Figure 8.13 Mechanical circuit (example 8.7)

$b_1 = 40$ N.s/m, determine (a) the damping required to dissipate maximum power in b_2 and (b) the maximum power dissipated.

The analogous circuit is shown in Fig. 8.13(b) and the Norton equivalent of the fixed part of the circuit is obtained from Fig. 8.13(c), the values being:

$$F_e = \frac{b_1 F}{b_1 + m_1 s} = \frac{40F}{40 + j30} \quad \quad ..(8.88a)$$

$$Y_e = m_2 s + \frac{1}{1/m_1 s + 1/b_1} = m_2 s + \frac{bm_1 s}{m_1 s + b_1}$$

$$= j9.6 + \frac{j(40)(30)}{40 + j30}$$

Sinusoidal Response

$$= 14.4 + j28.8 \qquad \qquad ..(8.88b)$$

For maximum power dissipation in the second damper, it must be set to the value

$$b_2 = \sqrt{14.4^2 + 28.8^2} = 32.2 \text{ N.s/m} \qquad ..(8.89)$$

The total admittance is the sum of b_2 and Y_e. Thus:

$$Y = 46.6 + j28.8 = 54.8 \underline{/31.7^\circ} \qquad ..(8.90)$$

If the amplitude of the driving force is 12.5 N:

$$F_e = \frac{500}{40 + j30} = 8 - j6 = 10 \underline{/-36.9^\circ} \qquad ..(8.91)$$

The amplitude of the velocity across the damper is:

$$V = \frac{|F_e|}{|Y|} = \frac{10}{54.8} = 0.182 \text{ m/s} \qquad ..(8.92)$$

Thus the maximum power dissipated is:

$$P = V_{eff}^2 \, b_2 = \frac{V^2}{2} b_2 = 0.533 \text{ W} \qquad ..(8.93)$$

8.4 SIGNAL FILTERS

A filter is, by definition, a device for maximizing or minimizing the response to certain frequencies without altering the response to other frequencies. The extent to which a particular frequency is present or absent in the response is measured primarily by its amplitude, The performance of a filter is then indicated by the ratio of output amplitude to input amplitude at each frequency. However, in some cases, such as when filters are used in instrumentation, the phase shift is equally important.

Although the theory and application of electrical signal filters is the most well-known, the same principles are relevant for the other physical media. Thus there are mechanical circuits to filter mechanical signals (for example, to reduce the transmission of vibrations) and fluid circuits to filter fluid signals.

We consider only passive filters here. These are circuits that are constructed with the passive components (resistance, capacitance and inductance).

8.4.1 Low-pass Filters

A low-pass filter, as the name implies, passes low frequency signals and blocks high frequency signals. Fig. 8.14 shows a schematic representation of the characteristics of low-pass filters. For the ideal low-pass filter, there is a sharp cutoff between the pass region and the block region. This means that an

Sinusoidal Response

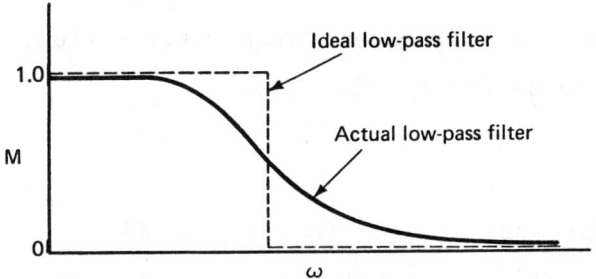

Figure 8.14 Low-pass filter characteristics

input signal in the low frequency pass region produces an output signal of the same amplitude and, of course, the same frequency. In other words, the signal passes through the filter. On the other hand, an input signal in the high frequency block region yields no output signal. Actual low-pass filters do not have as sharp a cutoff as ideal low-pass filters. Some frequencies that are desired eliminated are passed with greatly reduced amplitudes. Other frequencies that are to be retained are also passed out with only slightly reduced amplitudes.

The transfer function for typical low-pass passive filters may be described by:

$$G(s) = \frac{K}{D(s)} \qquad ..(8.94)$$

where K is a constant such that G(0) is less than or

equal to unity, and D(s) is a function of s. The distinguishing feature of the low-pass filter is that the numerator of the transfer function is not a function of s. The following examples will illustrate some passive low-pass filter circuits.

Example 8.8

In the mass-damper circuit of Fig. 8.15(a), the input velocity is V_i sin ωt. This circuit may be considered as a mechanical filter where the output is the velocity of the mass. In another mass-damper circuit of Fig. 8.15(b), there are two mass-damper combinations in tandem. This circuit has the same input velocity as the single mass-damper circuit. The output, however, is the velocity of the second mass component. Determine the magnitude-frequency response for both the circuits and discuss their behavior as mechanical filters.

The analogous circuit for the single mass-damper combination is shown in Fig. 8.15(c). The transfer function $G_1(s)$ is:

$$G_1(s) = \frac{V_o}{V_i} = \frac{1}{1 + ms/b} \qquad ..(8.95)$$

The magnitude-frequency response will be

$$M_1 = |G_1(j\omega)| = \left|\frac{1}{1 + j\omega m/b}\right| = \frac{1}{\left[1 + (\omega m/b)^2\right]^{1/2}} \qquad ..(8.96)$$

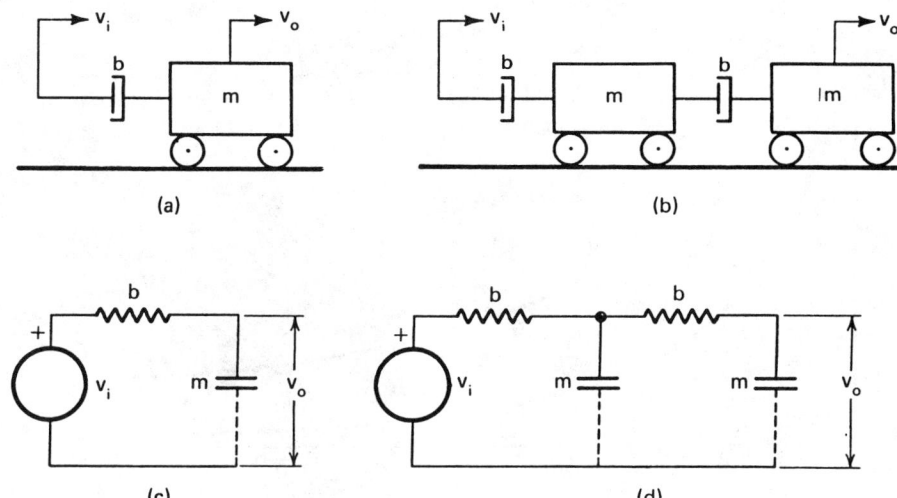

Figure 8.15 Mechanical circuits (example 8.8)

The analogous circuit for the two-section mass-damper arrangement is shown in Fig. 8.15(d). The transfer function, $G_2(s)$, is:

$$G_2(s) = \frac{V_o}{V_i} = \frac{1}{1 + (3ms/b) + (ms/b)^2} \quad \ldots(8.97)$$

The magnitude-frequency response is, therefore,

$$M_2 = |G_2(j\omega)| = \left| \frac{1}{[1 - (\omega m/b)^2] + j3\omega m/b} \right|$$

$$= \frac{1}{\left[\left\{1 - (\omega m/b)^2\right\}^2 + 9(\omega m/b)^2 \right]^{1/2}} \quad \ldots(8.98)$$

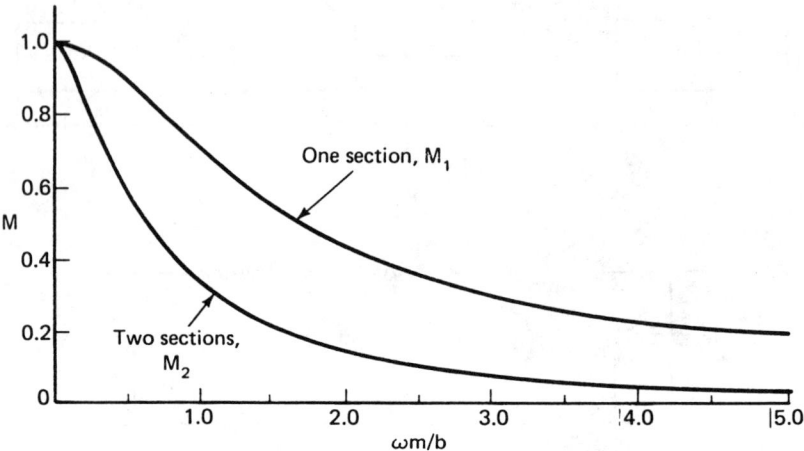

Figure 8.16 Magnitude-frequency curves (example 8.8)

Fig. 8.16 shows the magnitude-frequency curves for M_1 and M_2. Both circuits may be classified as low-pass filters. The one section circuit passes larger amplitude output signals at all frequencies than the two-section circuit. This is a desirable feature in the low-frequency region but is undesirable at high frequencies. For example, suppose the desired cutoff frequency is $\omega = b/m$. At frequencies above cutoff ($\omega m/b > 1$), the two-section circuit provides considerably more attenuation than the one-section circuit. In fact, the two-section arrangement practically blocks out all frequencies above $\omega m/b = 3$. However, below cutoff ($\omega m/b < 1$), the two-section circuit still attenuates signals although no attenuation is desired.

When additional mass-damper sections are employed,

the magnitude-frequency curves have sharper cutoffs but the cutoff frequency decreases.

Example 8.9

The electric circuit shown in Fig. 8.17 may be used as a low-pass filter, if the RLC components are

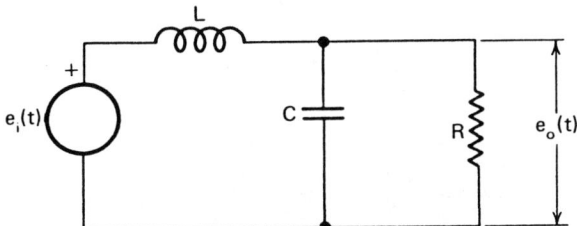

Figure 8.17 Electrical circuit (example 8.9)

selected judiciously. The characteristics of the filter depend on the damping factor $\delta = \sqrt{L}/(2R\sqrt{C})$ and $\omega_N = 1/\sqrt{LC}$. Investigate the filter characteristics of this circuit by plotting magnitude-frequency curves for various values of damping factor.

The transfer function, $G(s)$, of the electric circuit shown in Fig. 8.17 is:

$$G(s) = \frac{E_o}{E_i} = \frac{1}{LCs^2 + \frac{L}{R}s + 1} \qquad ..(8.99)$$

from which we may develop the magnitude as:

$$M = |G(j\omega)| = \left|\frac{1}{(1 - \omega^2 LC) + j\omega L/R}\right| \quad ..(8.100)$$

Now, in terms of the normalizing natural frequency, ω_N and damping factor, δ,

$$M = \left|\frac{1}{(1 - \omega^2/\omega_N^2) + j\, 2\delta(\omega/\omega_N)}\right|$$

$$= \frac{1}{\left[\left\{1 - (\omega/\omega_N)^2\right\}^2 + 4\delta^2(\omega/\omega_N)^2\right]^{1/2}} \quad ..(8.101)$$

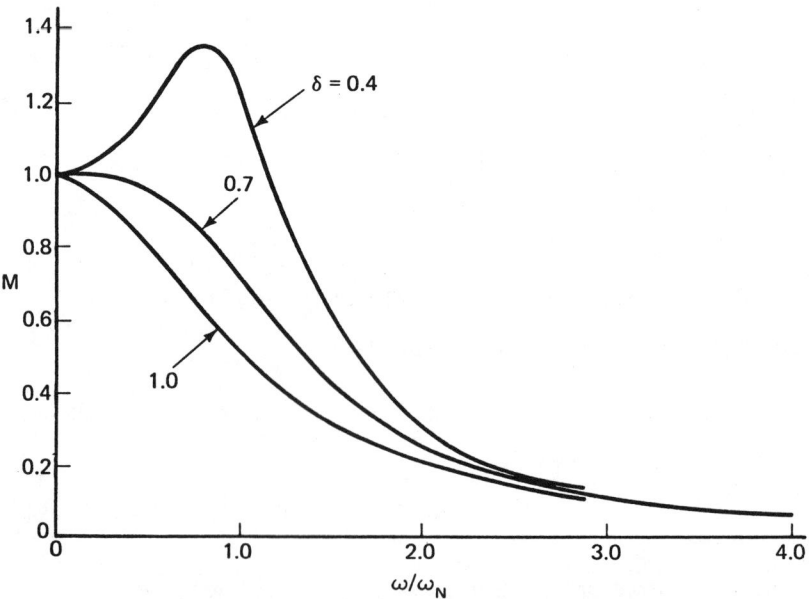

Figure 8.18 Magnitude-frequency curves (example 8.9)

Fig. 8.18 shows a plot of magnitude versus non-dimensional frequency ω/ω_N for various values of δ. The circuit attenuates high frequency (i.e., $\omega/\omega_N > 2$) and may be classed as a low-pass filter. However, when the damping factor is small, the circuit may magnify some signals at lower frequencies. A compromise must be made between sharp cutoff with low-frequency magnification and gradual cutoff without magnification. Damping factors of approximately 0.6 often provide a good trade-off.

8.4.2 High-Pass Filters

A high-pass filter has characteristics opposite to those of a low-pass filter. It passes high frequency signals and blocks low frequency signals. Fig. 8.19 shows the characteristic of an ideal high-pass filter and a typical actual high-pass filter. The ideal filter has a sharp discontinuity in its characteristics. For frequencies higher than the discontinuity, the magnitude of output and input are equal. The output has no magnitude when the frequencies are lower than the cutoff frequency. In the actual filter, the very low frequency and very high frequency behavior is the same as the ideal filter. However, at intermediate frequencies, there is a gradual magnitude change rather than the abrupt change desired.

The transfer function for the typical high-pass passive filters has the form:

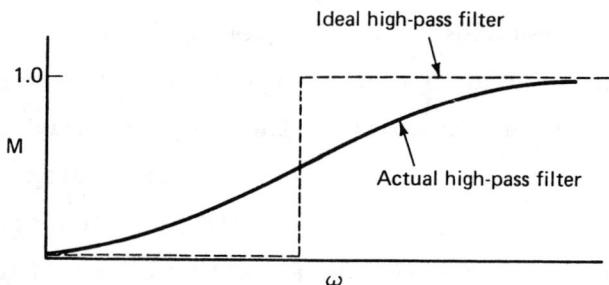

Figure 8.19 High-pass filter characteristics

$$G(s) = \frac{A_1 s^m}{A_1 s^m + A_2 s^{m-1} + \ldots + A_{m-1} s + A_m} \quad \ldots (8.102)$$

where A_1, A_2, . . . , A_m are the coefficients of the polynomial in s. The distinguishing feature of the high-pass transfer function is that the numerator term has the same order as the highest order term in the denominator. Examples 8.10 and 8.11 which follow will describe some high-pass filter circuits.

Example 8.10

Fig. 8.20(a) and (b) show one-section and two-section spring-damper combinations. The values of the damping, b, and the spring constant, k, are the same in each mechanical circuit. The velocity input to the damper is V_i sinωt in both circuits. The location of the output velocity is indicated in Figs. 8.20(a) and (b). Determine the magnitude-frequency characteristics of these circuits and discuss the characteristics in terms of signal filters.

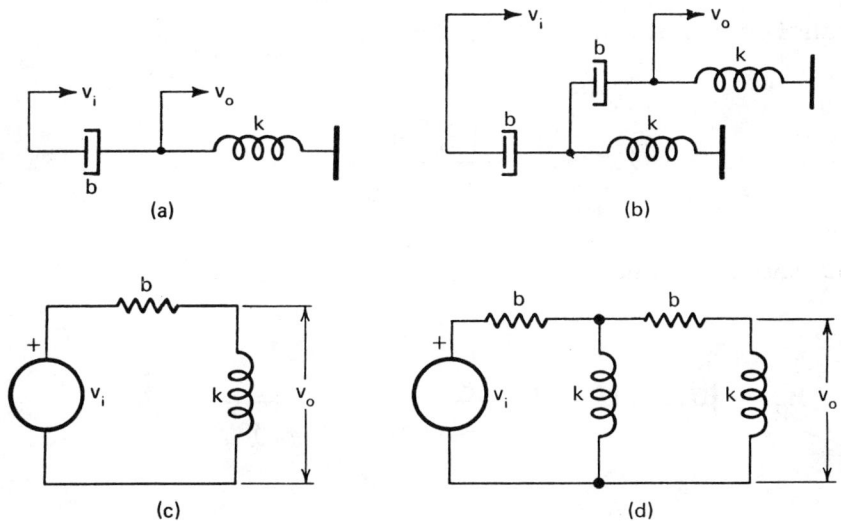

Figure 8.20 Mechanical circuit (example 8.10)

The analogous circuit diagrams for the one-section and two-section mechanical circuits are shown in Figs. 8.20(c) and (d). The transfer function $G_1(s)$ of the single spring-damper circuit is:

$$G_1(s) = \frac{bs/k}{bs/k + 1} \qquad ..(8.103)$$

The magnitude-frequency response is:

$$M_1 = |G_1(j\omega)| = \left|\frac{(\omega b/k)}{1 + j(\omega b/k)}\right|$$

$$= \frac{(\omega b/k)}{\left[1 + (\omega b/k)^2\right]^{1/2}} \qquad ..(8.104)$$

For the two-section spring-damper circuit, the transfer function, $G_2(s)$, is:

$$G_2(s) = \frac{b^2 s^2/k^2}{b^2 s^2/k^2 + 3bs/k + 1} \qquad ..(8.105)$$

and the frequency response is:

$$M_2 = |G_2(j\omega)| = \left| \frac{-(\omega b/k)^2}{[1 - (\omega b/k)^2] + j3\omega b/k} \right|$$

$$= \frac{(\omega b/k)^2}{\left[\{1 - (\omega b/k)^2\}^2 + 9(\omega b/k)^2\right]^{1/2}} \qquad ..(8.106)$$

Fig. 8.21 shows the magnitude-frequency plots for M_1 and M_2. The non-dimensional frequency parameter for this circuit is $\omega b/k$. The results indicate that both circuits pass very high frequency signals and block very low frequency signals. Thus, we may classify these circuits as types of high-pass filter. The two-section circuit does a better job at eliminating the low frequencies. However, it attenuates the high frequencies in the process. Once again, we have to make

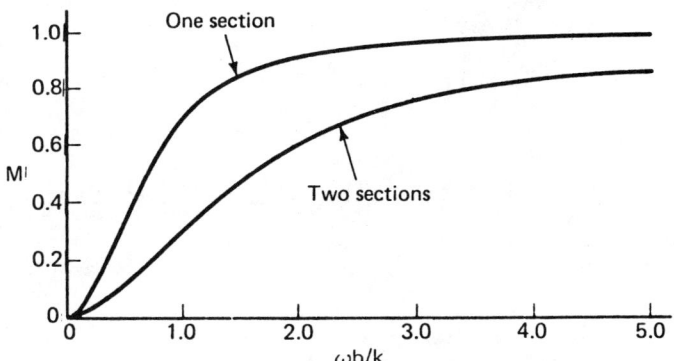
Figure 8.21 Magnitude-frequency curves (example 8.10)

a trade-off to obtain the most satisfactory result.

Example 8.11

Determine the magnitude-frequency characteristics of the LC electrical circuit shown in Fig. 8.22. The input is a sine wave and the output is the voltage across the inductance.

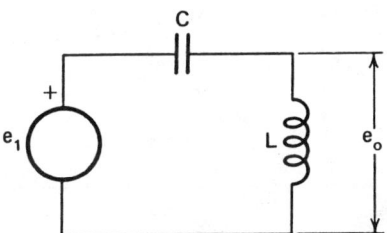
Figure 8.22 Electric circuit (example 8.11)

The transfer function of the LC circuit is:

$$G(s) = \frac{LCs^2}{LCs^2 + 1} \qquad \qquad ..(8.107)$$

The magnitude is therefore:

$$M = |G(j\omega)| = \left|\frac{\omega^2 LC}{1 - \omega^2 LC}\right| \quad \quad ..(8.108)$$

If we normalize by using the natural frequency $\omega_N = \sqrt{LC}$, then

$$M = \frac{(\omega/\omega_N)^2}{1 - (\omega/\omega_N)^2} \quad \quad ..(8.109)$$

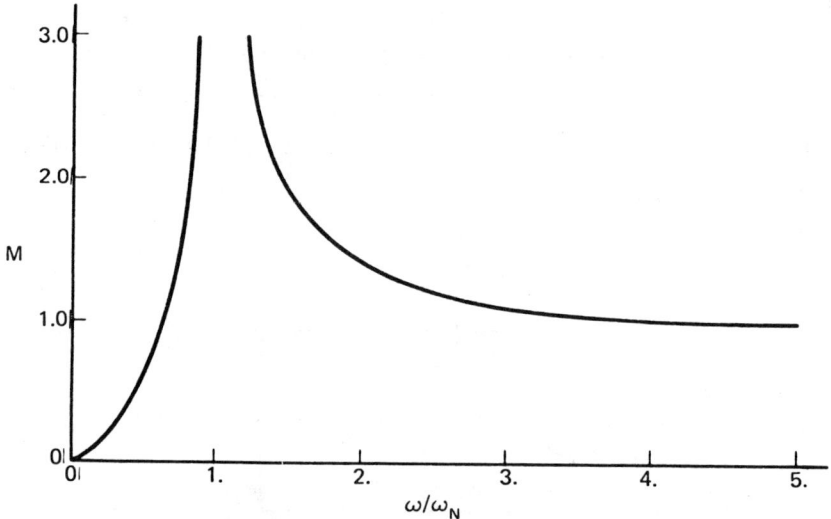

Figure 8.23 Magnitude-frequency curve (example 8.11)

Fig. 8.23 shows the magnitude-frequency response for this circuit. We may observe at once that the circuit passes high frequency signals and blocks low frequency signals. Thus the characteristics are the same in this respect as a high-pass filter. There is a dissimilarity with an ideal high-pass filter, however, for applied signals near the natural frequency ($\omega/\omega_N = 1$). For the circuit under consideration, the output increases without bound in this vicinity. If the frequency band between $\omega/\omega_N = 0.8$ and $\omega/\omega_N = 1.7$ is never present in the input signal or can otherwise be avoided, the circuit behaves well as a high-pass filter.

8.4.3 Band-Pass and Notch Filters

The combination of a low-pass and a high-pass filter results in a band-pass filter. Fig. 8.24(a) shows the ideal band-pass characteristics. At very low and very high frequencies, there is no output amplitude. Only frequencies in a band around some intermediate frequency produce an output signal. The ideal characteristics, of course, are just approximated very roughly by passive circuits. Active filters are required to approach the ideal characteristics more closely.

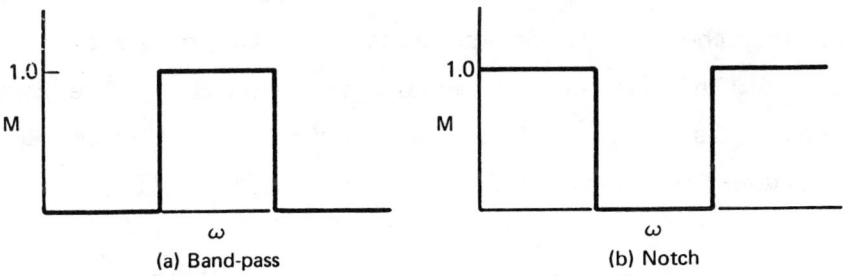

Figure 8.24 Ideal band-pass and notch filter characteristics

A low-pass and a high-pass filter can also be constructed to reject a specific frequency band and to pass all other frequencies. This type of filter is often called a notch filter. The characteristics of an ideal notch filter are shown in Fig. 8.24(b). In this filter, the low and the high frequency signals are passed. A narrow band in the intermediate frequency range is blocked. Here again, passive circuits only produce a crude approximation to the ideal characteristics. Sometimes this is sufficient for practical purposes. Many measuring instruments are disturbed by 60 Hz signals. A notch filter tuned to 60 Hz may be used to eliminate these undesirable signals. However, care must be exercised in filter selection to ensure that the filter does not block desired frequencies.

The following examples demonstrate some of the crudeness of passive band-pass and notch filters.

Example 8.12

Fig. 8.25 shows a bridged-T electric circuit. The circuit has two resistors and two capacitors. In this example, the resistors are equal and the capacitors are equal. (This is not necessary in general). The input signal is a sine wave. Investigate the magnitude-frequency characteristics of the circuit.

We may use the nodal circuit analysis to determine the transfer function for this circuit. The result is:

Sinusoidal Response

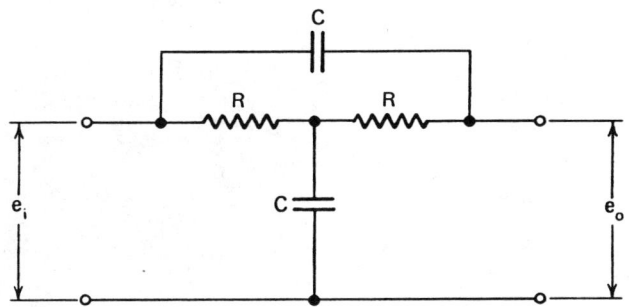

Figure 8.25 Bridged-T circuit (example 8.12)

$$G(s) = \frac{E_o}{E_i} = \frac{(RC)^2 s^2 + 2(RC)s + 1}{(RC)^2 s^2 + 3(RC)s + 1} \qquad ..(8.110)$$

The magnitude equals $|G(j\omega)|$ so that:

$$M = \left| \frac{1 - (\omega RC)^2 + j(2\omega RC)}{1 - (\omega RC)^2 + j(3\omega RC)} \right|$$

$$= \left[\frac{(1 - (\omega RC)^2)^2 + 4(\omega RC)^2}{(1 - (\omega RC)^2)^2 + 9(\omega RC)^2} \right] \qquad ..(8.111)$$

Fig. 8.26 shows the magnitude-frequency characteristics for the bridged-T circuit. At very low and very high frequency, the input signal is not attenuated. In the vicinity of $\omega RC = 1$, the magnitude is reduced. However, the circuit does not block any

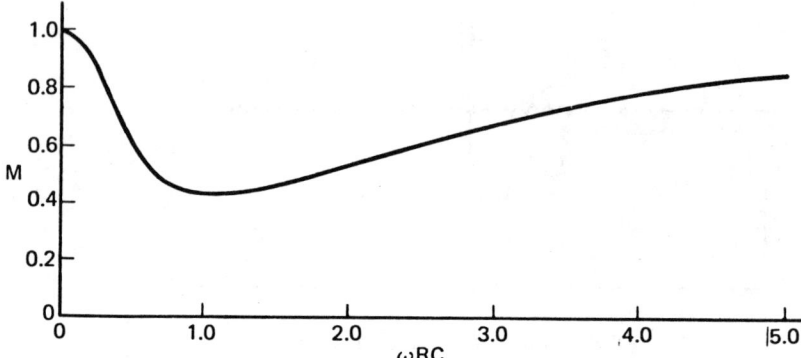

Figure 8.26 Magnitude-frequency characteristics (example 8.12)

frequency completely. The degree of blockage may be improved by using two different resistors or two different capacitors. We may consider the bridged-T circuit as a crude type of notch filter.

Example 8.13

Determine the magnitude-frequency characteristics of the electric LC circuit shown in Fig. 8.27.

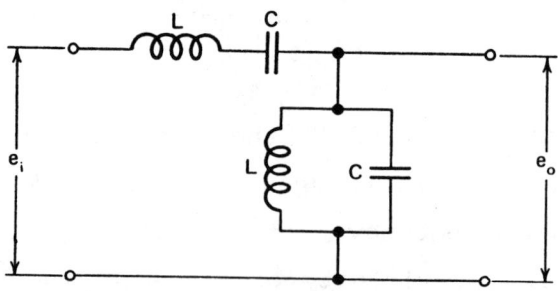

Fig. 8.27 Electric circuit (example 8.13)

Sinusoidal Response

The transfer function of the circuit is:

$$G(s) = \frac{LCs^2}{(LC)^2 s^4 + 3LCs^2 + 1} \qquad ..(8.112)$$

and in the frequency domain:

$$G(j\omega) = \frac{-\omega^2 LC}{\omega^4 (LC)^2 - 3\omega^2 LC + 1} \qquad ..(8.113)$$

If we define

$$\omega_N = 1/\sqrt{LC} \qquad ..(8.114a)$$

the magnitude of the transfer function becomes:

$$M = \frac{(\omega/\omega_N)^2}{1 + (\omega/\omega_N)^4 - 3(\omega/\omega_N)^2} \qquad ..(8.114b)$$

The above equation is shown plotted in Fig. 8.28. These characteristics indicate that the magnitude is zero at low and high frequencies. Thus the circuit is an effective block at the frequency extremes. However, at intermediate frequencies, signals will pass. Unfortunately, there are two frequencies (ω/ω_N = 0.62 and 1.62) where the amplitude is unbounded. These frequencies would have to be avoided. In general, we would classify this circuit as a band-pass filter.

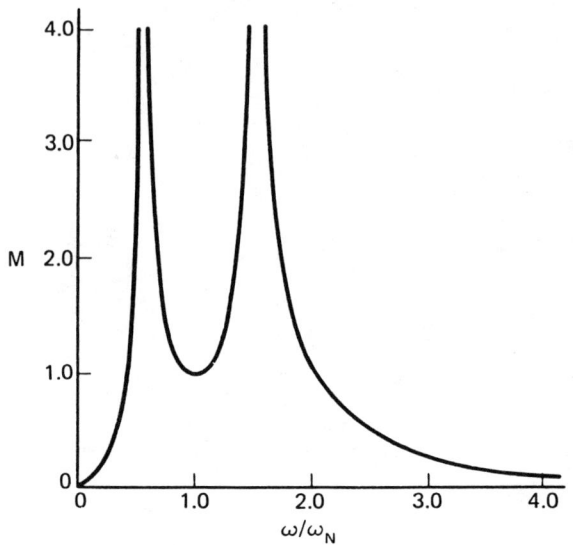

Figure 8.28 Magnitude-frequency characteristics (example 8.13)

8.5 RESONANCE

Resonance is a condition that occurs in circuits with reactive or energy storage components, when the frequency response is maximum. If the transfer function, $G(j\omega)$, represents an impedance or admittance, resonance comes about when there is no phase shift. The condition for resonance in such a circuit is therefore:

$$\angle G(j\omega_r) = 0 \qquad \ldots (8.115)$$

where ω_r is the frequency which satisfies the condition given in (8.115) and is called the resonant frequency.

In many cases, resonant circuits are deliberately sought to perform a highly selective frequency discrimination. These resonant circuits may be considered as a special type of band-pass filter, such as would be required in tuners for radio signals. On the other hand, however, resonance may be an extremely undesirable condition and we may try to avoid it. This would be the case in mechanical structural circuits where excessive strain could produce a catastrophic mechanical failure. In either case, we should know how to analyze resonant circuits. Then we may design the circuit to use resonance or to avoid it.

There are basically two types of resonant circuits: (a) series resonance and (b) parallel resonance. We will discuss these circuits in turn.

8.5.1 Series Resonance

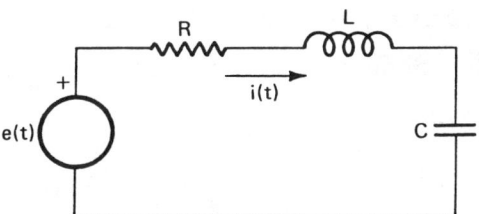

Figure 8.29 Series RLC electric circuit

Fig. 8.29 shows a series RLC electric circuit. The input voltage, e(t), is assumed to be a sine wave of the form E sinωt. The output signal is the current,

$i(t)$. The relation between the output and the input for exponential inputs is called the admittance, $Y(s)$, and may be expressed for this circuits as:

$$Y(s) = \frac{I}{E} = \frac{Cs}{LCs^2 + RCs + 1} \qquad ..(8.116)$$

In the frequency domain, (8.116) becomes:

$$Y(j\omega) = \frac{j\omega C}{1 - \omega^2 LC + j(\omega RC)} \qquad ..(8.117)$$

and in terms of magnitude and phase:

$$Y(j\omega) = \frac{\omega C}{\left[(1 - \omega^2 LC)^2 + (\omega RC)^2\right]^{1/2}} \; \angle 90° - \tan^{-1}\frac{\omega RC}{1-\omega^2 LC}$$

$$..(8.118)$$

The resonant condition of (8.115) may now be applied to the phase portion of (8.118). To obtain a zero phase, the denominator of the inverse tangent function must be zero so that

$$\omega_r = 1/\sqrt{LC} \qquad ..(8.119)$$

Thus the resonant frequency is equivalent to the natural frequency of the circuit (ω_N) for the case of series resonance. In terms of the resonant frequency, the magnitude portion of (8.118) may be written as:

Sinusoidal Response

$$M = |Y(j\omega)| = \frac{\frac{1}{R}[\frac{1}{Q}\frac{\omega}{\omega_r}]}{\left[\left\{1 - (\frac{\omega}{\omega_r})^2\right\}^2 + \left\{\frac{1}{Q}\frac{\omega}{\omega_r}\right\}^2\right]^{1/2}} \quad ..(8.120)$$

where Q is the quality factor of resonator selectivity and has the form

$$Q = \frac{\sqrt{L}}{R\sqrt{C}} \quad ..(8.121a)$$

This Q factor is inversely related to the damping ratio of the circuit:

$$Q = \frac{1}{2\delta} \quad ..(8.121b)$$

Eq. (8.120) is shown plotted in Fig. 8.30 for Q = 1.25 and 2.50. These values of Q were obtained by changing the value of R from 2 ohms to 1 ohm with L and C at fixed values. The magnitude in Fig. 8.30 represents admittance. Thus we observe that for series resonance, the driving point admittance is maximized and the driving point impedance is minimized.

For tuners, Q values of about 20,000 are commonly used. This increases the selectivity of the circuit so that it, in effect, passes a very narrow band of frequencies.

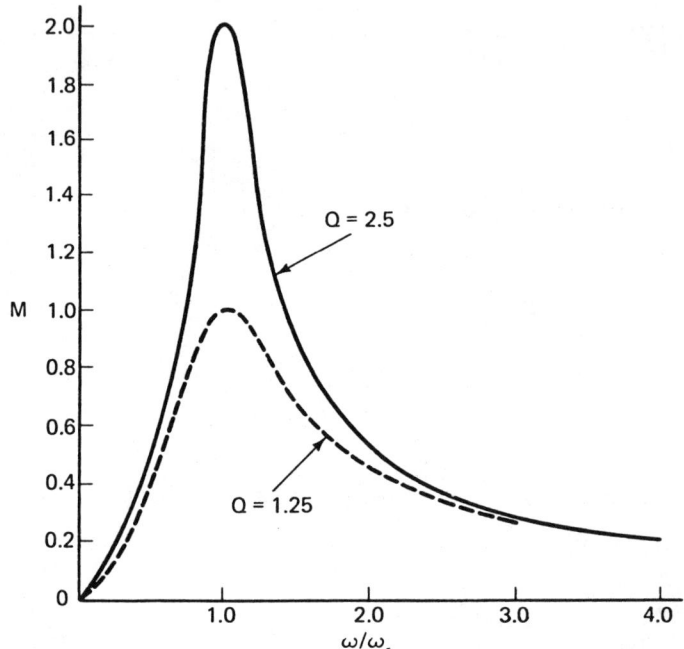

Figure 8.30 Effect of Q on magnitude-frequency characteristic of series RLC circuit

A fluid circuit with the same analogous circuit diagram is known as the Helmholtz resonator [Fig. 8.31].

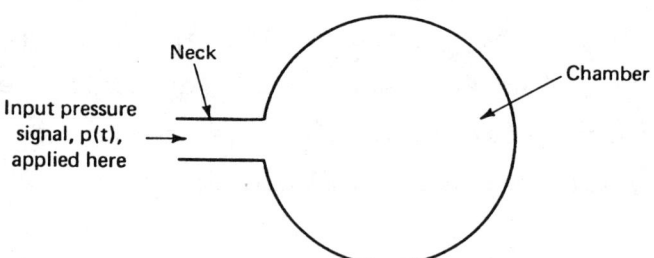

Figure 8.31 Helmholtz resonator

This device consists of a chamber with a short neck

entrance passage. The R and L portion of the circuit are due to the neck. In this case, the R is really nonlinear and we would have to use a linear approximation to analyze the circuit. The chamber, of course, represents the capacitance.

8.5.2 Parallel Resonance

Fig. 8.32 shows an electric circuit which exhibits parallel resonance.

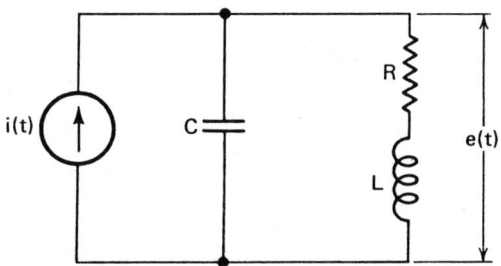

Figure 8.32 Electric circuit for parallel resonance

The circuit has a capacitor in parallel with a series RL combination. The input is the current i(t) and the output is the voltage across the capacitor e(t). In this circuit, the relation between the output and the input for exponential inputs is the driving point impedance Z(s), and appears as:

$$Z(s) = \frac{E}{I} = \frac{R + Ls}{LCs^2 + RCs + 1} \qquad ..(8.122)$$

and in the frequency domain:

$$Z(j\omega) = \frac{R + j\omega L}{1 - \omega^2 LC + j(\omega RC)} \qquad ..(8.123a)$$

$$= \frac{R(1-\omega^2 LC) + \omega^2 RLC + j[\omega(L-R^2C) - \omega^3 L^2 C]}{[(1-\omega^2 LC)^2 + (\omega RC)^2]}$$

$$..(8.123b)$$

The resonant condition is that the reactive portion of (8.123b) must equal zero. Thus:

$$\omega_r(L - R^2 C) - \omega_r^3 L^2 C = 0 \qquad ..(8.124)$$

and

$$\omega_r = \frac{1}{\sqrt{LC}} \sqrt{1 - \frac{R^2 C}{L}} \qquad ..(8.125a)$$

or

$$\omega_r = \omega_N \sqrt{1 - 1/Q^2} \qquad ..(8.125b)$$

When R is small (high Q), the resonant frequency for parallel and series resonance is the same. The magnitude for the parallel resonant circuit, [Eq. (8.123b)], may be expressed as:

$$M = \frac{R\left[1 + \left\{Q\frac{\omega}{\omega_N}(1 - \frac{\omega^2}{\omega_N^2}) - \frac{1}{Q}\frac{\omega}{\omega_N}\right\}^2\right]^{1/2}}{\left[1 - (\frac{\omega}{\omega_N})^2\right]^2 + \left[\frac{1}{Q}\frac{\omega}{\omega_N}\right]^2} \qquad ..(8.126)$$

Sinusoidal Response 561

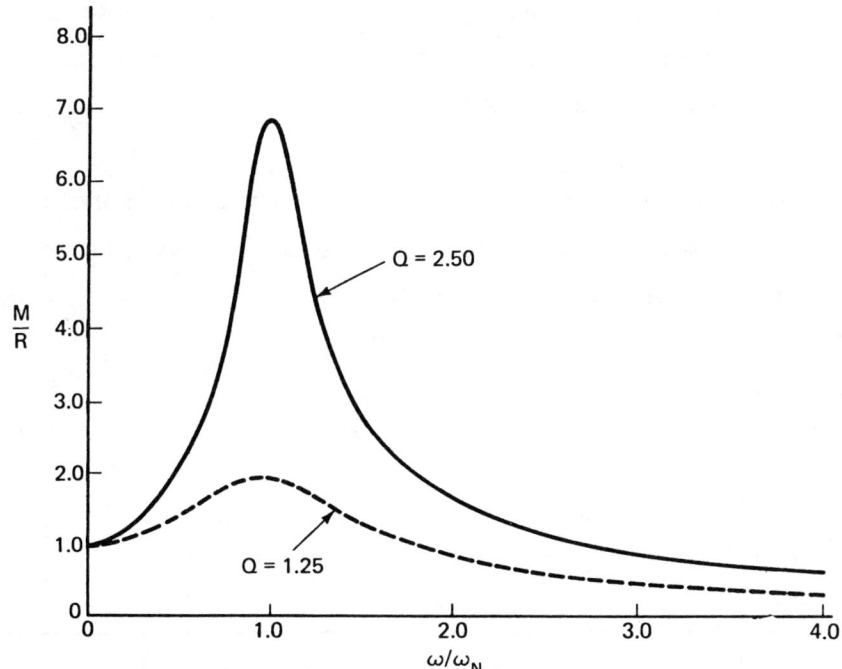

Figure 8.33 Magnitude-frequency response for parallel resonance

Fig. 8.33 shows the magnitude-frequency response for parallel resonance. In this case, the driving point impedance is maximum at resonance. Recall that for the series circuit, the driving point impedance is minimum at resonance.

8.6 BODE PLOT

With the advent of computers, especially PC's, it is easy to obtain the magnitude and phase responses of any function. However, it is required to know what to expect in such plots and this ensures the accuracy of the results. In order to confirm these results, it is only necessary to know the approximate shape of the response (magnitude and/or phase) and also some spot values. One of the convenient methods to accomplish the same is by the use of Bode plot.

The starting point is the given rational function, where the numerator and the denominator are available in the factored form as given in (8.127).

$$G(s) = \frac{N(s)}{D(s)} =$$

$$\frac{A\, s^{\eta_o} \left[\prod_{i=1}^{n_1} \left(\frac{s}{a_i} + 1\right)^{\alpha_i} \right]^{\alpha_{n_1}} \prod_{j=1}^{n_2} \left[\left(\frac{s}{\omega_{Nj}}\right)^2 + 2\delta_{Nj}\left(\frac{s}{\omega_{Nj}}\right) + 1 \right]^{\beta_j}^{\beta_{n_2}}}{\left[\prod_{k=1}^{m_1} \left(\frac{s}{b_k} + 1\right)^{\gamma_k} \right]^{\gamma_{m_1}} \prod_{\ell=1}^{m_2} \left[\left(\frac{s}{\omega_{D\ell}}\right)^2 + 2\delta_{D\ell}\left(\frac{s}{\omega_{D\ell}}\right) + 1 \right]^{\nu_\ell}^{\nu_{m_2}}}$$

..(8.127)

where $\qquad |\delta_{Nj}| < 1 \qquad$ ..(8.128a)

and $\qquad |\delta_{D\ell}| < 1 \qquad$ ..(8.128b)

There is no loss of generality in (8.127). Any polynomial can be factorized and it is known that the roots are either real or complex conjugates. Also, any root can be simple or multiple. If the root is simple,

Sinusoidal Response   563

the corresponding multiplicity is unity. The factor corresponding to complex conjugate roots can be expressed in different ways. But, in (8.127), the same has been expressed in the form used in the discussion of second order systems. Further, it should be noted that for each term, the constant term is unity.

A general transfer function is given here, even though in the text so far, only first order and second order transfer functions are discussed. As will be seen at the end of the discussion, the magnitude and the phase plots of any given transfer function will be obtained as the sum of the corresponding first and second order responses. It is now required to evaluate the magnitude and phase of $G(j\omega)$. We shall consider the magnitude and the phase plots separately.

8.6.1 MAGNITUDE PLOT

When $|G(j\omega)|$ is evaluated directly, it will be a product of several terms. Instead, the magnitude may be evaluated as a sum of several terms, if the logarithm of $|G(j\omega)|$ is considered. This permits us to obtain the magnitude in decibels (abbreviated dB), which is

$$H(\omega) = 20 \log_{10} |G(j\omega)| \text{ dB} \qquad \qquad ..(8.129)$$

Before writing the complete expansion of $H(\omega)$, we shall first determine the contribution of each term.

(a) The constant term A contributes

$$[20 \log_{10} A] \text{ dB} \qquad ..(8.130)$$

to $H(\omega)$ and this does not vary with frequency.

(b) The term s is considered next. This contributes

$$20 \log_{10} |(j\omega)| = 20 \log_{10} \omega \text{ dB} \qquad ..(8.131)$$

At $\omega = 0$, this becomes $-\infty$ and hence, it is preferable that zero frequency is not considered in such a case. If we consider a semi-logarithmic plot, the horizontal axis will have a logarithmic scale and hence, in general, the point $\omega = 0$ is not considered, while obtaining the Bode plot. If the term s is absent in the transfer function, a linear scale can be used for the frequency for purposes of representation. It is preferable to use semi-logarithmic plot, as it is easier to obtain Bode plots.

In (8.131), if the frequency is doubled, the magnitude is increased by 6 dB. This is expressed as a slope of 6 dB/octave. Instead, if the frequency is increased by a factor of ten, the magnitude is increased by 20 dB. This is expressed as a slope of 20 dB/decade. This provides the sketch of the contribution of the factor s to the magnitude.

Sinusoidal Response

This can be generalized further, when a factor of the type s^{η_o} is considered. The contribution will be

$$H_o(\omega) = \eta_o (20 \log_{10}\omega) \text{ dB} \qquad ..(8.132)$$

When the frequency is doubled, the magnitude is increased by $6\eta_o$ dB. When the frequency is increased by ten times, the magnitude is increased by $20\ \eta_o$ dB. Fig. 8.34 shows the sketch of $H_o(\omega)$ for two cases, $\eta_o = 1$ and $\eta_o = 3$.

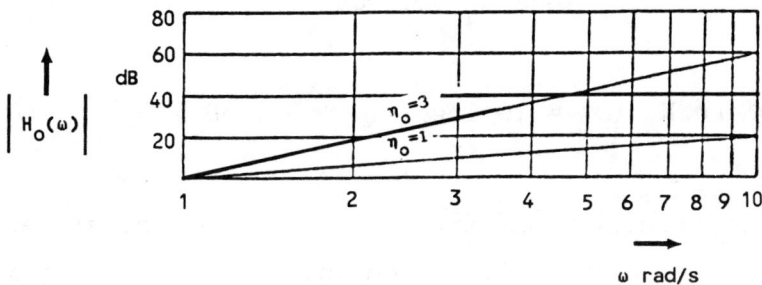

Figure 8.34 The contribution of $H_o(\omega)$ for $\eta_o=1$, $\eta_o=3$

It is noted that the term is contained only in the numerator of (8.127). If such a term occurs in the denominator, it is seen that its contribution to magnitude will be

$$-\eta_o (20 \log_{10}\omega) \text{ dB} \qquad ..(8.133)$$

In other words, the slopes of the lines in Fig. 8.34

will be negative.

(c) The term of the type $(\frac{s}{c} + 1)^\xi$ is considered next. This is a typical term occurring in either the numerator or the denominator or both. We shall first obtain the contribution of this term to the magnitude, when it is contained in the numerator. If it is contained in the denominator, the same discussion applies with a multiplication factor of (-1). The contribution to the magnitude will be

$$H_1(\omega) = \xi \left[10 \log_{10} (1 + \frac{\omega^2}{c^2})\right] \text{ dB}$$

$$= \xi H_{a_1}(\omega) \text{ dB (say)} \qquad ..(8.134)$$

where $H_{a_1}(\omega) = 10 \log_{10} (1 + \frac{\omega^2}{c^2}) \text{ dB} \qquad ..(8.135)$

The quantity 'c' may be positive or negative. In both cases, the magnitude remains the same. Some of the properties of $H_{a_1}(\omega)$ are given below:

(i) At $\omega = 0$, $H_{a_1} = 0$ dB.

(ii) When $\frac{\omega}{c} \ll 1$, $(1 + \frac{\omega^2}{c^2})$ can be approximated to be equal to unity and hence

$$H_{a_1} = 0 \text{ dB} \qquad ..(8.136)$$

(iii) when $\frac{\omega}{c} \gg 1$, $(1 + \frac{\omega^2}{c^2})$ can be approximated to be equal to $\frac{\omega^2}{c^2}$ and hence

$$H_{a_1} = 20 \log_{10} (\frac{\omega}{c}) \text{ dB} \qquad \qquad ..(8.137)$$

This will have the same properties as that of the term s and hence the slope of the line will be 6 dB/octave or 20 dB/decade.

It is this seen that (8.136) and (8.137) form the asymptotes for the magnitude curve of $H_{a_1}(\omega)$. This permits us to draw the two asymptotes which meet at $\frac{\omega}{c} = 1$. [See Fig. 8.35]. The point of intersection of the two asymptotes is known as the **break point**.

(iv) When $\frac{\omega}{c} = 1$,

$$H_{a_1}(\omega) = 3 \text{ dB} \qquad \qquad ..(8.138)$$

This will be the error and hence the actual sketch will be the curve shown as (a) in Fig. 8.35. The two straight lines serve as asymptotes. Further refinement of the actual curve is possible, if required.

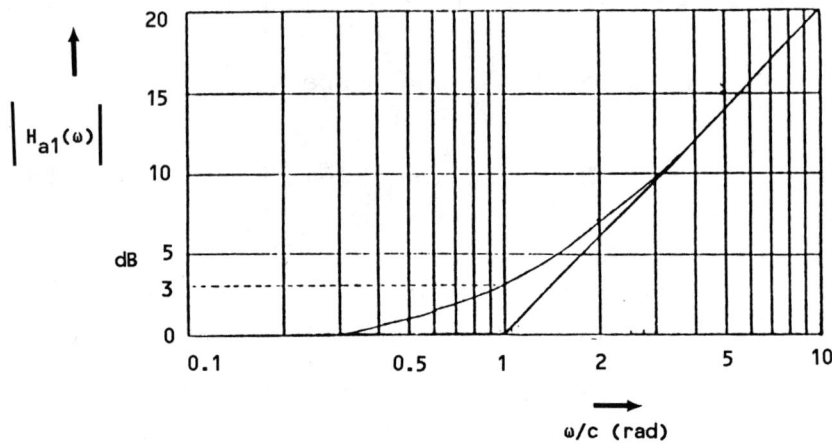

Figure 8.35 The plot of the term $\left|1 + s/c\right|_{s=j\omega}$

(v) $\dfrac{\omega}{c} = 2$, the approximate curve gives a value of 6 dB. The actual value of $H_{a_1}(\omega)$ will be

$$\left. H_{a_1}(\omega) \right|_{\frac{\omega}{c} = 1} = 20 \log_{10} \left|(1 + j2)\right| \text{ dB} = 7 \text{ dB}$$

.. (8.139)

There is an error of 1 dB between the approximate and the actual values.

(vi) When $\dfrac{\omega}{c} = \dfrac{1}{2}$, the approximate curve gives a value of 0 dB. The actual value of $H_{a_1}(\omega)$ will be

$$H_{a_1}(\omega)\bigg|_{\frac{\omega}{c} = \frac{1}{2}} = 20 \log_{10} |(1 + j\tfrac{1}{2})| = 1 \text{ dB}$$

..(8.140)

In this case also, there is an error of 1 dB.

This discussion permits us to obtain the properties of $H_1(\omega)$ which are as follows:

(i) When $\frac{\omega}{c} \ll 1$. $H_1(\omega) = 0$

(ii) When $\frac{\omega}{c} \gg 1$, the slope of $H_1(\omega)$ will be 6ξ dB/octave or 20ξ dB/decade.

(iii) When $\frac{\omega}{c} = 1$, $H_1(\omega) = 3\xi$ dB.

(iv) When $\frac{\omega}{c} = 2$, $H_1(\omega) = 7\xi$ dB.

(v) When $\frac{\omega}{c} = \frac{1}{2}$, $H_1(\omega) = \xi$ dB.

The curve can be easily drawn.

If the term $(\frac{s}{c} + 1)^{\xi}$ occurs in the denominator,

$$H_1(\omega) = -\xi H_{a_1}(\omega) \qquad ..(8.141)$$

This means that, in the properties of $H_1(\omega)$ discussed above, ξ is replaced by $-\xi$.

(d) The second-order term is considered now. A typical term is of the type

$$g(s) = \left[\left(\frac{s}{\omega_N}\right)^2 + 2\delta \left(\frac{s}{\omega_N}\right) + 1\right] \quad \ldots (8.142)$$

From the previous discussion, it is relatively easy to obtain the contribution, if the multiplicity is greater than unity. Also, the values of $\delta \geq 1$ need not be considered, because the factors contain real roots. Therefore, only values of $\delta < 1$ are considered here. The value of δ may be negative in which case the roots of (8.142) are contained in the right-half plane. In both cases, the magnitude is positive.

The contribution of $g(s)$ to the magnitude will be:

$$H_2(\omega) = 10 \log_{10}\left[1 + \left(\frac{\omega}{\omega_N}\right)^2(4\delta^2 - 2) + \left(\frac{\omega}{\omega_N}\right)^4\right]$$

$$= 10 \log_{10}\left[f\left(\frac{\omega^2}{\omega_N^2}\right)\right] \quad \ldots (8.143)$$

Some properties of $H_2(\omega)$ are obtained as follows:

(i) When $\omega = 0$, $H_2(\omega) = 0$ dB $\quad \ldots (8.144)$

(ii) When $\frac{\omega}{\omega_N} = 1$, $H_2(\omega) = 10 \log_{10}(4\delta^2)$ dB $\ldots (8.145)$

This serves as the **break point** in the second order case. However, particular care should be taken to evaluate this quantity when $\delta = 0$.

(iii) Let $u = (\frac{\omega}{\omega_N})^2$ ..(8.146a)

Then $f(u) = \left[u^2 + u(4\delta^2 - 1) + 1\right]$ ..(8.146b)

Equating $f'(u) = 0$, we get

$$u = (1 - 2\delta^2)$$ ..(8.146c)

Under this condition, $H_2(\omega)$ will have a minimum value given by

$$10 \log_{10}\left[4\delta^2(1 - \delta^2)\right]$$ ..(8.147)

This occurs only when $\delta < 0.707$.

(iv) When $\frac{\omega}{\omega_N} = 2$,

$$H_2(\omega)\bigg|_{\frac{\omega}{\omega_N} = 2} = 10 \log_{10}(9 + 16\delta^2) \text{ dB}$$ ..(8.148)

We can conclude this part of the discussion by saying that, for a given value of δ, a curve for $H_2(\omega)$ can be drawn. Fig. 8.36 shows the curves of $1/H_2(\omega)$ vs $(\frac{\omega}{\omega_N})$ for some representative values of δ.

If $\left[g(s)\right]^\mu$ is considered, (that is, if the multiplicity is greater than unity), $H_2(\omega)$ is multiplied by the factor μ.

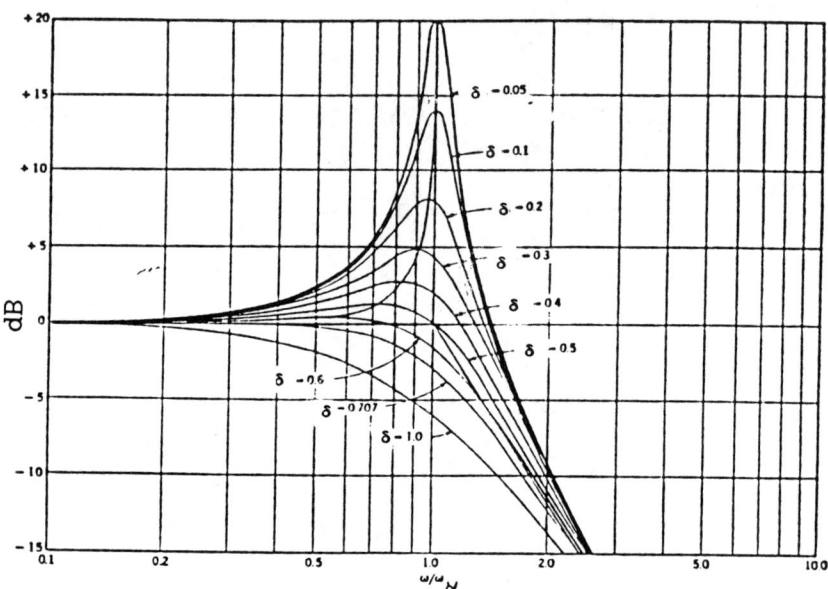

Figure 8.36 Magnitude versus frequency of

$$\frac{1}{H_2(\omega)} = \left| \frac{1}{(s/\omega_N)^2 + 2\delta(s/\omega_N) + 1} \right|_{s=j\omega}$$

(M. E. Van Valkenburg, INTRODUCTION TO MODERN NETWORK SYNTHESIS, copyright ©1960 by John Wiley and Sons, figure 9.18. Adapted and reprinted with permission from John Wiley and Sons, Inc.)

From the foregoing discussion, the magnitude of $G(j\omega)$ can be written as follows:

$$20 \log_{10}|G(j\omega)| \text{ dB} = 20 \log_{10}A + \eta_o 20 \log_{10}\omega$$
$$+ \sum_{i=1}^{n_1} \alpha_i \ 10 \log_{10}(1 + \frac{\omega^2}{a_i^2}) +$$

$$+ \sum_{j=1}^{\beta_{n_2}} \beta_j \; 10 \; \log_{10}\left[1 + (\frac{\omega}{\omega_{Nj}})^2 (4\delta_{Nj}^2 - 2) + (\frac{\omega}{\omega_{Nj}})^4\right]$$

$$- \sum_{k=1}^{\gamma_{m_1}} \gamma_k \; 10 \; \log_{10}(1 + \frac{\omega^2}{b_k^2})$$

$$- \sum_{l=1}^{\nu_{m_2}} \gamma_l \; 10 \; \log_{10}\left[1 + (\frac{\omega}{\omega_{Dl}})^2 (4\delta_{Dl}^2 - 2) + (\frac{\omega}{\omega_{Dl}})^4\right]$$

..(8.149)

As remarked earlier, it is seen that the magnitude will be the sum of several terms. The magnitude plot can be obtained as follows:

(a) Select the break points.

(b) For every break point, draw the asymptotes which have the following properties:

    (i) The slope of the asymptote will be $\pm 6x_o$ dB/octave, if the break point corresponds to a first order factor. The positive sign shall be used, if the term is contained in the numerator. Otherwise, the negative sign shall be used. The quantity x_o denotes the multiplicity of the term.

    (ii) The slope of the asymptote is $\pm 12y_o$ dB/octave, if the break point corresponds to a second order factor. The quantity y_o denotes the multiplicity of the term. In this case, both the asymptotes are drawn. Depending on the value of δ, the appropriate curve is drawn.

(iii) Locate the points corresponding to the various values of ω taking into account the contribution made by each asymptote. Apply the corrections at the break points. Draw a smooth curve.

8.6.2 PHASE PLOT

In order to obtain the phase plot of a transfer function, we can consider the first order and the second order factors separately and then add up the various values of phase angles obtained.

(a) The constant A does not contribute to the phase. Its phase angle is always $0°$. Hence, this factor need not be considered in the phase plot.

(b) The term s is considered next. The phase of $s = j\omega$ is always $90°$ and is independent of the frequency. As a consequence, the contribution of s^{η_o} to the phase is $\eta_o \cdot 90°$.

(c) The term of the type $(\frac{s}{c} + 1)$ is considered next. The phase is given by

$$\theta_1(\omega) = \tan^{-1} \frac{\omega}{c} \qquad \qquad ..(8.150)$$

Sinusoidal Response

This gives the following three corner points:

$\frac{\omega}{c}$	$\theta_1(\omega)$
0	0
1	$45°$
∞	$90°$

Two points can be joined by a straight line and hence can serve as the asymptote. On a logarithmic scale, neither $\omega = 0$ nor $\omega = \infty$ will be available. Therefore, $\frac{\omega}{c}$ is chosen as a small quantity equal to 0.1 for which $\theta_1(\omega)$ will be $5.7°$. This point and another point $\frac{\omega}{c} = 1$ for which $\theta_1(\omega) = 45°$ can be joined by a straight line which serves as the asymptote.

Noting that $\tan^{-1}(1/2) = 26.6°$ (octave below the reference frequency of $\frac{\omega}{c} = 1$) and $\tan^{-1}2 = 63.4°$ (octave above the frequency of $\frac{\omega}{c} = 2$), the asymptote and the actual curve can be drawn and are as shown in Fig. 8.37. This means that the actual phase curve reaches $90°$ asymptotically. It can also be verified that the maximum error in the range $0 \le \frac{\omega}{c} \le 1$ is $4.1°$. It is noted that the actual curve lies below the asymptote when $0 < \frac{\omega}{c} < 1$ and lies above the asymptote when $1 < \frac{\omega}{c} < 1$. Also shown in Fig. 8.37 is the case when c is negative. In such a case, the phase is negative.

The contribution of $(\frac{s}{c} + 1)^{\xi}$ to the phase is $\xi \tan^{-1}(\frac{\omega}{c})$, and this can be obtained easily.

(c) The second order term is considered. A typical term is of the type given in (8.142). Its contribution to phase is:

$$\theta_2(\omega) = \tan^{-1} \frac{2\delta \frac{\omega}{\omega_N}}{1 - (\frac{\omega}{\omega_N})^2} \quad \quad ..(8.151)$$

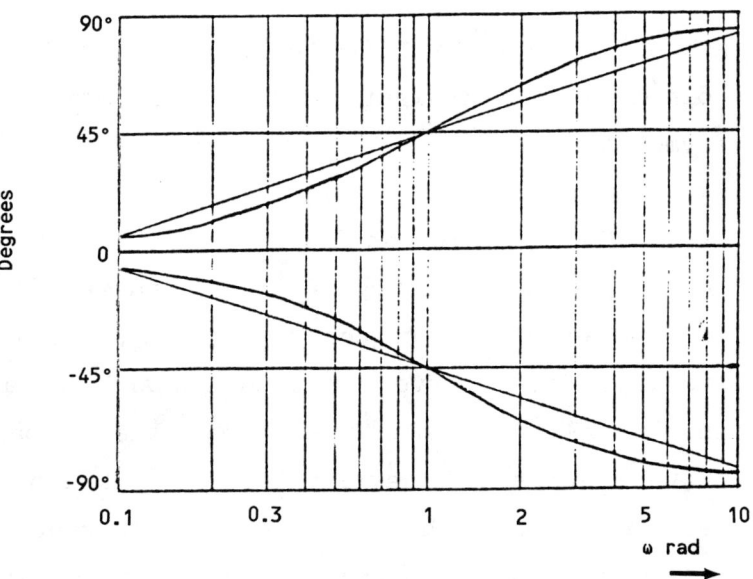

Figure 8.37 The phase plot of the term (1 + jω/c)

We obtain a family of curves for different values of δ. Some representative curves are shown in Fig. 8.38. It is noted that all the curves pass through the point of $90°$, when $\frac{\omega}{\omega_N} = 1$. The curves shown in Fig. 8.38 are for values of positive δ. If δ is negative, the phase is negative, and the same curves are valid with the numbers on the vertical axis being multiplied by (-1).

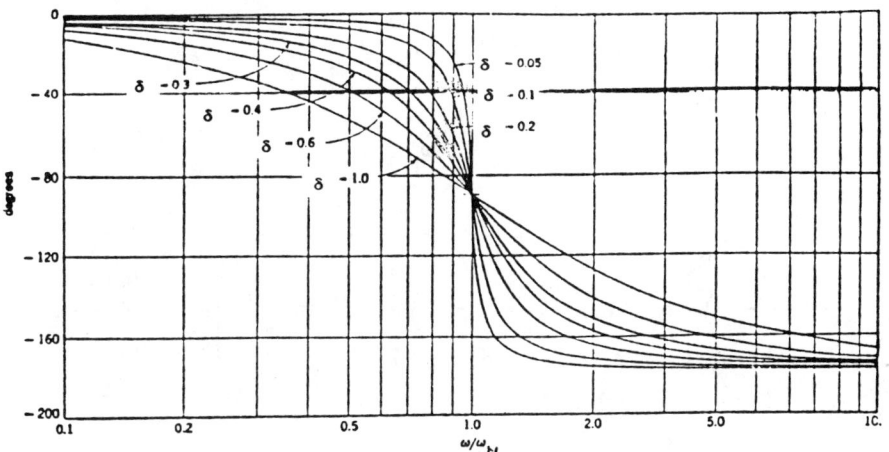

Figure 8.38 Phase angle versus frequency of

$$\frac{1}{H_2(\omega)} = \frac{1}{(s/\omega_N)^2 + 2\delta(s/\omega_N) + 1}$$

(M. E. Van Valkenburg, INTRODUCTION TO MODERN NETWORK SYNTHESIS, copyright ⊕1960 by John Wiley and Sons, figure 9.19. Adapted and reprinted with permission from John Wiley and Sons, Inc.)

When a factor of the type $[g(s)]^\mu$ is considered, the phase is also multiplied by the factor μ.

Regarding the plotting of the phase function,

simple rules as enunciated in the case of the magnitude plot cannot be given. The points shall be located at appropriate values of ω and a smooth curve shall be drawn. It is noted that:

$$\angle G(j\omega) = \eta_o \cdot 90° + \sum_{i=1}^{\alpha_{n_i}} \alpha_i \tan^{-1}\left(\frac{\omega}{a_i}\right) +$$

$$+ \sum_{j=1}^{\beta_{n_2}} \beta_j \tan^{-1}\left[\frac{2\delta_{Nj}\left(\frac{\omega}{\omega_{Nj}}\right)}{1 - \left(\frac{\omega}{\omega_{Nj}}\right)^2}\right]$$

$$- \sum_{k=1}^{\gamma_{m_1}} \gamma_k \tan^{-1}\left(\frac{\omega}{b_k}\right) - \sum_{l=1}^{\nu_{m_2}} \nu_l \tan^{-1}\left[\frac{2\delta_{Dl}\left(\frac{\omega}{\omega_{Dl}}\right)}{1 - \left(\frac{\omega}{\omega_{Dl}}\right)^2}\right]$$

..(8.152)

An example illustrates the procedure.

Example 8.14

It is required to obtain the Bode plot of

$$G(s) = \frac{(s + 10)}{(s + 0.5)(s + 4)} \qquad ..(8.153)$$

Before proceeding further, (8.153) is rearranged as

Sinusoidal Response

$$G(s) = \frac{5(\frac{s}{10} + 1)}{(\frac{s}{0.5} + 1)(\frac{s}{4} + 1)} \quad \quad ..(8.154)$$

We shall consider the magnitude plot first. Accordingly, we have:

$$20 \log_{10}|(j\omega)| \text{ dB} = g_1 + g_2 - g_3 - g_4 \quad \quad ..(8.155)$$

where $g_1 = 20 \log_{10} 5$ ..(8.156a)

$$g_2 = 10 \log_{10}\left[1 + (\frac{\omega}{10})^2\right] \quad \quad ..(8.156b)$$

$$g_3 = 10 \log_{10}\left[1 + (\frac{\omega}{0.5})^2\right] \quad \quad ..(8.156c)$$

$$g_4 = 10 \log_{10}\left[1 + (\frac{\omega}{4})^2\right] \quad \quad ..(8.156d)$$

The break points are $\omega = 0.5$, 4 and 10. The various asymptotes are obtained as follows:

(a) The first interval is between $\omega = 0.1$ and 0.5. In this interval, the contribution by the various terms to the slope will be zero. The overall magnitude has a constant value equal to $20 \log_{10} 5 = 14$ dB.

(b) The second interval is between $\omega = 0.5$ and 4. In this interval, g_2 and g_4 do not contribute to the slope. Hence the slope of the magnitude function will be -20 dB/decade. The asymptote starts at $\omega = 0.5$ and ends at $\omega = 4$. At $\omega = 0.5$, the value will be 14 dB and at $\omega = 4$, the value will be -4 dB.

(c) The third interval is between $\omega = 4$ and 10. In this interval, g_2 does not contribute to the slope. Hence the slope of the magnitude function will be -40 dB/decade. The asymptote starts at -4 dB at $\omega = 4$ and ends at $\omega = 10$, where the value will be -20 dB.

(d) The last interval will start at $\omega = 10$. In this interval, all the three terms g_2, g_3 and g_4 contribute to the slope. Hence the slope of the magnitude function will be -20 dB/decade. The asymptote starts at $\omega = 10$ at a value of -20 dB.

The asymptotes and the actual curve of $|G(j\omega)|$ after the corrections are incorporated are shown in Fig. 8.39. The actual curve is obtained by applying the required corrections, as described earlier.

Regarding the phase curve, the phase of $G(j\omega)$ at the break points and at some other appropriate points are obtained and a smooth curve joining these points is drawn. The curve is also shown in Fig. 8.39.

Only one example is given in the text. The students should do well to work more problems in order to get familiarity with the methods.

8.7 MODELING BASED ON SINUSOIDAL RESPONSE

In Chapter III, the principle of modeling has been formulated based on the terminal behaviour of elements.

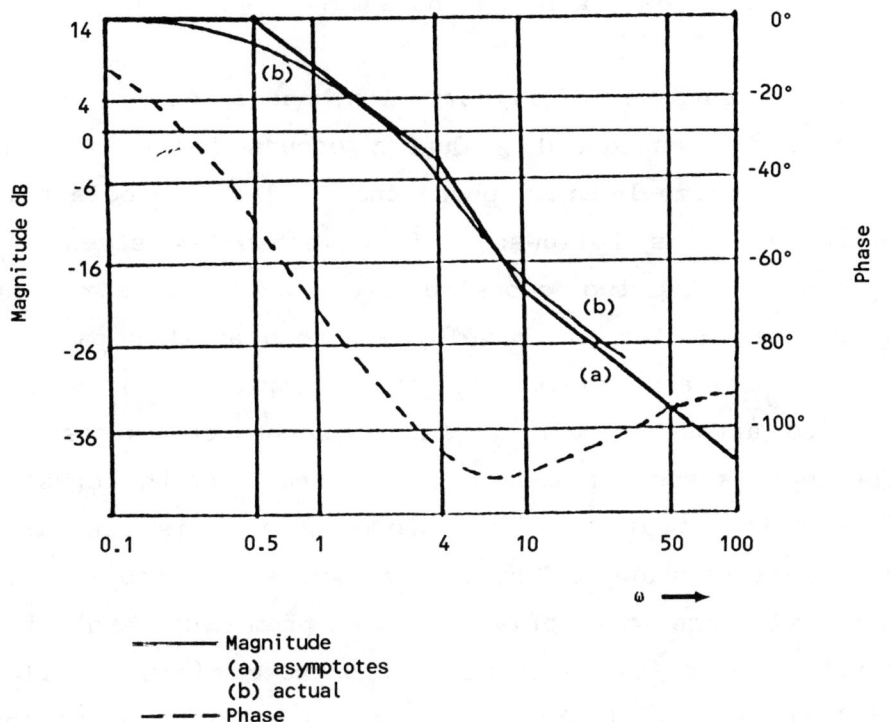

Figure 8.39 Magnitude and phase plots of $G(j\omega)$ (example 8.14)

Only one element was considered. The across and through variables were functions of time and no excitation was specified. The same principle can be extended further by considering the sinusoidal behaviour across two terminals of a network or a system. The problem can be defined as follows: given the sinusoidal response across two terminals of a network, it is required to determine the elements contained in the network so that the sinusoidal behaviour coincides or closely approximates the given response. It is obvious at the outset that the problem is a difficult one. However, the intention of this section is to illustrate some basic principles regarding the same.

For purposes of illustration, modeling of a quartz crystal is considered. Quartz occurs in nature and exhibits **piezoelectric** phenomena. The piezoelectric property is as follows: If a mechanical stress is applied across two opposite faces of a crystal, an electrical voltage is developed across another pair of faces. Alternatively, if an electrical voltage is applied across a pair of faces, mechanical stress is developed across another pair of faces and the crystal is set in vibration. The quartz crystal is generally hexagonal in shape. A number of synthetic piezoelectric materials are available. Some prominent synthetic crystals are: (i) Aluminium Phosphate (also called Berlinite), (ii) Lithium Niobate and Lithium Tantalate, (iii) Bismuth Germanium Oxide and Bismuth Silicium Oxide, (iv) Piezoelectric ceramics and polymers. Each material has its own characteristic crystalline structure.

Depending on the orientation of the axes, crystals can be cut into thin wafers by slicing sections. These sections are then ground, lapped and etched so that the crystal vibrates at a given frequency. Two of the typical crystal cuts are shown in Fig. 8.40. There are many other crystal cuts, each cut exhibiting different

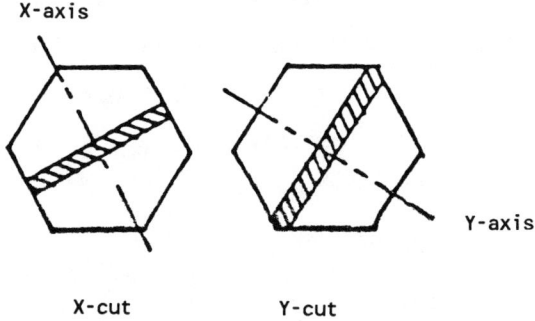

Figure 8.40   X-cut and Y-cut quartz crystals

properties. However, the resonant frequency of the crystal is a function of the ambient temperature, in addition to its dependence on the mode of vibration. A crystal is mounted between two metal electrodes and the entire unit is evacuated so that the air inside does not influence the frequency of oscillations. A crystal is represented as shown in Fig. 8.41.

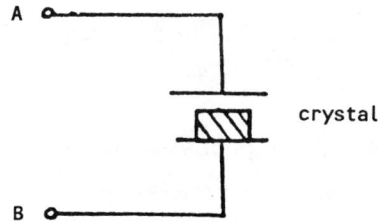

Figure 8.41   Representation of a crystal

What is now required is to obtain an electrical equivalent circuit that represents the behaviour of the crystal across the two-terminals A and B. In order to accomplish this, the reactance curve of a crystal is required. A typical reactance curve of a crystal is shown in Fig. 8.42. Restricting ourselves to the range

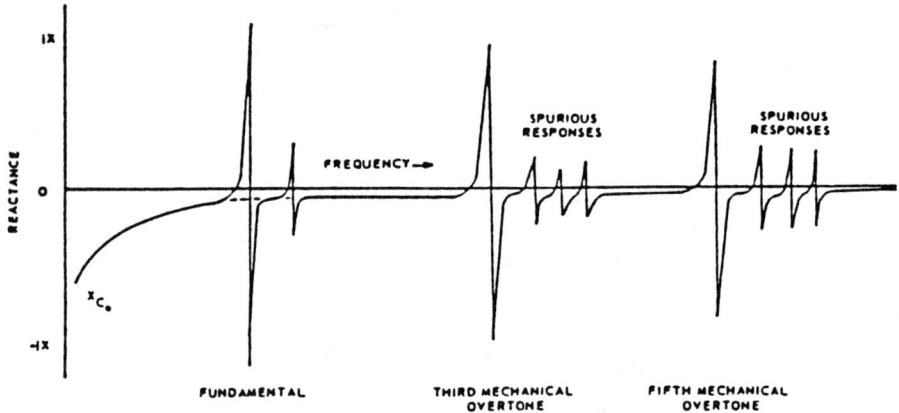

Figure 8.42 Typical reactance curve of a quartz crystal

(M. E. Frerking, CRYSTAL OSCILLATOR DESIGN AND TEMPERATURE COMPENSATION, copyright ©1978, figure 5.8. Adapted and reprinted with permission from Van Nostrand Reinhold, Inc.)

up to the fundamental frequency only, the following properties are noted:

(a) At zero frequency, the reactance is $-\infty$. This means that direct current cannot flow between the two terminals A and B.

(b) There will be one series resonant frequency ω_s and one parallel resonant frequency ω_r.

Based on these two observations, the equivalent circuit will be as shown in Fig. 8.43(a) and its reactance curve is given in Fig. 8.43(b). It is easily observed that the reactance curve of Fig. 8.43(b) has the same shape as the reactance curve of Fig. 8.42 up to the range of the fundamental frequency.

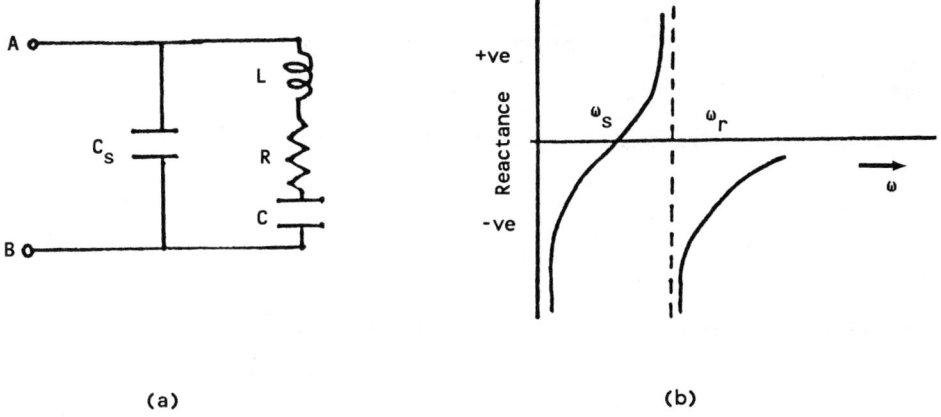

Figure 8.43 (a) Equivalent circuit of a quartz crystal
(b) Its reactance curve

From physical considerations also, the equivalent circuit can be justified. The capacitance C_s represents the parallel combination of the shunt capacitance between the two electrodes and the inherent capacitance of the holder. The inductance L is associated with the stiffness of the quartz material and the capacitance C is associated with the mass. The resistor R is present because of the losses in the crystal. The various parameters L, R and C are obtained by the mechanical properties of the crystal. For a typical quartz crystal resonant at 474 kc/s, the various values are L = 3.66 H,

$R = 1130\ \Omega$, $C = 0.0316$ pF and $C_s = 5.76$ pF. In practice these are measured by crystal impedance meters.

Analysis of the electrical equivalent circuit of Fig. 8.42(a) yields:

$$Z_{AB} = \frac{(1 - \omega^2 LC) + j\omega CR}{-\omega^2 CC_s R + j[\omega(C + C_s) - \omega^3 LCC_s]} \quad ..(8.157)$$

For the range of frequencies used, $\omega^2 CC_s R$ is a very small quantity and hence can be neglected. It is obvious that the series resonant frequency is given by

$$\omega_s = \frac{1}{\sqrt{LC}} \quad ..(8.158)$$

and the parallel resonant frequency is given by

$$\omega_r = \frac{1}{\sqrt{L\dfrac{CC_s}{C + C_s}}} \quad ..(8.159)$$

Generally, ω_o, the frequency of oscillations of the crystal, is set between ω_s and ω_r.

The Q-factor of the crystal can be readily calculated as:

$$Q = \frac{\omega_r L}{R}$$
$$= 9641.4 \quad ..(8.154)$$

It is worth noting that the Q-factor of crystal can

exceed 5×10^6.

The above discussion should make clear how to associate a network configuration given the sinusoidal response. The realization of networks approximating a given response involves another branch of mathematical theory and is beyond the scope of the present text. This branch of study has been responsible for the synthesis of electrical filters and in fact, modern filter theory is based on this study.

8.8 SUMMARY AND DISCUSSIONS

In this chapter, a sinusoidal or cosinusoidal excitation of a circuit is considered. It is shown that the response in any branch is sinusoidal, after the initial transient (which exists only for a short duration of time) has died down. The analytical method of obtaining the response has been discussed in detail. The response consists of two parts, namely magnitude and phase. The approximate methods of obtaining the sketches of magnitude and phase have been enunciated. The resonant properties of circuits have been discussed. In the case of series resonance, through variable becomes a maximum; and in the case of parallel resonance, across variable becomes a maximum. It is also indicated how to obtain a network when a sinusoidal response is given.

8.9 PROBLEMS

8.1 In the electrical RC circuit shown in Fig. P8.1, R = 1,000 Ω and C = 5 μF. The input is a voltage source e(t) = 100 cos 100t, V and the output is the voltage across the capacitor, $e_c(t)$. Find and plot the sinusoidal steady state response, $e_c(t)$.

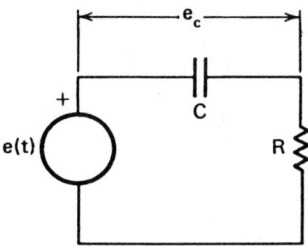

Figure P 8.1

8.2 The input impedance to many electronic measuring instruments is often modeled as a resistance of 2 MΩ in parallel with a capacitance of 50 pF. Determine the series equivalent circuit, when the frequency is: (a) 15.92 Hz; (b) 1.592 kHz; (c) 159.2 kHz.

8.3 In a series RLC electric circuit, R = 200 Ω, L = 0.15 H and C = 60 μF. The voltage source provides a signal with a frequency of 200 rad/s and an amplitude of 60 V. Find the sinusoidal steady state response of the current in the circuit.

8.4 A fluid line has a source of pressure connected to one end and is open to atmosphere at the other end.

The pressure source may be described by p(t) = 40 cos 0.1t, Pa. If the line may be modeled as R = $8(10^5)$ N.s/m^5 and L = $8(10^6)$ N.s^2/m^5 in series, find the steady-state response for the flow in the line.

8.5 A plate of negligible mass is connected to a wall by a mechanical spring and viscous damper as shown in Fig. P8.5. The input force applied to the plate is f(t) = F sin ωt and k = 18 N/m and b = 1.6 N.s/m. If the velocity of the plate is considered as the output, find the frequency at which the magnitude ratio M = 1/2. Also find the corresponding value of the phase angle.

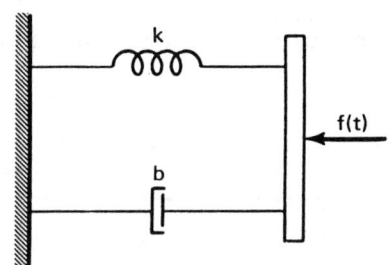

Figure P 8.5

8.6 A damper of 2 N.s/m is connected in parallel with a spring of 5 N/m. Find the series equivalent circuit, when the frequency is: (a) 0.04 Hz; (b) 0.4 Hz; (c) 4 Hz.

8.7 In the mass-damper system shown in Fig. P8.7, a velocity is applied to the free end of the damper so that v(t) = 100 sin 0.5t, m/s. The mass m = 20 kg.

The magnitude response for the velocity of the mass is M = 0.25. Find the value of the damping.

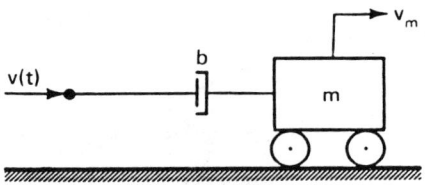

Figure P 8.7

8.8 In the mechanical system shown in Fig. P8.8, mass m_1 is subjected to a velocity input $v_1(t) = V_1 \, \varepsilon^{j\omega t}$.
(a) Obtain the sytem function $G(j\omega) = V_2(j\omega)/V_1(j\omega)$.
(b) If b = 4 N.s/m, k = 6 N/m, m_1 = 1 kg and m_2 = 1 kg, sketch the magnitude and phase of $G(j\omega)$.
(c) Determine the driving point admittance $Y_{dp}(j50)$.

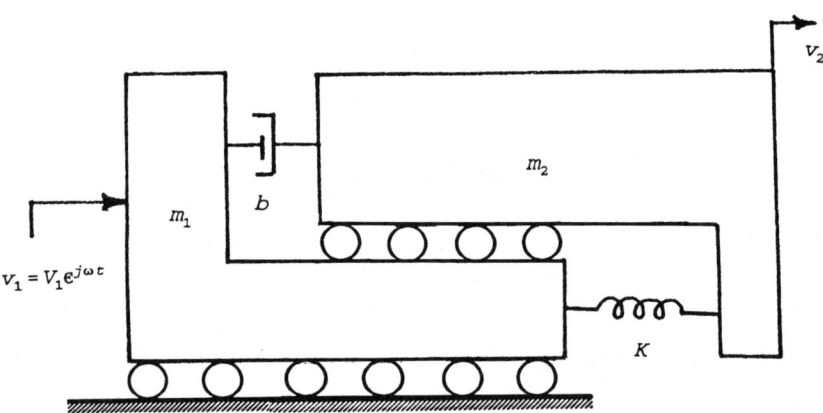

Figure P 8.8

8.9 The input torque to the rotational flywheel-damper system shown in Fig. P8.9 is described by T(t) = T cos ωt. If the angular velocity of the flywheel is considered as the output, find the complete response of the system. Sketch the frequency response.

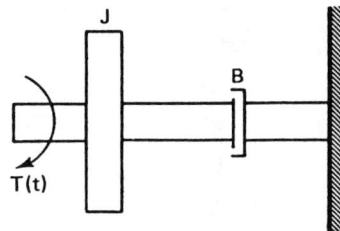

Figure P 8.9

8.10 In the rotational system shown in Fig. P8.10, $J = 4$ kg.m^2, $K = 20$ N.m and $B = 2$ N.m.s. The torque input, $T(t)$, is $10 \sin 5t$, N.m. Find the sinusoidal steady state response for the angular velocity of the flywheel.

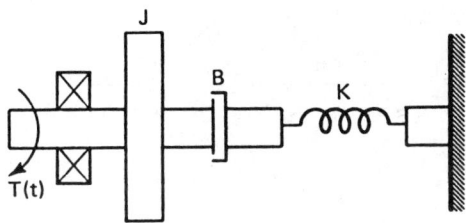

Figure P 8.10

8.11 Fig. P8.11 shows a simplified version of a pneumatic control system component known as a phase-lag compensator. If $R_{f1} = 22(10^6)$ N.s/m^5, $R_{f2} = 3(10^6)$ N.s/m^5 and $C_f = 2(10^{-7})$ m^5/N, obtain the system function $G(j\omega) = P_o(j\omega)/P_i(j\omega)$. Sketch the magnitude and phase of $G(j\omega)$.

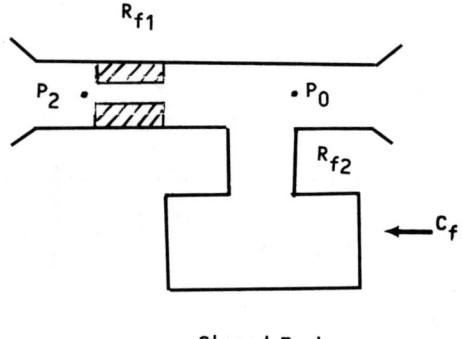

Closed Tank

Figure P 8.11

8.12 Obtain the Bode magnitude and phase plots of the following functions:

(a) $G(s) = \dfrac{s(s + 100)}{(s + 1)(s + 10)(s + 500)}$

(b) $G(s) = \dfrac{(s^2 - 2s + 4)}{(s + 4)(s^2 + 2s + 10)}$

CHAPTER 9

LAPLACE TRANSFORMS

Laplace transform is a powerful tool in the analysis of linear systems. It is not the purpose of this book to discuss all the properties of Laplace transforms and the consequences. Attention will be centred on some basic properties of Laplace transforms and their usefulness in the solution of linear differential equations and in the analysis of circuits.

9.1 DEFINITION OF LAPLACE TRANSFORM

The function F(s), which is the Laplace transform of another function f(t), is defined by

$$F(s) = \mathbb{L}\{f(t)\} = \int_0^\infty f(t)\, \varepsilon^{-st}\, dt, \quad \text{Re}(s) > 0 \quad \quad ..(9.1)$$

where $s = \sigma + j\omega$ is a complex variable.

The independent variable 't' shall not be confused with the time. The Laplace transform can be defined for any function having an independent variable, say 'x'. It should also be noted that the interval of 't' is between 0 and ∞. This implies that f(t) does not exist when t < 0. This is known as **one-sided Laplace transform**. There exists two-sided Laplace transform also but this topic is not considered here.

In order that f(t) has Laplace transform, it must satisfy the following two properties:

(a) f(t) shall be bounded and has a finite number of maxima and minima and a finite number of finite discontinuities in any interval of the type $0 \leq t_1 \leq t \leq t_2$.

(b) f(t) shall be of **exponential order**, that is, there exist real and positive constants α and M such that

$$\varepsilon^{-\alpha t} |f(t)| < M \quad \text{for } t \geq 0 \qquad ..(9.2)$$

Condition (a) can be easily verified. The importance of condition (b) is illustrated by an example. (The actual constants need not be determined. It is required to establish that such constants exist).

Example 9.1:

Consider $\quad f(t) = \varepsilon^{t^n} \qquad ..(9.3a)$

Then $\quad f_1(t) = \varepsilon^{-\alpha t} \varepsilon^{t^n} = \varepsilon^{t(t^{n-1} - \alpha)} \qquad ..(9.3b)$

Laplace Transforms

The constant α is finite. There are two conditions to be considered.

(a) $n > 1$: In this case, there is no way a finite positive constant M can be found that such that $f_1(t)$ is bounded.

(b) $n = 1$: It is obvious that as $t \to \infty$, $f_1(t) \to 0$, with $\alpha > 1$.

This means that ε^{t^n} is a function of exponential order only when $n = 1$.

Example 9.2

It is required to determine $\mathcal{L}\{\varepsilon^{at} \cdot U_s(t)\}$, where $a = a_1 + ja_2$.

By (1), $F(s) = \int_0^\infty \varepsilon^{at} \varepsilon^{-st} dt = \int_0^\infty \varepsilon^{-(s-a)t} dt$

$$= \left[\frac{1}{s-a} \, \varepsilon^{-(s-a)t}\right]_0^\infty = \frac{1}{s-a} \quad \ldots (9.4)$$

provided $\varepsilon^{-(s-a)t} \to 0$ as $t \to \infty$.
This is possible when $\text{Re}(s) > \text{Re}(a)$
or $\text{Re}(s) > a_1$. $\quad \ldots (9.5)$

The quantity a_1 is called the **abscissa of convergence,** because the Laplace transform holds only when the inequality (9.5) holds. This is illustrated in Fig. 9.1 on the s-plane.

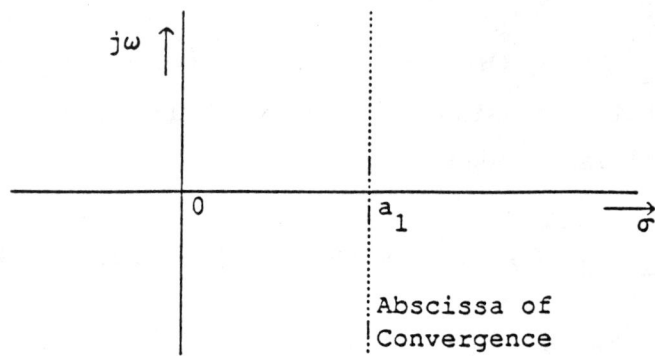

Figure 9.1 Diagram showing the abscissa of convergence of example 9.2

Example 9.3

It can be easily shown that

$$\mathcal{L}\{U_s(t)\} = \frac{1}{s}, \qquad \qquad ..(9.7)$$

In fact, this follows from (9.4) when a = 0. The abscissa of convergence is the imaginary axis, that is, $\sigma = 0$. [The students are encouraged to derive (9.7) starting from (9.1).]

Henceforward, $f(t) \cdot U_s(t)$ will be written as $f(t)$ only [that is, $U_s(t)$ will be omitted], because the Laplace transform is valid only when $t \geq 0$.

9.2 INVERSE LAPLACE TRANSFORM

The inverse Laplace transform is defined by

Laplace Transforms

$$\mathcal{L}^{-1}\{F(s)\} = \frac{1}{2\pi j} \int_{c-j\infty}^{c+j\infty} F(s)\, \varepsilon^{st}\, ds \qquad ..(9.8)$$

This is a contour integral. In many cases, it is not required to perform this integration. We can build up Laplace transform pairs, because the inverse Laplace transform is unique.

Henceforth, the Laplace transform pair is designated as

$$f(t) \longleftrightarrow F(s) \qquad ..(9.9)$$

9.3 PROPERTIES OF LAPLACE TRANSFORMS

Now, we shall discuss some properties of Laplace transforms.

Theorem 9.1

If $\qquad f_1(t) \longleftrightarrow F_1(s) \qquad ..(9.10a)$

and $\qquad f_2(t) \longleftrightarrow F_2(s), \qquad ..(9.10b)$

then $\qquad \mathcal{L}\{c_1 f_1(t) + c_2 f_2(t)\} = c_1 F_1(s) + c_2 F_2(s)$

$$..(9.11)$$

Proof: This theorem is proved starting from the definition (9.1).
We have

$$\mathcal{L}\{c_1 f_1(t) + c_2 f_2(t)\} = \int_0^\infty \{c_1 f_1(t) + c_2 f_2(t)\} \varepsilon^{-st}\, dt$$

$$= \int_0^\infty c_1 f_1(t)\, \varepsilon^{-st}\, dt + \int_0^\infty c_2 f_2(t)\, \varepsilon^{-st}\, dt$$

$$= c_1 F_1(s) + c_2 F_2(s) \qquad \ldots (9.12)$$

The consequence of this theorem is that **Superposition** holds. We can obtain both the Laplace and inverse transforms by the application of this theorem. First, we shall illustrate the use of this theorem in evaluating the Laplace transform of some functions. In these cases, the transform can be obtained starting from the definition and then evaluating the integral. Students are encouraged to verify the results by the direct evaluation of the integral.

Example 9.4

It is required to determine $\mathcal{L}\{\cos kt\}$.

We know that
$$\cos kt = \frac{1}{2}\{\varepsilon^{jkt} + \varepsilon^{-jkt}\} \qquad \ldots (9.13a)$$

Therefore,
$$\mathcal{L}\{\cos kt\} = \frac{1}{2}\{\mathcal{L}[\varepsilon^{jkt}] + \mathcal{L}[\varepsilon^{-jkt}]\}$$

$$= \frac{1}{2}\left\{\frac{1}{s - jk} + \frac{1}{s + jk}\right\}$$

$$= \frac{s}{s^2 + k^2} \qquad \ldots (9.13b)$$

Laplace Transforms

The abscissa of convergence is obviously the imaginary-axis.

Example 9.5

It is required to determine $\mathcal{L}\{\sin kt\}$.

We know that

$$\sin kt = \frac{1}{2j}\left\{\varepsilon^{jkt} - \varepsilon^{-jkt}\right\} \qquad ..(9.14)$$

Therefore, $\mathcal{L}\{\sin kt\} = \dfrac{1}{2j}\left\{\dfrac{1}{s-jk} - \dfrac{1}{s+jk}\right\}$

$$= \frac{k}{s^2 + k^2} \qquad ..(9.15)$$

The abscissa of convergence is the imaginary axis.

Theorem 9.2

If $\qquad f(t) \longleftrightarrow F(s),$

then (a) $\mathcal{L}\{f'(t)\} = sF(s) - f(0) \qquad ..(9.16a)$

and (b) $\mathcal{L}\{f''(t)\} = s^2 F(s) - sf(0) - f'(0) \qquad ..(9.16b)$

where $\qquad f'(t) = \dfrac{df(t)}{dt} \qquad ..(9.17a)$

and $\qquad f''(t) = \dfrac{d^2 f(t)}{dt^2} \qquad ..(9.17b)$

Proof: (a) The relationship (9.16a) is proved starting

from the definition (9.1). This gives

$$\mathcal{L}\{f'(t)\} = \int_0^\infty f'(t)\, \varepsilon^{-st}\, dt \qquad \ldots (9.18)$$

Integrating by parts, (9.18) becomes

$$= \left[f(t)\, \varepsilon^{-st} \right]_0^\infty - \int_0^\infty (-s)\, \varepsilon^{-st} f(t)\, dt$$

$$= sF(s) - f(0)$$

which is (9.17a).

In this theorem, $f(0)$ and $f'(0)$ are the initial conditions of $f(t)$. These are evaluated at $t = 0^+$. But the notation $t = 0^+$ is generally replaced by $t = 0$.

(b) The relationship (9.16b) can be proved similarly. From the definition, we have

$$\mathcal{L}\{f''(t)\} = \int_0^\infty f''(t)\, \varepsilon^{-st}\, dt \qquad \ldots (9.19)$$

Integrating by parts, (9.19) becomes

$$= \left[f'(t)\, \varepsilon^{-st} \right]_0^\infty - \int_0^\infty (-s)\, \varepsilon^{-st} f'(t)\, dt$$

$$= -f'(0) + s \int_0^\infty f'(t)\, \varepsilon^{-st}\, dt \qquad \ldots (9.20)$$

Laplace Transforms

By the substitution of (9.18), (9.20) becomes

$$s^2 F(s) - s f(0) - f'(0)$$

which is (16b).

By proceeding on similar lines, it can be proved that $\mathcal{L}\{f^{(n)}(t)\}$, the Laplace transform of the nth derivative of $f(t)$ is given by

$$\mathcal{L}\{f^{(n)}(t)\} = s^n F(s) - s^{n-1} f(0) - s^{n-2} f'(0) -$$
$$\ldots\ldots -f^{n-1}(0) \qquad \ldots (9.21)$$

This is left as an exercise for the student.

One of the obvious applications of this theorem is in the solution of differential equations, which will be discussed later. Another application is to obtain the Laplace transform of a function which is the derivative of another function whose Laplace transform is known.

Example 9.6

It is required to obtain $\mathcal{L}\{\cos kt\}$ by the application of Theorem 9.2.

It is known that

$$\mathcal{L}\{\sin kt\} = \frac{k}{s^2 + k^2} \qquad \qquad ..(9.22a)$$

Then
$$\mathcal{L}\{\cos kt\} = \mathcal{L}\left\{\frac{1}{k}\frac{d}{dt}(\sin kt)\right\}$$

$$= \frac{1}{k}\left\{s \cdot \frac{k}{s^2 + k^2} - \sin(0)\right\}$$

$$= \frac{s}{s^2 + k^2} \qquad \qquad ..(9.22b)$$

Example 9.7

It is required to determine the Laplace transform of an impulse function.

Since
$$U_i(t) = \frac{d}{dt}\{U_s(t)\},$$

we have
$$\mathcal{L}\{U_i(t)\} = s\left\{\frac{1}{s}\right\} - U_s(0) = 1 \qquad ..(9.23)$$

where $U_s(0)$ is taken to be zero.

In the next theorem, we shall discuss the property of the derivative of F(s) with respect to s, designated as F'(s).

Theorem 9.3

If
$$f(t) \longleftrightarrow F(s),$$

then
$$t \cdot f(t) \longleftrightarrow -F'(s) \qquad \qquad ..(9.24)$$

Laplace Transforms

Proof: Differentiating (9.1) with respect to s, we get

$$F'(s) = \frac{d}{ds} \int_0^\infty f(t)\, \varepsilon^{-st}\, dt$$

$$= \int_0^\infty \frac{d}{ds} \left\{ f(t)\, \varepsilon^{-st}\, dt \right\}$$

$$= \int_0^\infty -t\, f(t)\, \varepsilon^{-st}\, dt$$

$$= -\, \mathcal{L}\left\{ t\, f(t) \right\} \qquad \ldots (9.25)$$

As illustrations of this theorem, we shall obtain the Laplace transform of the Unit Ramp and the Unit Parabolic functions.

Example 9.8

It is required to determine (a) $\mathcal{L}\{U_r(t)\}$ and (b) $\mathcal{L}\{U_p(t)\}$.

(a) It is known that

$$U_r(t) = t \cdot U_s(t), \qquad \ldots (9.26a)$$

where 't' arises from integrating $U_s(t)$.

Therefore, $\mathcal{L}\{U_r(t)\} = -\dfrac{d}{ds}\left[\mathcal{L}\{U_s(t)\}\right]$

$$= -\frac{d}{ds}\left\{\frac{1}{s}\right\} = \frac{1}{s^2} \quad \quad ..(9.26b)$$

(b) It is known that

$$U_p(t) = \frac{t}{2}\left\{U_r(t)\right\} \quad \quad ..(9.27a)$$

Therefore, $\mathcal{L}\left\{U_p(t)\right\} = -\frac{d}{ds}\left\{\frac{1}{2s^2}\right\} = \frac{1}{s^3} \quad ..(9.27b)$

In the next theorem, we shall study the Laplace transform of the function $\varepsilon^{-at}.f(t)$.

Theorem 9.4
If $\quad\quad\quad f(t) \longleftrightarrow F(s),$

then $\quad \varepsilon^{-at}.f(t) \longleftrightarrow F(s+a) \quad\quad ..(9.28)$

Proof: Starting from the definition, we have

$$\mathcal{L}\left\{\varepsilon^{-at}.f(t)\right\} = \int_0^\infty \varepsilon^{-at}.f(t)\,\varepsilon^{-st}\,dt$$

$$= \int_0^\infty f(t)\,\varepsilon^{-(s+a)t}\,dt$$

$$= F(s+a)$$

which is (9.28).

As an illustration of this theorem, we shall find the Laplace transform of a damped sinusoidal function $\varepsilon^{-at}.\sin kt$, which is an underdamped response of a

Laplace Transforms

second order system.

Example 9.10

It is required to determine the Laplace transform of $\varepsilon^{-at}.\sin kt$.

From Theorem 9.4 and Example 9.5, we have

$$\mathcal{L}\left\{\varepsilon^{-at}.\sin kt\right\} = \frac{k}{(s+a)^2 + k^2} \quad ..(9.29)$$

Similarly, it is seen that

$$\mathcal{L}\left\{\varepsilon^{-at}\cos kt\right\} = \frac{(s+a)}{(s+a)^2 + k^2} \quad ..(9.30)$$

It is concluded that if $f(t)$ is multiplied by ε^{-at}, the corresponding Laplace transform will become $F(s+a)$.

In the next theorem, we shall discuss shifting in the t-plane.

Theorem 9.5

If $\quad f(t) \longleftrightarrow F(s)$,

then $\quad \left\{f(t-\tau).U_s(t-\tau)\right\} \longleftrightarrow \varepsilon^{-\tau s}.F(s) \quad ..(9.31)$

Proof: Starting from the definition, we have

$$\mathcal{L}\{f(t-\tau).U_s(t-\tau)\} = \int_0^\infty f(t-\tau).U_s(t-\tau).\varepsilon^{-st} dt$$

$$..(9.32)$$

The lower limit of integration can be changed to τ, since $f(t-\tau).U_s(t-\tau)$ is zero within the interval $0 \le t \le \tau$. As a consequence, (9.32) becomes

$$\int_\tau^\infty f(t-\tau).U_s(t-\tau).\varepsilon^{-st} dt \qquad ..(9.33)$$

We can make the substitution

$$x = t - \tau \qquad ..(9.34)$$

in which case, (9.33) becomes

$$\int_0^\infty f(x)\, \varepsilon^{-s(x+\tau)} dx$$

$$= \varepsilon^{-s\tau}.F(s)$$

which is (9.31).

This theorem will be helpful in determining the Laplace transform of composite functions, which are decomposed using the singularity functions, as discussed in Chapter II.

Example 9.11

It is required to obtain the Laplace transform of

Laplace Transforms

$$g(t) = U_p(t) + 2U_s(t-2) - 2U_r(t-4) \qquad ..(9.35)$$

From Theorem 9.5, it is seen that

$$G(s) = \mathcal{L}\{g(t)\} = \frac{1}{s^3} + \frac{2\varepsilon^{-2s}}{s} - \frac{2\varepsilon^{-4s}}{s^2} \qquad ..(9.36)$$

In the next theorem, we shall discuss the inverse Laplace transform of the product of two functions.

Theorem 9.6

If
$$f_1(t) \longleftrightarrow F_1(s) \qquad ..(9.37a)$$
and
$$f_2(t) \longleftrightarrow F_2(s) \qquad ..(9.37b)$$

then
$$\mathcal{L}^{-1}\{F_1(s).F_2(s)\} = \int_0^t f_1(\tau) f_2(t-\tau) \, d\tau$$

$$= \int_0^t f_1(t-\tau) f_2(\tau) \, d\tau \quad ..(9.38)$$

This theorem will not be proved here. The integral on the right hand side of (9.38) is known as **Convolution** integral and is represented by $\{f_1(t) * f_2(t)\}$.

Example 9.11

It is required to determine

$$\mathcal{L}^{-1}\left\{\frac{1}{(s+a)(s+b)}\right\}, \quad a \neq b \qquad ..(9.39)$$

by Theorem 9.6.

It is known that

$$\mathcal{L}^{-1}\left\{\frac{1}{s+a}\right\} = \varepsilon^{-at} \qquad ..(9.40a)$$

and $\quad \mathcal{L}^{-1}\left\{\frac{1}{s+b}\right\} = \varepsilon^{-bt} \qquad ..(9.40b)$

From (9.38), (9.39) can be evaluated as

$$\int_0^\infty \varepsilon^{-a\tau} \varepsilon^{-b(t-\tau)} d\tau = \varepsilon^{-bt} \int_0^\infty \varepsilon^{(b-a)\tau} d\tau$$

$$= \varepsilon^{-bt} \left[\frac{\varepsilon^{(b-a)\tau}}{b-a}\right]_0^t = \frac{\varepsilon^{-at} - \varepsilon^{-bt}}{(b-a)} \qquad (9.41)$$

An important consequence of this theorem is that the Laplace transform of a product of two functions is **not equal** to the product of the Laplace transforms of the two functions.

Table 9.1 gives the Laplace transform pairs of some commonly used functions.

TABLE OF SELECTED LAPLACE TRANSFORM PAIRS

	$f(t)$	$F(s)$
(9.1)	$U_i(t)$	1
(9.2)	$U_s(t)$	$\dfrac{1}{s}$
(9.3)	$U_r(t)$	$\dfrac{1}{s^2}$
(9.4)	$\varepsilon^{-at} U_s(t)$	$\dfrac{1}{s+a}$
(9.5)	$t\varepsilon^{-at} U_s(t)$	$\dfrac{1}{(s+a)^2}$
(9.6)	$\sin kt\, U_s(t)$	$\dfrac{k}{s^2+k^2}$
(9.7)	$\cos kt\, U_s(t)$	$\dfrac{s}{s^2+k^2}$
(9.8)	$f(t-a)$	$\varepsilon^{-as} F(s)$
(9.9)	$\varepsilon^{-at} \sin kt\, U_s(t)$	$\dfrac{k}{(s+a)^2+k^2}$
(9.10)	$\varepsilon^{-at} \cos kt\, U_s(t)$	$\dfrac{(s+a)}{(s+a)^2+k^2}$

We shall now discuss two more theorems which are known as **Initial Value** and **Final Value Theorems**.

Theorem 9.7 (Initial Value Theorem)

If $\quad f(t) \longleftrightarrow F(s),$

then $\quad \lim_{s \to \infty} sF(s) = \lim_{t \to 0} f(t) = f(0^+) \quad ..(9.42)$

Proof

From Theorem 9.2, it is known that

$$\mathcal{L}\{f'(t)\} = s F(s) - f(0^+) \quad ..(9.43)$$

Therefore,

$$\lim_{s \to \infty} \mathcal{L}\{f'(t)\} = \lim_{s \to \infty} sF(s) - f(0^+) \quad ..(9.44)$$

Since $f(t)$ is a function of exponential order, it can be shown that

$$\lim_{s \to \infty} \mathcal{L}\{f'(t)\} = 0 \quad ..(9.45)$$

Hence, it follows

$$\lim_{s \to \infty} \mathcal{L}\{f'(t)\} = f(0^+) \quad ..(9.46)$$

This theorem can be effectively used to obtain the initial condition of the function $f(t)$ starting from $F(s)$ without performing the inverse Laplace transform.

Laplace Transforms

Theorem 9.8 (Final Value Theorem)

If $f(t) \longleftrightarrow F(s)$,

then
$$\lim_{s \to 0} sF(s) = \lim_{s \to \infty} f(t) \qquad \ldots (9.47)$$

Proof

Starting from (9.43), we can write

$$\lim_{s \to 0} sF(s) = \lim_{s \to 0} \mathcal{L}\{f'(t)\} + f(0^+) \qquad \ldots (9.48)$$

By definition,

$$\lim_{s \to 0} \mathcal{L}\{f'(t)\} = \lim_{s \to 0} \int_0^\infty \varepsilon^{-st} f'(t) \, dt$$

$$= \int_0^\infty \lim_{s \to \infty} \varepsilon^{-st} f'(t) \, dt$$

$$= \Big[f(t)\Big]_0^\infty$$

$$= \lim_{t \to \infty} f(t) - f(0^+) \qquad \ldots (9.49)$$

Combination of (9.48) and (9.49) proves the theorem.

This theorem can be effectively used to determine the final condition of the function $f(t)$ starting from $F(s)$ without performing the inverse Laplace transform.

9.4 SOLUTIONS OF DIFFERENTIAL EQUATIONS

One of the applications of Laplace transform is in the solution of a linear differential equation. We shall discuss this by considering a second order differential equation (SDE). Let the SDE be

$$a_2 y'' + a_1 y' + a_0 y = g(t) \qquad \qquad ..(9.50)$$

where a_2, a_1 and a_0 are constants,

$$y'' = \frac{d^2 y(t)}{dt^2}, \qquad \qquad ..(9.51a)$$

and $\qquad y' = \frac{dy}{dt} \qquad \qquad ..(9.51b)$

Along with the SDE, the initial conditions $y(0)$ and $y'(0)$ are known.

The steps to be followed are:
(i)   Apply the Laplace transform to both sides of the differential equation.
(ii) Rearrange such that $Y(s)$ is obtained as a ratio of two polynomials.
(iii) Expand the right hand side into partial fractions. (This will be explained when examples are solved).
(iv) By means of Table I and the various theorems, obtain $y(t)$.

This will be illustrated by some examples.

Example 9.12

Consider

Laplace Transforms

$$y'' + 12y' + 35y = \varepsilon^{-\alpha t} \qquad ..(9.52a)$$

with $\qquad y(0) = 1$ and $y'(0) = -2 \qquad ..(9.52b)$

Taking the Laplace transform of (9.52a) on both sides, we have:

$$\{s^2 Y(s) - s\, y(0) - y'(0)\} + 12\{s\, Y(s) - y(0)\}$$
$$+ 35\, Y(s) = \frac{1}{s + \alpha} \qquad ..(9.53)$$

Rearranging we have,

$$Y(s) = \frac{(s + 10)(s + \alpha) + 1}{(s + \alpha)(s + 5)(s + 7)} \qquad ..(9.54)$$

Depending on the value of α, two different cases arise.

<u>Case (a)</u>: Let $\alpha = 1$. In this case, (9.54) becomes

$$Y(s) = \frac{s^2 + 11s + 11}{(s + 1)(s + 5)(s + 7)} \qquad ..(9.55)$$

This can be expanded into partial fractions and rewritten as

$$\frac{s^2 + 11s + 11}{(s + 1)(s + 5)(s + 7)} \equiv \frac{A_1}{s + 1} + \frac{A_2}{s + 5} + \frac{A_3}{s + 7} \qquad ..(9.56)$$

By multiplying both sides of (9.56) by the denominator polynomial, we get

$$s^2 + 11s + 11 \equiv A_1(s+5)(s+7) + A_2(s+1)(s+7)$$
$$+ A_3(s+1)(s+5) \qquad ..(9.57)$$

This is an identity and is valid for all values of s. This permits us to substitute certain values of s to obtain the constants A_1, A_2 and A_3.

Put s = -1. We have $1 = A_1(4)(6)$

$$\text{or } A_1 = \frac{1}{24} \qquad ..(9.58a)$$

Put s = -5. We have $-19 = A_2(-4)(2)$

$$\text{or } A_2 = \frac{19}{8} \qquad ..(9.58b)$$

Put s = -7. We have $-17 = A_3(-6)(-2)$

$$\text{or } A_3 = \frac{-17}{12} \qquad ..(9.58c)$$

Therefore,

$$Y(s) = \frac{\frac{1}{24}}{s+1} + \frac{\frac{19}{8}}{s+5} + \frac{\frac{-17}{12}}{s+7} \qquad ..(9.59)$$

from which we get

$$y(t) = \frac{1}{24}\varepsilon^{-t} + \frac{19}{8}\varepsilon^{-5t} - \frac{17}{12}\varepsilon^{-7t} \qquad ..(9.60)$$

Laplace Transforms

In general, the form of the partial fraction expansion when the denominator polynomial contains only simple roots will be

$$\sum_{i=1} \frac{A_i}{s + a_i} \qquad \ldots(9.61)$$

and the various A_i's are evaluated as illustrated above.

Case (b): Let $\alpha = 5$. (α can be equal to 7. In such a case or when the denominator contains roots of multiplicity greater than unity, the following procedure applies).

The function $Y(s)$ becomes

$$Y(s) = \frac{s^2 + 15s + 51}{(s+5)^2(s+7)} \qquad \ldots(9.62)$$

The partial fraction expansion of RHS (right hand side) of (9.62) shall be written as

$$\frac{s^2 + 15s + 51}{(s+5)^2(s+7)} = \frac{B_1}{s+5} + \frac{B_2}{(s+5)^2} + \frac{B_3}{s+7} \qquad \ldots(9.63)$$

(The students shall note that this is the only form possible).

By multiplying both sides of (9.63) by the denominator polynomial, we get

$$s^2 + 15s + 51 \equiv B_1(s+5)(s+7)$$
$$+ B_2(s+7) + B_3(s+5)^2 \quad ..(9.64)$$

The constant B_3 is evaluated by the substitution of $s = -7$. This gives

$$B_3 = \frac{-5}{4} \quad ..(9.65a)$$

The constant B_2 is evaluated by the substitution of $s = -5$. This gives

$$B_2 = \frac{1}{2} \quad ..(9.65b)$$

The constant B_1 cannot be evaluated by the substitution of either $s = -5$ or $s = -7$. One way to obtain B_1 is to substitute $s = 0$ in (9.64). This gives

$$51 = 35B_1 + 7B_2 + 25B_3 \quad \text{or} \quad B_1 = \frac{9}{4} \quad ..(9.65c)$$

An alternative way (in fact, the general way) of evaluating B_1 is to differentiate (9.64) once with respect to s and then substituting $s = -5$. Accordingly, differentiation of (9.64) once gives

$$2s + 15 = B_1(2s + 12) + B_2 + B_3 \cdot 2(s+5) \quad ..(9.66a)$$

Substitution of $s = -5$ gives

$$5 = 2B_1 + B_2 \quad \text{or} \quad B_1 = \frac{9}{4} \quad ..(9.66b)$$

Laplace Transforms

Therefore

$$Y(s) = \frac{\frac{9}{4}}{s+5} + \frac{\frac{1}{2}}{(s+5)^2} + \frac{-\frac{5}{4}}{s+7} \qquad ..(9.67)$$

from which $y(t)$ is obtained as

$$y(t) = \frac{9}{4}\varepsilon^{-5t} + \frac{1}{2}t\,\varepsilon^{-5t} - \frac{5}{4}\varepsilon^{-7t} \qquad ..(9.68)$$

In general, the partial fraction expansion of a function whose denominator contains multiple roots is given by

$$\frac{N(s)}{(s+\beta)^q D_1(s)} = \frac{A_1}{(s+\beta)} + \frac{A_2}{(s+\beta)^2} + \ldots$$

$$\ldots + \frac{A_q}{(s+\beta)^q} + \frac{N_1(s)}{D_1(s)} \qquad ..(9.69)$$

By multiplying both sides of (9.69) by the denominator polynomial $(s+\beta)^q D_1(s)$, we have

$$\left\{ A_1(s+\beta)^{q-1} + A_2(s+\beta)^{q-2} + \ldots \right.$$
$$\left. + A_{q-1}(s+\beta) + A_q \right\} D_1(s) + N_1(s)(s+\beta)^q = N(s)$$
$$..(9.70)$$

Substitute $s = -\beta$; A_q is obtained.

Differentiate (9.70) once and substitute $s = -\beta$.

The constant A_{q-1} is obtained. The procedure is continued till the various constants are evaluated.

Example 9.13

Consider

$$y'' + 6y' + 45y = \varepsilon^{-2t} \quad \quad ..(9.71)$$

with $y(0) = 1$ and $y'(0) = 0$.

Taking Laplace transforms on both sides, we have

$$\{s^2 Y(s) - s y(0) - y'(0)\} + 6\{s Y(s) - y(0)\}$$
$$+ 45 Y(s) = \frac{1}{s+2} \quad ..(9.72)$$

Rearranging, we have

$$Y(s) = \frac{s^2 + 8s + 13}{(s+2)(s^2 + 6s + 45)} \quad ..(9.73)$$

The RHS can be expanded into partial fractions as

$$\frac{s^2 + 8s + 13}{(s+2)(s^2 + 6s + 45)} = \frac{A_1}{s+2} + \frac{A_2}{s+3+j6} + \frac{A_2^*}{s+3-j6} \quad ..(9.74)$$

where A_2^* is the complex conjugate of A_2.

Laplace Transforms

It is to be noted that evaluation of A_2 and A_2^* will be cumbersome. It is preferable to combine the last two terms on the right hand side of (9.74) which can be rewritten as

$$\frac{s^2 + 8s + 13}{(s+2)(s^2 + 6s + 45)} = \frac{A_1}{s+2} + \frac{A_4 s + A_5}{(s+3)^2 + 6^2} \quad ..(9.75)$$

Multiplying both sides of (9.75) by the denominator, we have

$$A_1 \left\{ (s+3)^2 + 6^2 \right\} + (A_4 s + A_5)(s+2) = (s^2 + 8s + 13) \quad ..(9.76)$$

Substitution of s = −2 yields

$$A_1 = \frac{1}{37} \quad ..(9.77)$$

By substituting s = 0 in (9.76), we have

$$45 A_1 + 2 A_5 = 13$$

or

$$A_5 = \frac{218}{37} \quad ..(9.78)$$

The constant A_4 can be evaluated by substituting any other value of s. By substituting s = −1 in (9.76), we have

$$40 A_1 - A_4 + A_5 = 6$$

or
$$A_4 = \frac{36}{37} \qquad ..(9.79)$$

As an alternative, we can equate the corresponding coefficients in (9.76) and obtain

$$A_1 + A_4 = 1 \qquad ..(9.80)$$

where $(A_1 + A_4)$ is the coefficient of s^2. This shows that by equating the corresponding coefficients on both sides of (9.80), a set of simultaneous equations will be obtained. This set can be solved to obtain A_1, A_4 and A_5. This method will not be discussed further.

Therefore,

$$Y(s) = \frac{\frac{1}{37}}{(s+2)} + \frac{\frac{36}{37}s + \frac{218}{37}}{(s+3)^2 + 6^2} \qquad ..(9.81)$$

It is known that

$$\mathcal{L}^{-1}\left\{\frac{\frac{1}{37}}{s+2}\right\} = \frac{1}{37}\varepsilon^{-2t} \qquad ..(9.82)$$

In order to obtain the inverse Laplace transform of the second term, it has to be rearranged in order to fit into one or more of the pairs in Table 9.1.

Laplace Transforms

Accordingly, in order to match trnsform pairs (9.9) and (9.10), we have

$$\frac{\frac{36}{37} s + \frac{218}{37}}{(s + 3)^2 + 6^2} = \frac{\frac{36}{37}(s + 3)}{(s + 3)^2 + 6^2} + \frac{\frac{110}{37} \cdot \frac{1}{6} \cdot 6}{(s + 3)^2 + 6^2} \quad ..(9.83)$$

We have

$$\mathcal{L}^{-1}\left\{\frac{\frac{36}{37}(s + 3)}{(s + 3)^2 + 6^2}\right\} = \frac{36}{37} \varepsilon^{-3t} \cos 6t \quad ..(9.84a)$$

and $\quad \mathcal{L}^{-1}\left\{\dfrac{\frac{55}{111} \cdot 6}{(s + 3)^2 + 6^2}\right\} = \dfrac{55}{111} \varepsilon^{-3t} \sin 6t \quad ..(9.84b)$

Hence, $y(t) = \dfrac{1}{37} \varepsilon^{-2t} + \dfrac{36}{37} \varepsilon^{-3t} \cos 6t + \dfrac{55}{111} \varepsilon^{-3t} \sin 6t$

$$..(9.85)$$

The advantages of using Laplace transforms for obtaining solutions of linear differential equations are obvious and they are:

(i) The given differential equation is transformed into polynomials and this will permit one to obtain the function to be solved in the form of a rational function (a ratio of two polynomials) in a large number of cases.

(ii) The inverse Laplace transform can be obtained by algebraic manipulations in such cases.

(iii) There is no need to evaluate the homogeneous solution (complementary function) and the particular solution (integral) separately, which are then used to evaluate the various constants in conjunction with the initial conditions. All these are combined together in order to obtain the final solution.

9.5 APPLICATIONS IN THE ANALYSIS OF NETWORKS

The concept of Laplace transforms can be conveniently used in the analysis of networks. There are two methods: (a) In this method, the differential equation governing the network is first established. This is then solved using the techniques discussed earlier. (b) The network elements are modelled in the transform domain first. The various theorems are then used to solve for the required variable. This will be discussed in Section 9.6.

The first method will be discussed now and will be illustrated by examples. (These examples are taken from the earlier chapters of the book and appropriate references have been be quoted).

Example 9.14

Consider the network shown in Fig. 5.14 The

Laplace Transforms

governing equation is

$$v' + 0.2\, v = 0.1\, f_1(t) \qquad \ldots (9.86a)$$

where
$$v' = \frac{dv}{dt}, \qquad \ldots (9.86b)$$

$$f_1(t) = U_r(t) - U_r(t-2) \qquad \ldots (9.86c)$$

and
$$v(0) = 0 \qquad \ldots (9.86d)$$

Let
$$v(t) \longleftrightarrow V(s). \qquad \ldots (9.87)$$

Applying Laplace transforms to both sides of (9.86a) and using appropriate formulas in Table 9.1, we have

$$V(s) = \frac{0.1\,(1 - \varepsilon^{-2s})}{s^2\,(s + 0.2)} \qquad \ldots (9.88)$$

The factor $(1 - \varepsilon^{-2s})$ can be taken into account later. We can write the partial fraction expansion for the remaining portion as

$$\frac{0.1}{s^2\,(s + 0.2)} = \frac{A_1}{s^2} + \frac{A_2}{s} + \frac{A_3}{s + 0.2} \qquad \ldots (9.89)$$

Multiplying both sides by $s^2(s + 0.2)$, we get

$$0.1 = A_1(s + 0.2) + A_2 s(s + 0.2) + A_3 s^2 \qquad \ldots (9.90)$$

Put s = 0.   We get $A_1 = 0.5$ ..(9.91a)

Put s = -0.2.   We get $A_3 = 2.5$ ..(9.91b)

Equating the coefficient of s^2 on both sides, we have

$$0 = A_2 + A_3 \quad \text{or} \quad A_2 = -2.5 \quad ..(9.91c)$$

Hence, V(s) can be written as

$$V(s) = \left\{\frac{0.5}{s^2} - \frac{2.5}{s} + \frac{2.5}{s+0.2}\right\} - \varepsilon^{-2s}\left\{\frac{0.5}{s^2} - \frac{2.5}{s} + \frac{2.5}{s+0.2}\right\}$$
..(9.92)

The inverse transform of V(s) gives

$$v(t) = \left\{0.5t - 2.5 + 2.5\,\varepsilon^{-0.2t}\right\} U_s(t)$$

$$- \left\{0.5(t-2) - 2.5 + 2.5\,\varepsilon^{-0.2(t-2)}\right\} U_s(t-2)$$
..(9.93)

which is the same as the solution given earlier.

Example 9.15

Consider the network shown in Fig. 6.5(b). The differential equation is known to be

$$v'' + 3v' + 2v = 10\, U_i(t) \qquad ..(9.94)$$

with $\qquad v(0) = 0$ and $v'(0) = 0 \qquad ..(9.95)$

Let $\qquad v(t) \longleftrightarrow V(s).$

Applying the Laplace transform to both sides of (9.94) and rearranging, we get

$$V(s) = \frac{10}{s^2 + 3s + 2} = \frac{10}{s+1} - \frac{10}{s+2} \qquad ..(9.96)$$

The inverse Laplace transform yields

$$v(t) = 10\, (\varepsilon^{-t} - \varepsilon^{-2t})\, U_s(t) \qquad ..(9.97)$$

which is the same as the solution given earlier.

9.6 MODELS OF BASIC COMPONENTS IN THE TRANSFORM DOMAIN AND SOME OF THEIR USES

In this section, we shall develop models of the basic components in the Laplace transform domain. The mathematical relationships of the three basic elements are given by (3.1) and (7.37), and they will be used in developing these models.

(a) **Dissipative Elements**

The basic equation governing this type of element is given by

$$av(t) = K_1 \, tv(t) \qquad ..(9.98)$$

When (9.98) is subjected to Laplace transforms, we get

$$AV(s) = K_1 \, TV(s) \qquad ..(9.99)$$

Taking a resistor as an example, (9.99) can be represented by Fig. 9.2.

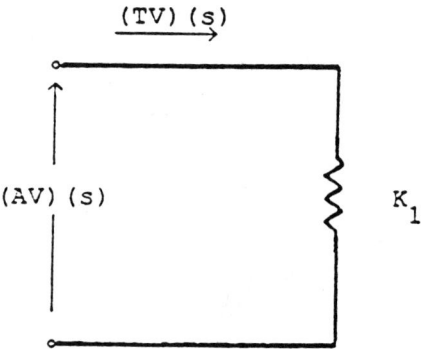

Figure 9.2 The resistor in the transformed domain

(b) **T-type Storage Elements**

The basic equation governing this type of element is given by

$$av(t) = K_2 \frac{d}{dt}\left\{tv(t)\right\} \qquad ..(9.100)$$

When (9.100) is subjected to Laplace transforms, we get

$$AV(s) = K_2 \left\{ s\ TV(s) - tv(0) \right\} \qquad ..(9.101)$$

Eq. (9.101) can be rearranged as

$$TV(s) = \frac{AV(s)}{sK} + \frac{tv(0)}{s} \qquad ..(9.102)$$

Taking an inductor as an example (in which case K_2 = L) and using KCL, (9.102) can be represented by the circuit shown in Fig. 9.3. The model is a parallel connection of a current generator tv(0)/s and an impedance sL.

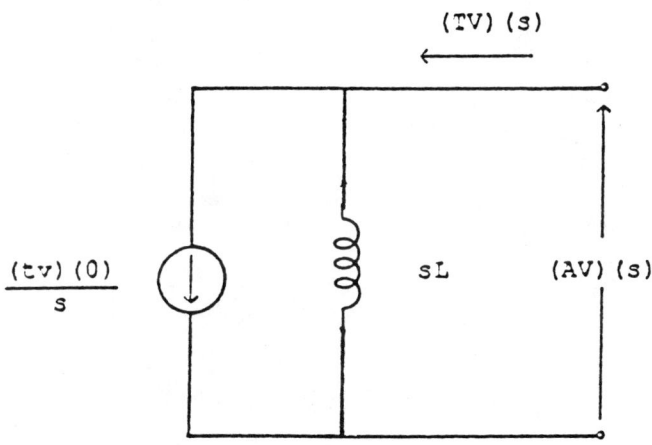

Figure 9.3 A network model of an inductor in the transformed domain

The circuit of Fig.9.3 can be compared with the

circuit model of an inductor given in Fig. 4.23. It is observed that the current generator in Fig. 9.3 is the Laplace transform of $tv(0).U_s(t)$ which is in parallel with an inductor which has its generalized impedance sL. (Note that this is not the actual inductor).

Another model is possible by the interpretation of (9.101) or by rewriting (9.100). This is not considered here.

(c) **A-Type Storage Elements**

The basic equation governing this type of element can be written as

$$tv(t) = \frac{1}{K_3} \frac{d}{dt}\{av(t)\} \qquad ..(9.103)$$

which is the inverse relation of (4.8).

When (9.103) is subjected to Laplace transforms, we get

$$TV(s) = \frac{1}{K_3}\{s\ AV(s) - av(0)\} \qquad ..(9.104)$$

Eq. (9.104) can be rearranged as

$$AV(s) = \frac{TV(s)}{sK_3} + \frac{av(0)}{s} \qquad ..(9.105)$$

Taking a capacitor as an example (in which case $\frac{1}{K_3}$ = C), and using KVL, (9.105) can be represented by the circuit shown in Fig. 9.4, which is a series connection of a voltage generator av(0)/s and an impedance (1/sC).

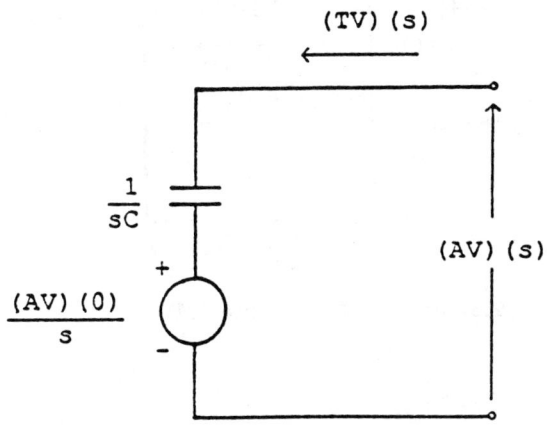

Figure 9.4 A network model of a capacitor in the transformed domain

The circuit of Fig. 9.4 can be compared with the circuit model of a capacitor given in Fig. 4.21. It is observed that the voltage generator in Fig. 9.4 is the Laplace transform of $(av)(0) \cdot U_s(t)$ and the capacitor C is replaced by its generalized impedance (1/sC) (Table 7.1).

In this case also, another model is possible by the proper interpretation of (9.104) or by rewriting (9.103).

Example 9.15

It is required to determine $v_C(t)$ and $i_L(t)$ in the network shown in Fig. 9.5, using Laplace transforms. The switch S is closed at t = 0.

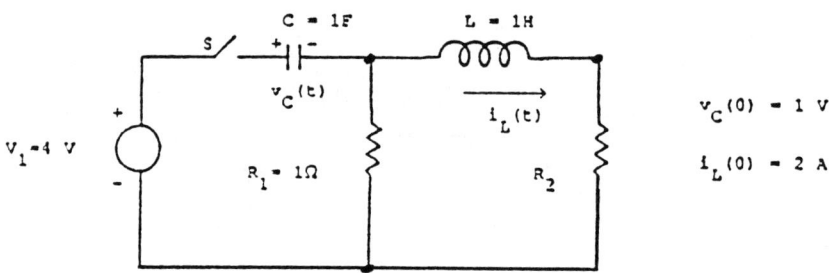

Figure 9.5 Network for example 9.15

In the Laplace transform domain, the network of Fig. 9.5 is transformed to that shown in Fig. 9.6. This network can be solved by any one of the known methods. We shall solve it by using mesh analysis. The mesh equations are

$$I_3(s) = \frac{i_L(0)}{s}$$

$$(R_1 + \frac{1}{sC}) I_1(s) - R_1 I_2(s) = \frac{V_1 - v_C(0)}{s} \qquad ..(9.106)$$

$$-R_1 I_1(s) + (R_1 + R_2 + sL) I_2(s) = L\, i_L(0)$$

Laplace Transforms

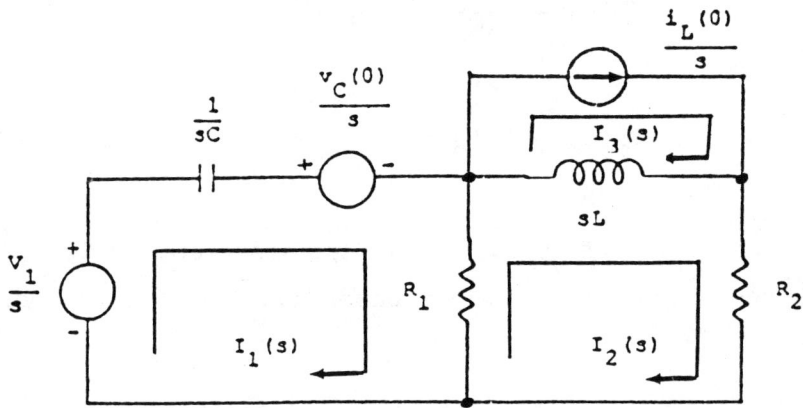

Figure 9.6 The network of figure 9.5 in the Laplace transform domain

The set of equations can be solved to get

$$I_1(s) = \frac{sL\{V_1 - v_c(0) + R_1 i_L(0)\} + (R_1 + R_2)\{V_1 - v_c(0)\}}{s^2 R_1 L + s(R_1 R_2 + \frac{L}{C}) + (R_1 + R_2)/C}$$

.. (9.107a)

$$I_2(s) = \frac{sR_1 L i_L(0) + \left\{\frac{L i_L(0)}{C} + R_1 (V_1 - v_c(0))\right\}}{s^2 R_1 L + s(R_1 R_2 + \frac{L}{C}) + (R_1 + R_2)/C}$$

.. (9.107b)

It is also known that

$$V_c(s) = \frac{I_1(s)}{sC} + \frac{v_c(0)}{s} \qquad ..(9.107c)$$

and $\qquad I_2(s) = I_L(s) \qquad ..(9.107d)$

Now, we can substitute the numbers and obtain the required quantities. Three cases will be considered.

<u>Case (i)</u>: Let $R_2 = 2\ \Omega$. From (9.107b), we get

$$I_2(s) = \frac{2s+5}{s^2+3s+3}$$

$$= \frac{2(s+1.5)}{(s+1.5)^2+(0.866)^2} + \frac{2(\frac{1}{0.866})(0.866)}{(s+1.5)^2+(0.866)^2}$$

$$..(9.108a)$$

Hence, $i_L(t) =$

$$2\ \varepsilon^{-1.5t} \cos(0.866t) + 2.3095\ \varepsilon^{-1.5t} \sin(0.866t)$$
$$..(9.108b)$$

Also, from (9.107a), we get

$$I_1(s) = \frac{5s+9}{s^2+3s+3} \qquad ..(9.109a)$$

from which we get

$$V_c(s) = \frac{I_1(s)}{sC} + \frac{1}{s} = \frac{5s + 9}{s(s^2 + 3s + 3)} + \frac{1}{s}$$

$$= \frac{4}{s} + \frac{-3s - 4}{s^2 + 3s + 3} = \frac{4}{s} + \frac{-3(s + 1.5) + 0.5}{(s + 1.5)^2 + (0.866)^2}$$

Hence, $v_c(t) = 4 - 3\varepsilon^{-1.5t} \cos(0.866t)$
$$+ 0.577\varepsilon^{-1.5t} \sin(0.866t) \quad ..(9.109c)$$

Case (ii): Let $R_2 = 3\,\Omega$. From (9.107b), we get

$$I_2(s) = \frac{2s + 5}{(s + 2)^2} = \frac{2}{s + 2} + \frac{1}{(s + 2)^2} \quad ..(9.110a)$$

Hence, $\quad i_L(t) = 2\varepsilon^{-2t} + t\varepsilon^{-2t} \quad ..(9.110b)$

From (9.107a) and (9.107c), we get

$$V_c(s) = \frac{5s + 12}{(s + 2)^2} + \frac{1}{s}$$

$$= \frac{4}{s} + \frac{-3}{s + 2} + \frac{-1}{s(s + 2)^2} \quad ..(9.111a)$$

Hence, $\quad v_c(t) = 4 - 3\varepsilon^{-2t} - t\varepsilon^{-2t} \quad ..(9.111b)$

Case (iii): Let $R_2 = 6\,\Omega$. From (9.107b), we get

$$I_2(s) = \frac{2s + 5}{s^2 + 7s + 7}$$

$$= \frac{0.5636}{s + 1.2087} + \frac{1.4364}{s + 5.7913} \quad ..(9.112a)$$

Hence, $i_L(t) = 0.5636 \, \varepsilon^{-1.2087t} + 1.4363 \, \varepsilon^{-5.7913t}$

$$..(9.112b)$$

From (9.107a) and (9.107c), we get

$$V_c(s) = \frac{5s + 21}{s(s^2 + 7s + 7)} + \frac{1}{s}$$

$$= \frac{4}{s} + \frac{-2.7}{s + 1.2087} + \frac{-0.3}{s + 5.7913} \quad ..(9.113a)$$

Hence, $v_c(t) = 4 - 2.7 \, \varepsilon^{-1.2087t} - 0.3 \, \varepsilon^{-5.7913t}$

$$..(9.113b)$$

9.7 SUMMARY AND DISCUSSIONS

In this chapter, we have discussed some of the properties of Laplace transforms. Their uses in the solution of linear differential equations and in the solution of networks have been illustrated. Particular mention has to be made of the fact that the generalized impedances of the basic network elements and the impedances in the transform domain are identical. Hence the other properties discussed in Chapters VII and VIII are valid here also.

9.8 PROBLEMS

9.1 Solve all the problems of Chapter V using Laplace transforms only.

9.2 Solve all the problems of Chapter VI using Laplace transforms only.

9.3 An electrical circuit is shown in Fig. P9.3.
   (a) Switches SW1 and SW2 are closed at t = 0. The circuit was not previously energized. Find $i_R(0^+)$ and $\dfrac{di_R}{dt}(0^+)$.

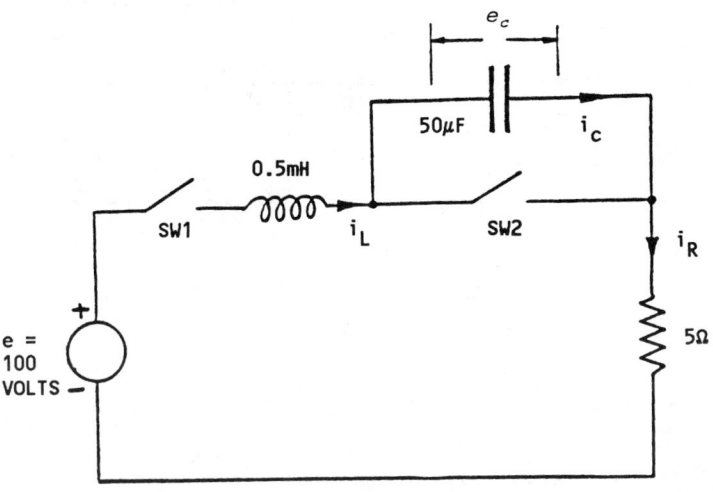

Figure P 9.3

(b) Switch SW2 is opened at t = 1 second. Find

(i) $e_c(t = 1^+ \text{ sec})$ and $\dfrac{de_c}{dt}(t = 1^+ \text{ sec})$,

(ii) $i_L(t = 1^- \text{ sec})$ and $i_L(t = 1^+ \text{ sec})$

(c) Write the differential equation for

(i) $0 \le t \le 1$ sec, (ii) $1 \le t \le \infty$ sec.

Solve them using Laplace transforms.

9.4 The mechanical system shown in Fig. P9.4(a) is initially at rest with the spring neither compressed

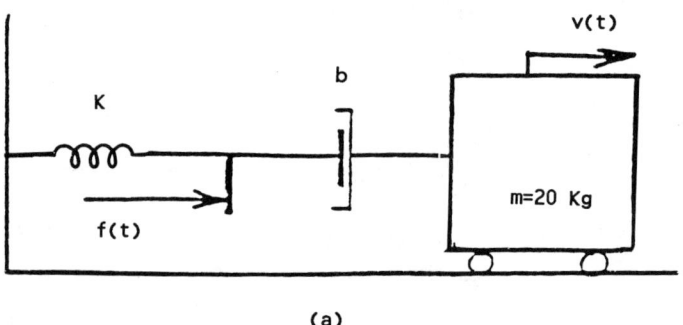

(a)

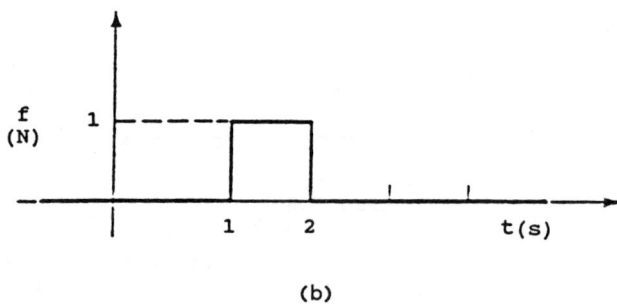

(b)

Figure P 9.4

nor extended. The mass is 20 kg, the spring stiffness is 30 N/m, and the damping is 5 N.s/m. The system is driven by applying a force f(t) as shown in Fig. P9.4(b).

(a) Draw the equivalent electrical circuit.
(b) Obtain the differential equation, if the velocity of the mass is the output.
(c) Obtain an expression for the resulting transfer function in time domain, using Laplace transforms.

CHAPTER 10

TWO-PORT NETWORKS AND MUTUALLY COUPLED COMPONENTS

Up to this point, we have dealt exclusively with two-terminal components. These components were described by a relation between one through variable and one across variable. As a result, each two terminal component was represented completely by a single impedance or admittance.

We now consider a class of components with two sets of signal variables. These components, which are generally referred to as two-terminal pairs, have two through variables and two across variables. Since there is a coupling between the variables of each pair and those of the other pair, it requires more than one impedance or one admittance to describe two-terminal pair networks.

10.1 TWO-TERMINAL PAIRS

A block diagram representation of a two-terminal pair network is shown in Fig. 10.1. We designate the left side as terminal-pair 1 and the right side as terminal-pair 2. The network (within the block) may consist of combinations of conventional two-terminal components or may be a mutually coupled component that we will describe in Sections 10.2 and 10.3. This network does not contain any independent sources. However, whatever the implementation of the two-terminal pair, its mathematical description has the same form.

Figure 10.1 General two-terminal-pair network

10.1.1 Open-Circuit Impedance Parameters

To obtain the standard form for two-terminal pair networks, we must recognize that an input (across or through variable type) at one terminal of the pair produces a response at that pair and also at the other pair. Suppose, for example, that the two-terminal pair is simulated by through variable sources TV_1 and TV_2 [Fig. 10.2(a)]. Then, by using superposition, the responses of the corresponding across variables (AV_1 and AV_2) are:

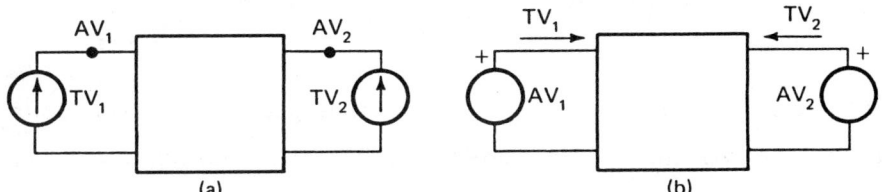

Figure 10.2 Two-terminal-pair networks

$$AV_1 = z_{11} TV_1 + z_{12} TV_2 \qquad (10.1a)$$

$$AV_2 = z_{21} TV_1 + z_{22} TV_2 \qquad (10.1b)$$

where z_{11} is the driving point impedance at terminal 1 under the condition that terminal 2 is open-circuited ($TV_2 = 0$). Similarly, z_{22} is the driving point impedance at terminal 2 with terminal 1 open-circuited ($TV_1 = 0$). The impedance z_{12} represents the effect that source TV_2 on AV_1 when $TV_1 = 0$. Correspondingly, z_{21} relates TV_1 and AV_2 when $TV_2 = 0$. In mathematical terms, these impedances may be written as:

$$z_{11} = \left.\frac{AV_1}{TV_1}\right|_{TV_2 = 0} \qquad z_{12} = \left.\frac{AV_1}{TV_2}\right|_{TV_1 = 0}$$

$$z_{21} = \left.\frac{AV_2}{TV_1}\right|_{TV_2 = 0} \qquad z_{22} = \left.\frac{AV_2}{TV_2}\right|_{TV_1 = 0}$$

..(10.2)

These are known as z-parameters or Open-Circuit

Impedance parameters, because they are defined under open-circuit conditions of the terminal pairs.

Eqs. (10.1) and (10.2) indicate that we need four impedances to describe two-terminal pair networks. It should also be emphasized that these impedances are defined under specific terminal conditions and hence should be used with great care. Also, if reciprocity holds, from the reciprocity theorem for linear networks, the effect that TV_1 has on AV_2 is the same as the effect that TV_2 has on AV_1. This means that $z_{12} = z_{21}$ and that we need only three impedances to describe the two-terminal pair network.

It is usually convenient to place the governing relationships for two-terminal pair networks in matrix form. Thus, (10.1) becomes:

$$\begin{bmatrix} AV_1 \\ AV_2 \end{bmatrix} = \begin{bmatrix} z_{11} & z_{12} \\ z_{21} & z_{22} \end{bmatrix} \begin{bmatrix} TV_1 \\ TV_2 \end{bmatrix} \qquad ..(10.3)$$

and can also be represented by

$$[AV] = [z][TV] \qquad ..(10.4)$$

where
$$[AV] = \begin{bmatrix} AV_1 \\ AV_2 \end{bmatrix} \quad \quad ..(10.5a)$$

$$[z] = \begin{bmatrix} z_{11} & z_{12} \\ z_{21} & z_{22} \end{bmatrix} \quad \quad ..(10.5b)$$

$$[TV] = \begin{bmatrix} TV_1 \\ TV_2 \end{bmatrix} \quad \quad ..(10.5c)$$

Eq.(10.3) is known as the impedance matrix representation of the two-terminal pair. There are other matrix forms that can also be used to describe the performance of the two-terminal pair.

10.1.2 Short-Circuit Admittance Parameters

Consider the two-terminal pair network with across variable sources AV_1 and AV_2 [Fig.10.2(b)]. The response of the corresponding through variables may be placed into the admittance matrix form:

$$\begin{bmatrix} TV_1 \\ TV_2 \end{bmatrix} = \begin{bmatrix} y_{11} & -y_{12} \\ -y_{21} & y_{22} \end{bmatrix} \begin{bmatrix} AV_1 \\ AV_2 \end{bmatrix} \quad \quad ..(10.6)$$

where

$$y_{11} = \left.\frac{TV_1}{AV_1}\right|_{AV_2 = 0} \qquad -y_{12} = \left.\frac{TV_1}{AV_2}\right|_{AV_1 = 0}$$

$$-y_{21} = \left.\frac{TV_2}{AV_1}\right|_{AV_2 = 0} \qquad y_{22} = \left.\frac{TV_2}{AV_2}\right|_{AV_1 = 0}$$

..(10.7)

and $-y_{12} = -y_{21}$ by reciprocity, when the components are linear and bilateral.

The admittances in (10.7), however, are not the reciprocals of the impedances in (10.2) [e.g., $y_{11} \neq 1/z_{11}$]. The admittances are defined under short-circuit conditions (AV_1 or $AV_2 = 0$) whereas the impedances are defined under open-circuit conditions (TV_1 or $TV_2 = 0$). This point has been discussed in greater detail in Chapter VII.

Eq. (10.5) can also be represented by

$$[TV] = [y] [AV] \qquad ..(10.8)$$

where the individual matrices have the same meaning as given in (10.5). As can be seen, (10.8) can be obtained by (10.4) or vice versa. Hence, we have

$$[y] = [z]^{-1} \qquad \ldots (10.9a)$$

or
$$[z] = [y]^{-1} \qquad \ldots (10.9b)$$

provided the required inverse exists.

At this stage, it has to be pointed out that, for a given network, z-parameters or y-parameters may or may not exist. The following examples illustrate this point clearly.

Example 10.1

It is required to determine z- and y-parameters of the series impedance Z_s, shown in Fig. 10.3.

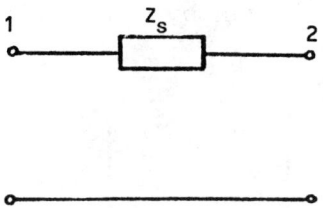

Figure 10.3 Series impedance of example 10.1

It is easily verified that z-parameters do not exist. The Y-parameters are given by

$$[y]_s = \begin{bmatrix} Y_s & -Y_s \\ -Y_s & Y_s \end{bmatrix} \qquad \ldots (10.10a)$$

where $\quad Y_s = 1/Z_s \quad$ ..(10.10b)

Example 10.2

It is required to determine z- and y-parameters of the shunt admittance shown in Fig.10.4.

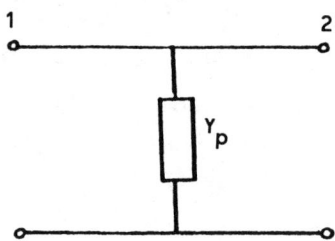

Figure 10.4 Shunt admittance of example 10.2

It is verified that y-parameters do not exist. The z-parameters are given by

$$[z]_s = \begin{bmatrix} Z_p & Z_p \\ Z_p & Z_p \end{bmatrix} \quad \text{..(10.11a)}$$

where $\quad Z_p = 1/Y_p \quad$ ..(10.11b)

Starting from the definitions, it is possible to determine z- and y-parameters of a network. This is illustrated by the following two examples.

Example 10.3

It is required to determine z- and y-parameters of

the T-network shown in Fig. 10.5.

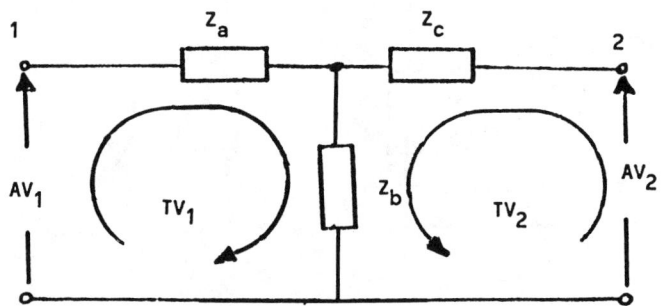

Figure 10.5 The T-network of example 10.3

The z-parameters are determined starting from their definitions given in (10.2). In order to determine z_{11}, TV_2 has to be made equal to zero. In this case, the network reduces to a single mesh circuit as shown in Fig. 10.6(a). This gives

$$z_{11} = \frac{AV_1}{TV_1}\bigg|_{TV_2 = 0} = Z_a + Z_b \quad \quad ..(10.12)$$

Similarly, in order to determine z_{22}, TV_1 has to be made equal to zero. In this case, the network reduces to a single mesh circuit as shown in Fig. 10.6(b). This gives

$$z_{22} = \frac{AV_2}{TV_2}\bigg|_{TV_1 = 0} = Z_b + Z_c \quad \quad ..(10.13)$$

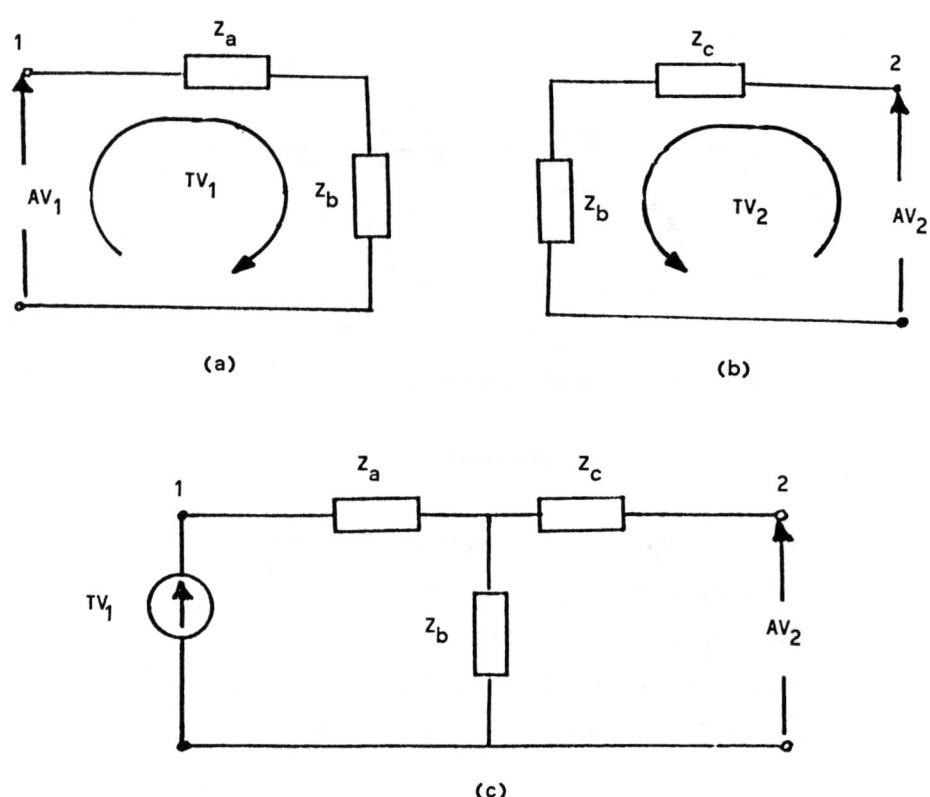

Figure 10.6 The equivalent circuits obtained for the determination of z-parameters of the T-network of figure 10.5

It is noted that the network is reciprocal and hence $z_{12} = z_{21}$. To determine z_{12}, a through variable generator TV_1 is connected to the terminal pair 1 and AV_2, the across variable at the terminal pair 2 is determined under open-circuit conditions. This is as shown in Fig. 10.6(c). This gives

Two-port Networks and Mutually Coupled Components   649

$$z_{12} = \left.\frac{AV_2}{TV_1}\right|_{TV_2 = 0} = \frac{(Z_b)(TV_1)}{(TV_1)} = Z_b \qquad ..(10.14)$$

All the three results obtained above can be put together to obtain the overall z-parameters as

$$[z]_T = \begin{bmatrix} Z_a + Z_b & Z_b \\ \\ Z_b & Z_b + Z_c \end{bmatrix} \qquad ..(10.15)$$

From (10.15) and (10.9), y-parameters are obtained as

$$[y]_T = [z]_T^{-1} = \frac{1}{\Delta_z} \begin{bmatrix} Z_b + Z_c & -Z_b \\ \\ -Z_b & Z_a + Z_b \end{bmatrix} \qquad ..(10.16a)$$

where $\quad \Delta_z = Z_a Z_b + Z_b Z_c + Z_c Z_a \qquad ..(10.16b)$

Example 10.4

It is required to determine z- and y-parameters of the π-network shown in Fig. 10.7.

In this case, y-parameters are obtained starting from the basic definitions given in (10.7). In order to determine y_{11}, AV_2 has to be made equal to zero. The resulting network is shown in Fig. 10.8(a). It is

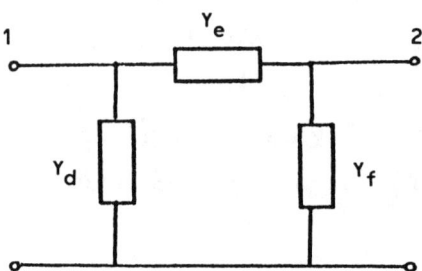

Figure 10.7 π-network of example 10.4

observed that the admittances y_d and y_e are in parallel and hence

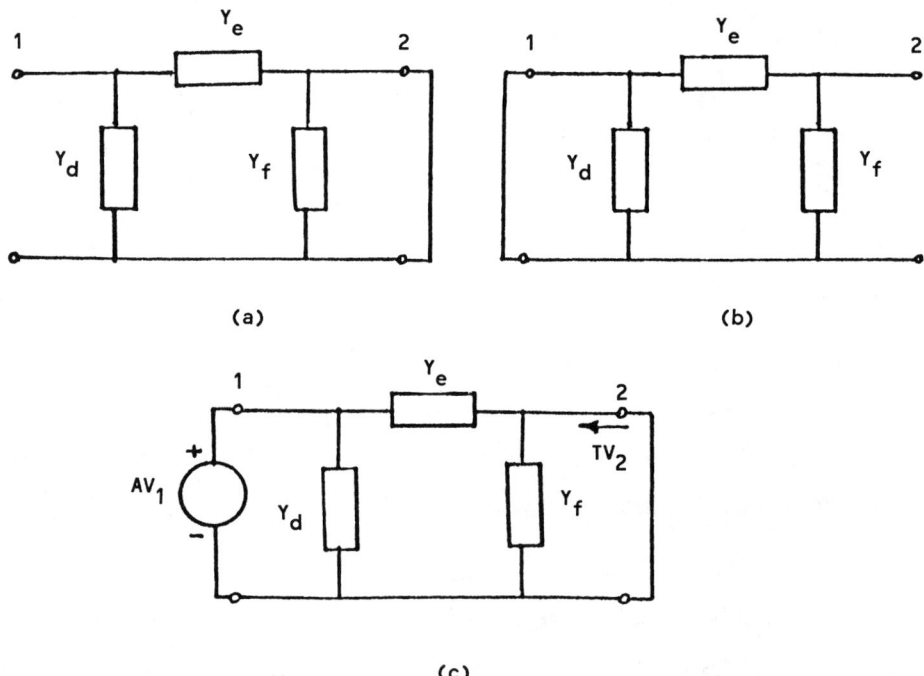

Figure 10.8 The equivalent circuits obtained for the determination of the y-parameters of the π-network of figure 10.7

Two-port Networks and Mutually Coupled Components 651

$$y_{11} = Y_d + Y_e \qquad \ldots (10.17)$$

Similarly, in order to determine y_{22}, AV_1 has to be made zero. In this case, the equivalent circuit will be as shown in Fig. 10.7(b). It is observed that Y_e and Y_f are in parallel and hence

$$y_{22} = Y_e + Y_f \qquad \ldots (10.18)$$

It is noted that the network is reciprocal and hence $-y_{12} = -y_{21}$. To determine $-y_{12}$, an across variable generator AV_1 is connected to the terminal pair 1 and it is required to determine TV_2, the through variable, flowing through the short-circuit placed across the terminal pair 2. The equivalent circuit is as shown in Fig. 10.7(c). This gives

$$-y_{12} = \left. \frac{TV_2}{AV_1} \right|_{AV_2 = 0} = \frac{(Y_e)(-AV_1)}{(AV_1)} = -Y_e \qquad \ldots (10.19)$$

All the three results obtained above can be put together to obtain the overall y-parameters as

$$[y]_\pi = \begin{bmatrix} Y_d + Y_e & -Y_e \\ -Y_e & Y_e + Y_f \end{bmatrix} \qquad \ldots (10.20)$$

From (10.20) and (10.9), z-parameters are obtained as

$$[z]_\pi = [y]_\pi^{-1} = \frac{1}{\Delta_y} \begin{bmatrix} Y_e + Y_f & Y_e \\ Y_e & Y_d + Y_e \end{bmatrix} \quad \ldots (10.21a)$$

where
$$\Delta_y = Y_d Y_e + Y_e Y_f + Y_f Y_d \quad \ldots (10.21b)$$

10.1.3 Hybrid Parameters

Another representation of the two-terminal pair network is known as the hybrid matrix. This may be derived by considering a through variable source at terminal - pair 1 and an across variable at terminal - pair 2. Thus:

$$\begin{bmatrix} AV_1 \\ TV_2 \end{bmatrix} = \begin{bmatrix} h_{11} & h_{12} \\ h_{21} & h_{22} \end{bmatrix} \begin{bmatrix} TV_1 \\ AV_2 \end{bmatrix} \quad \ldots (10.22)$$

where the values of h_{11}, h_{12}, h_{21} and h_{22} can also be defined similar to z- and y-parameters. They can also be expressed from (10.3) or (10.6) in terms of transfer impedances or admittances. The result is:

$$h_{11} = \frac{z_{11}z_{22} - z_{12}z_{21}}{z_{22}} = \frac{1}{y_{11}}$$

$$h_{12} = \frac{z_{12}}{z_{22}} = -\frac{y_{12}}{y_{11}}$$

..(10.23)

$$h_{21} = -\frac{z_{21}}{z_{22}} = \frac{y_{21}}{y_{11}}$$

$$h_{22} = \frac{1}{z_{22}} = \frac{y_{11}y_{22} - y_{12}y_{21}}{y_{11}}$$

Eq. (10.23) indicates that, in the hybrid matrix, $h_{12} = -h_{21}$, if reciprocity holds. It also shows clearly that the elements in the impedance and admittance matrices are not reciprocals.

As stated previously, the two-terminal pair components may be physically realized from a network of two-terminal components or from a mutually coupled component. The matrix representation given in (10.3), (10.6) and (10.22) are valid for either physical realization. However, there is a reduced form of these matrices in which some of the matrix elements are equal to zero and which can only occur through the use of mutually coupled components. The reduced form is called an **ideal coupler**. It occurs when the conditions

$$z_{11}z_{22} - z_{12}z_{21} = 0 \qquad ..(10.24a)$$

and
$$y_{11}y_{22} - y_{12}y_{21} = 0 \qquad \ldots (10.24b)$$

are fulfilled. Under these conditions, the hybrid matrix elements h_{11} and h_{22} equal to zero. Eq. (10.22) then becomes the matrix representation of the ideal coupler and takes the form:

$$\begin{bmatrix} AV_1 \\ TV_2 \end{bmatrix} = \begin{bmatrix} 0 & n \\ -n & 0 \end{bmatrix} \begin{bmatrix} TV_1 \\ AV_2 \end{bmatrix} \qquad \ldots (10.25)$$

where n is called the ideal coupling or transformation ratio.

Since our emphasis in this chapter is mainly on ideal couplers, we restrict our consideration to those two-terminal networks that are realized by mutually coupled components. Specifically, we examine the most well-known mutually coupled element, the transformer, and show the conditions under which approximates an ideal coupler.

10.2 TRANSFORMERS

10.2.1 Mutually Coupled Coils (Electrical Transformers)

Eq. (3.9) has expressed the relation that in a pure inductor (conducting coil), the flux linkage, λ_{12}, is directly proportional to the current. Faraday's law was then used to develop the elemental equation of the inductor. This law states that the voltage induced in a coil is equal to the rate of change of flux linkage. Thus, the result developed in (3.11) was that the induced coil voltage is proportional to the rate of change of current.

We may extend these basic concepts to the case where two separate coils are in close proximity to each other. In this arrangement, the change of current in one of the coils induces voltages in both the coil itself and also in the nearby coil. These coils are therefore designated as **mutually coupled** and the phenomenon is known as **mutual inductance**.

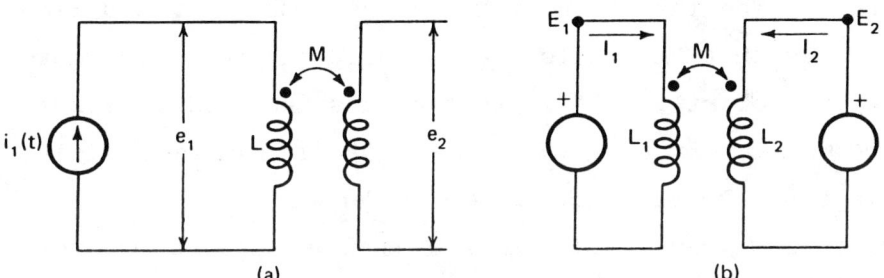

Figure 10.9 Mutually coupled coils

To describe the mutual inductance effect as a relation between signal variables, consider the mutually coupled coils shown in Fig. 10.9(a). Suppose coils 1 and 2 are mutually coupled and the current $i_1(t)$ flows in coil 1. If we designate the voltage induced in coil 1 as e_1 and the voltage induced in coil 2 as e_2, we may express these voltages as:

$$e_1 = L \frac{di_1}{dt} \qquad ..(10.26a)$$

$$e_2 = M \frac{di_1}{dt} \qquad ..(10.26b)$$

where L is the self-inductance of coil 1 and M is the mutual inductance between coil 1 and 2. The units for mutual inductance and self-inductance are the same (henrys).

Fig. 10.9(b) shows a two-terminal pair that consists of mutually coupled coils and two voltage sources. The inclusion of M and the double-arrowed line indicates that the coils are mutually coupled. The dots show the polarity. If current enters one coil at a dot, with a positive rate of change, it induces a positive voltage on the dot side of the other coil. If the loop equations (refer to Section 7.5) are applied to this arrangement, the result is:

$$\text{loop 1} \longrightarrow -E_1 + L_1 s\, I_1 + Ms\, I_2 = 0 \qquad ..(10.27a)$$

$$\text{loop 2} \longrightarrow -E_2 + L_2 s\, I_2 + Ms\, I_1 = 0 \qquad ..(10.27b)$$

Eq. (10.27) may be placed in the two-terminal pair matrix forms given in (10.3), (10.6) and (10.22). The impedance matrix representation [Eq. (10.3)] is:

$$\begin{bmatrix} E_1 \\ E_2 \end{bmatrix} = \begin{bmatrix} L_1 s & Ms \\ Ms & L_2 s \end{bmatrix} \begin{bmatrix} I_1 \\ I_2 \end{bmatrix} \quad ..(10.28)$$

The corresponding admittance matrix [Eq. (10.6)] for the mutually coupled coils is:

$$\begin{bmatrix} I_1 \\ I_2 \end{bmatrix} = \begin{bmatrix} \dfrac{L_2}{(L_1 L_2 - M^2)s} & \dfrac{-M}{(L_1 L_2 - M^2)s} \\ \dfrac{-M}{(L_1 L_2 - M^2)s} & \dfrac{L_1}{(L_1 L_2 - M^2)s} \end{bmatrix} \begin{bmatrix} E_1 \\ E_2 \end{bmatrix} \quad ..(10.29)$$

Note again that the corresponding elements in the impedance matrix [Eq. (10.28)] and the admittance matrix [Eq. (10.29)] are not reciprocals. We may also put (10.27) into the hybrid matrix form [Eq. (10.22)] with the result:

$$\begin{bmatrix} E_1 \\ I_2 \end{bmatrix} = \begin{bmatrix} \dfrac{(L_1 L_2 - M^2)s}{L_2} & \dfrac{M}{L_2} \\ -\dfrac{M}{L_2} & \dfrac{1}{L_2 s} \end{bmatrix} \begin{bmatrix} I_1 \\ E_2 \end{bmatrix} \quad ..(10.30)$$

Under special conditions, the hybrid matrix for the mutually coupled coils [Eq.(10.30)] may reduce to the form for the ideal coupler [Eq.(10.25)]. One of the conditions is that all the flux from one coil must be linked with the other coil. When this condition is met,

$$M = \sqrt{L_1 L_2} \qquad \qquad ..(10.31)$$

The other condition is that virtually no current is required to produce the magnetic flux linking the coils. This means that L_1, L_2 and M are very large (approaching infinite values) and therefore $1/L_2$ approaches zero. However, the ratio M/L_2 is determinate and may be expressed as:

$$\frac{M}{L_2} = \frac{\sqrt{L_1 L_2}}{L_2} = \sqrt{\frac{L_1}{L_2}} = \frac{N_1}{N_2} \qquad ..(10.32)$$

since, in general, the self-inductance of the coil is proportional to the square of the number of turns [e.g., see Eq.(3.12)]. If these conditions are satisfied, (10.30) reduces to:

$$\begin{bmatrix} E_1 \\ I_2 \end{bmatrix} = \begin{bmatrix} 0 & N_1/N_2 \\ -N_1/N_2 & 0 \end{bmatrix} \begin{bmatrix} I_1 \\ E_2 \end{bmatrix} \qquad ..(10.33)$$

Note that none of the terms in (10.33) includes the complex frequency, s. However, it does not follow that in the sinusoidal steady state (s = jω) the electrical transformer will behave approximately as an ideal transformer at all frequencies. This becomes evident, when we consider the term $1/j\omega L_2$ [h_{22} in (10.30)] which clearly cannot be zero when ω = 0. Indeed, if we recall that, when ω = 0, there is zero rate of change of the currents, we must conclude that the transformer cannot function under such a condition. Thus, a pair of coupled coils may be represented by an ideal transformer only when the two conditions considered above are met. In general, these conditions cannot be realized, even approximately, at low frequencies.

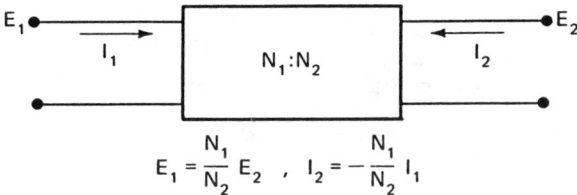

Figure 10.10 Representation of ideal electrical transformer

We use a block with the turns ratio ($N_1 : N_2$) indicated within as the circuit symbol for the ideal transformer. This is shown in Fig. 10.10. For convenience, one side of a transformer is designated as **primary** and the other side as **secondary**. The choice is somewhat arbitrary, but if there is only one source, it is common to consider as primary the side of the transformer connected to the source. For this discussion, we shall arbitrarily specify the left side as primary and the right side as secondary. The number of turns on the primary is N_1 and in general, the subscript 1 refers to the primary. Similarly, N_2 is the number of secondary turns, the subscript 2 being used for all secondary quantities. The ideal transformer relations may be simply expressed from (10.33) as:

$$E_1 = (N_1/N_2) E_2 \qquad \qquad ..(10.34a)$$

$$I_2 = - (N_1/N_2) I_1 \qquad \qquad ..(10.34b)$$

Note that the voltage is always larger on the side having the coil with the larger number of turns. The current is smaller on this side. The ideal transformer neither stores nor dissipates energy. It functions only to transform the variables as described in (10.34).

10.2.2 Mutually Coupled Mass (Mechanical Transformer)

Fig. 10.11 shows an unpivoted bar of length, l, and mass, m. When forces F_1 and F_2 are applied at each end, the bar has both translational and rotational motion. The resulting velocity at the left end is V_1 and that at the right end is V_2. The velocity at the centre of gravity of the bar is V_c. As in all the other components, the analysis must stem from physical laws. In this case, we use the basic laws of mechanics and

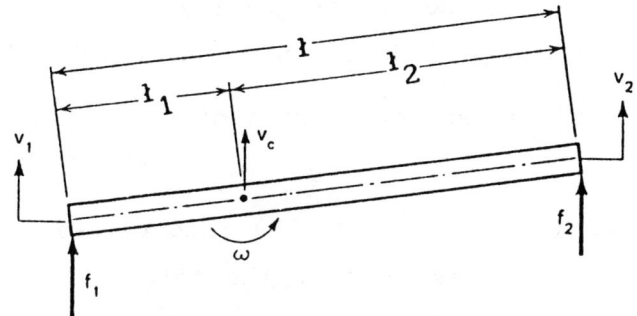

Figure 10.11 Unpivoted bar

then interpret the results in a circuit context. For the translational motion,

$$\sum f = (\text{mass})(\text{acceleration}) \qquad ..(10.35)$$

and in terms of exponential amplitudes, this may be expressed as:

$$F_1 + F_2 = ms V_c \qquad ..(10.36)$$

For the rotational motion,

$$\sum T = (\text{moment of inertia})(\text{angular acceleration}) \qquad ..(10.37)$$

This leads to:

$$F_2 l_2 - F_1 l_1 = J s \Omega \qquad ..(10.38)$$

where l_1 and l_2 = the distance from the ends to the centre of gravity,

Ω = the amplitude of the angular velocity

and J = the moment of inertia of the bar.

To place (10.37) and (10.38) into circuit format, we must separate the effects of the force at each end. The solution of these simultaneous equations yields:

$$m\, l_2\, s\, V_c - J\, s\, \Omega = F_1\, l \qquad \text{..(10.39a)}$$

$$m\, l_1\, s\, V_c + J\, s\, \Omega = F_2\, l \qquad \text{..(10.39b)}$$

Eqs. (10.39) represent the motion in terms of the angular velocity, Ω, and the velocity of the centre of gravity, V_c. For a two-terminal pair circuit representation, we require the motion to be in terms of V_1 and V_2, the velocities corresponding to the applied forces F_1 and F_2. We may relate these sets of variables (Ω, V_c and V_1, V_2) from the concept of relative motion. When the angle of rotation is small,

$$V_1 = V_c - l_1 \Omega \qquad \text{..(10.40a)}$$

and

$$V_2 = V_c + l_2 \Omega \qquad \text{..(10.40b)}$$

and thus:

$$V_c = \frac{V_1 l_2 + V_2 l_1}{l} \qquad ..(10.41a)$$

and
$$\Omega = \frac{V_2 - V_1}{l} \qquad ..(10.41b)$$

If the radius of gyration of the bar is a distance h from the centre of gravity, then the relation between the moment of inertia and mass is

$$J = mh^2 \qquad ..(10.42)$$

The elimination of Ω and V_c from (10.39) and (10.41) yields:

$$m \frac{(l_2^2 + h^2)}{l^2} s V_1 + m \frac{(l_1 l_2 - h^2)}{l^2} s V_2 = F_1 \qquad ..(10.43a)$$

$$m \frac{(l_1 l_2 - h)^2}{l^2} s V_1 + m \frac{(l_1^2 \mp h)^2}{l^2} s V_2 = F_2 \qquad ..(10.43b)$$

We may observe from (10.19) that the relation between V_1 and F_1 also depends on V_2. Similarly, the relation between V_2 and F_2 also depends on V_1. From a circuit viewpoint, this is equivalent to a mutual coupling effect. Fig. 10.12 shows a circuit representation of (10.43), where M_m represents this mutual coupling effect. From the nodal equations, we

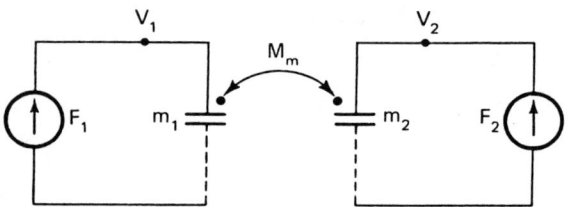

Figure 10.12 Mutually coupled masses

may analyze the circuit as:

$$\text{node 1} \longrightarrow m_1 s V_1 + M_m s V_2 = F_1 \quad \text{..(10.44a)}$$

$$\text{node 2} \longrightarrow M_m s V_1 + m_2 s V_2 = F_2 \quad \text{..(10.44b)}$$

Eq. (10.44) is identical to (10.43), when we define:

$$m_1 = m \frac{(l_2^2 + h^2)}{l^2} \quad \text{..(10.45a)}$$

$$m_2 = m \frac{(l_1^2 + h^2)}{l^2} \quad \text{..(10.45b)}$$

$$M_m = m \frac{(l_1 l_2 - h^2)}{l^2} \quad \text{..(10.45c)}$$

Eq. (10.43) for the unpivoted bar may be placed in hybrid matrix form [Eq. (10.30)] in terms of the analogous dual variables (that is, F_1 in place of E_1 and

V_2 in place of I_2, etc.,) with the result:

$$\begin{bmatrix} V_1 \\ F_2 \end{bmatrix} = \begin{bmatrix} \dfrac{l^2}{ms(h^2 + l_2^2)} & -\dfrac{l_1 l_2 - h^2}{h^2 + l_2^2} \\ \dfrac{l_1 l_2 - h^2}{l_1^2 + h^2} & \dfrac{smh^2}{(h^2 + l_2^2)} \end{bmatrix} \begin{bmatrix} F_1 \\ V_2 \end{bmatrix}$$

..(10.46)

Eq. (10.46) represents the mutually coupled masses in the format of a two-terminal pair network. As in the case of the mutually coupled coils, we may apply special conditions to reduce the components to an ideal coupler. One of the conditions is that the mass is concentrated at the centre of gravity ($h = 0$). The other condition is that the mass of the bar is large. With these conditions, (10.46) becomes:

$$\begin{bmatrix} F_1 \\ V_2 \end{bmatrix} = \begin{bmatrix} 0 & l_2/l_1 \\ -l_2/l_1 & 0 \end{bmatrix} \begin{bmatrix} V_1 \\ F_2 \end{bmatrix}$$

..(10.47)

The ideal unpivoted lever like the ideal electrical transformer requires a varying signal and does not operate down to zero frequency.

In dealing previously with two-terminal components, we have used the same circuit symbol in all the media. For example, mass and capacitance were represented similarly in circuit diagrams. We now extend the use of uniform symbols to the representation of ideal couplers. Thus, Fig. 10.13 shows the ideal mechanical transformer with the same symbol that was used for the ideal electrical transformer in Fig. 10.10. In this case, the numbers within the block ($l_1 : l_2$) refer to the ratio of the across variables, and in this case the distance from the pivot points to the ends. The larger velocity occurs on the side with the larger distance.

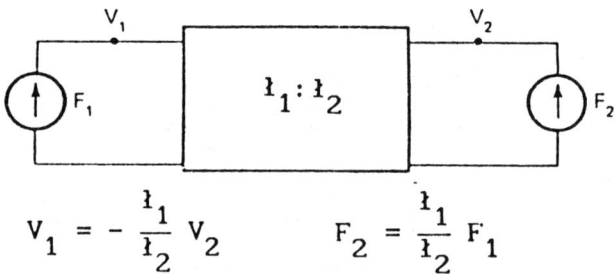

Figure 10.13 Representation of ideal mechanical transformer

10.3 IDEAL COUPLERS

There are other physical arrangements which perform the function of ideal couplers. These will be shown symbolically in the same way as in Figs. 10.10 and 10.13. In all cases, the larger number within the block is placed on the side with the larger value of the across variable. We emphasize this now because some of the couplers operate with the dual variable. As a result, a geometric parameter associated with one side may appear within the block on the opposite side. This will become clear when we describe the operation of gears and differential pistons.

10.3.1 Levers

In Section 10.2, we demonstrated that an unpivoted bar with a large mass concentrated at its centre of gravity acts as an ideal coupler. We can obtain another ideal coupling component by merely pivoting the bar about a fixed support and assuming that the mass of the bar is negligible. The component [Fig. 10.14] is then an ideal lever.

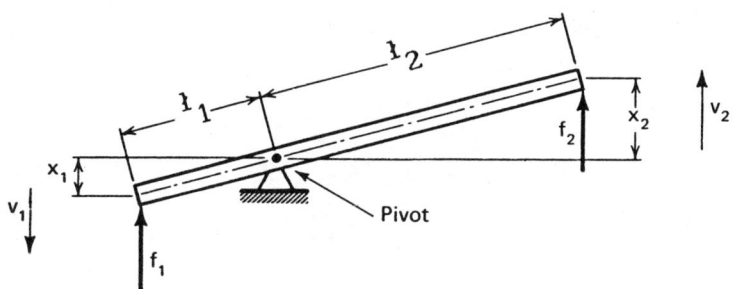

Figure 10.14 Pivoted lever

There are two relations that must be derived: one between v_1 and v_2 and the other between f_1 and f_2. We may relate the velocities from the geometry of the lever. For small angular displacements, the motion of the ends is approximately linear and thus:

$$\frac{x_1}{l_1} = \frac{x_2}{l_2} \qquad \ldots (10.48)$$

Since the lever arms l_1 and l_2 are fixed, we may differentiate (10.48) with respect to time and obtain:

$$v_1 = (l_1/l_2) \, v_2 \qquad \qquad ..(10.49)$$

It is important to notice that the directions of v_1 and v_2 are selected to eliminate negative signs in the relation between the velocities. Inspection of any particular lever arrangement will always indicate the proper choice of direction to avoid negatives.

The relation between f_1 and f_2 may be accomplished by equating the energy input on one side to the energy output of the opposite side. However, for a lever without appreciable mass, the moment of inertia is negligible. As a result, the sum of the moments about the pivot are approximately equal to zero and we may obtain directly that:

$$f_2 = (l_1/l_2) \, f_1 \qquad \qquad ..(10.50)$$

Eqs. (10.49) and (10.50) are the relations between the primary and secondary variables for the ideal lever. The transformation ratio is l_1/l_2. Note also that the pivoted lever operates at all frequencies including zero frequency.

Example 10.5

A mechanical arrangement which includes an ideal

lever, two masses, a spring and a damper is shown in Fig. 10.15. The source is a vertical force f(t) applied to the left end of the lever. Draw the analogous circuit diagram for this system. In addition, find the transformation ratio and the relation between the primary and the secondary variables.

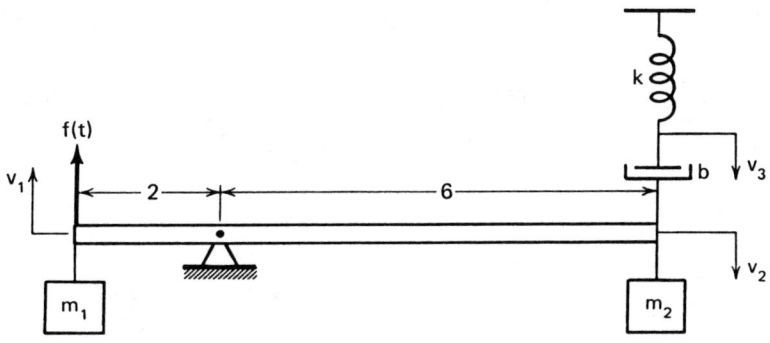

Figure 10.15 Mechanical arrangement for example 10.5

Fig. 10.16 shows the analogous circuit diagram. The primary and secondary sides are connected through an

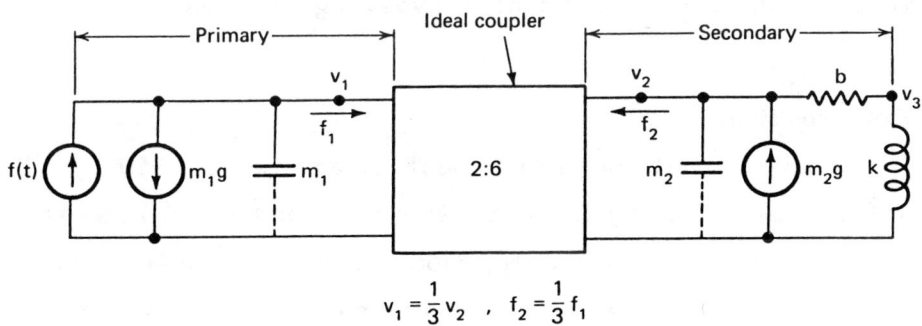

Figure 10.16 Analogous circuit for example 10.5

ideal coupler with the ratio 2 : 6. Since the velocity on the secondary side is larger than on the primary side, the larger number appears on the secondary side. The primary circuit consists of the source due to the applied force, the force source due to gravity and the mass, m_1. The secondary circuit has a mass m_2, a damper b, a spring k, and a force source due to gravity acting on m_2.

The transformation ratio is 1/3 and the relations between the primary and secondary variables are:

$$v_1 = (1/3)\ v_2 \quad \text{and} \quad f_2 = (1/3)\ f_1 \qquad \ldots (10.51)$$

10.3.2 Pulleys

Smooth pulleys of negligible mass may also be considered as ideal couplers. In this case, the transformation ratio depends on the number of lines supporting the pulley. The following examples demonstrate the determination of the analogous circuit for mechanical arrangements involving pulleys.

Example 10.6

In the pulley arrangement shown in Fig. 10.17(a), a force $f_1(t)$ is applied to the free end of a line that passes over a fixed pulley and under a movable pulley. The velocity of the free end is designated as v_1 in the direction of the force. The velocity of the movable pulley is v_2 and is assumed upward. Represent this

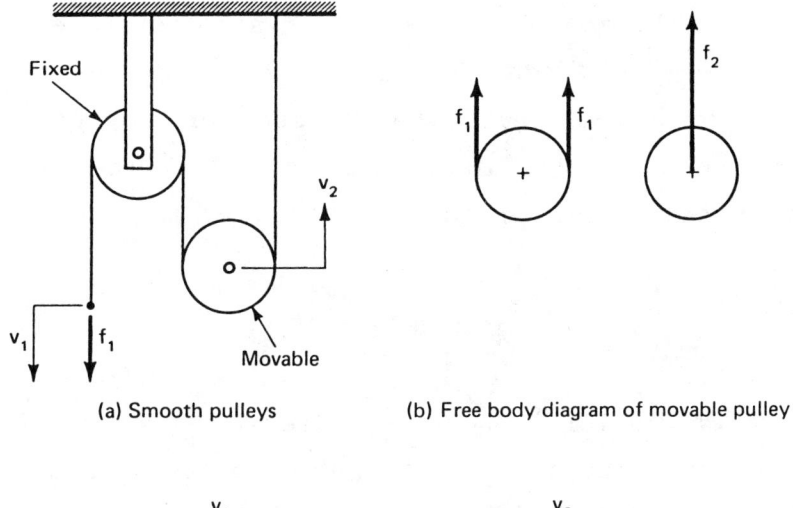

(a) Smooth pulleys (b) Free body diagram of movable pulley

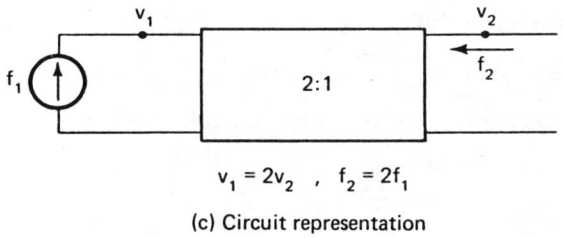

$v_1 = 2v_2$, $f_2 = 2f_1$

(c) Circuit representation

Figure 10.17 Pulley arrangement for example 10.6

mechanical device by an analogous circuit diagram, if the pulleys are assumed to be smooth. Find the transformation ratio and the relation between primary and secondary variables.

Fig. 10.16(b) shows a free body diagram of the movable pulley. The assumed smoothness of the pulley surfaces means that friction may be neglected. As a consequence, the applied force, f_1, is transmitted equally throughout the line. Thus, because of the loop in the line, the movable pulley is acted upon by two equal and parallel forces, f_1. The resultant force acting on the movable pulley, f_2, is therefore equal to $2f_1$. Geometric considerations show that the movable pulley moves only half the distance of the free end. Thus the transformation ratio is 2 and the relation between the variables is:

$$v_1 = 2v_2 \text{ and } f_2 = 2f_1 \qquad (10.52)$$

Fig. 10.16(c) shows the circuit representation of the pulley arrangement. The pulley behaves as an ideal coupler. Note that the number of lines supporting the movable pulley which is associated with the secondary variables is placed within the block on the primary side (2 : 1).

Example 10.7

Fig. 10.18(a) shows a hoist that has a force, $f(t)$, applied through a spring, k. The velocity of the free end of the hoist cable is v_1. The hoist has a fixed set of two smooth pulleys and a movable set of two smooth pulleys. The velocity associated with the movable set is v_2. The object of the device is to lift the mass, m. Represent the hoist by an analogous circuit. Determine the transformation ratio and the relation between the variables on the primary and secondary sides.

The tension in the hoist cable, f_1, is equal to

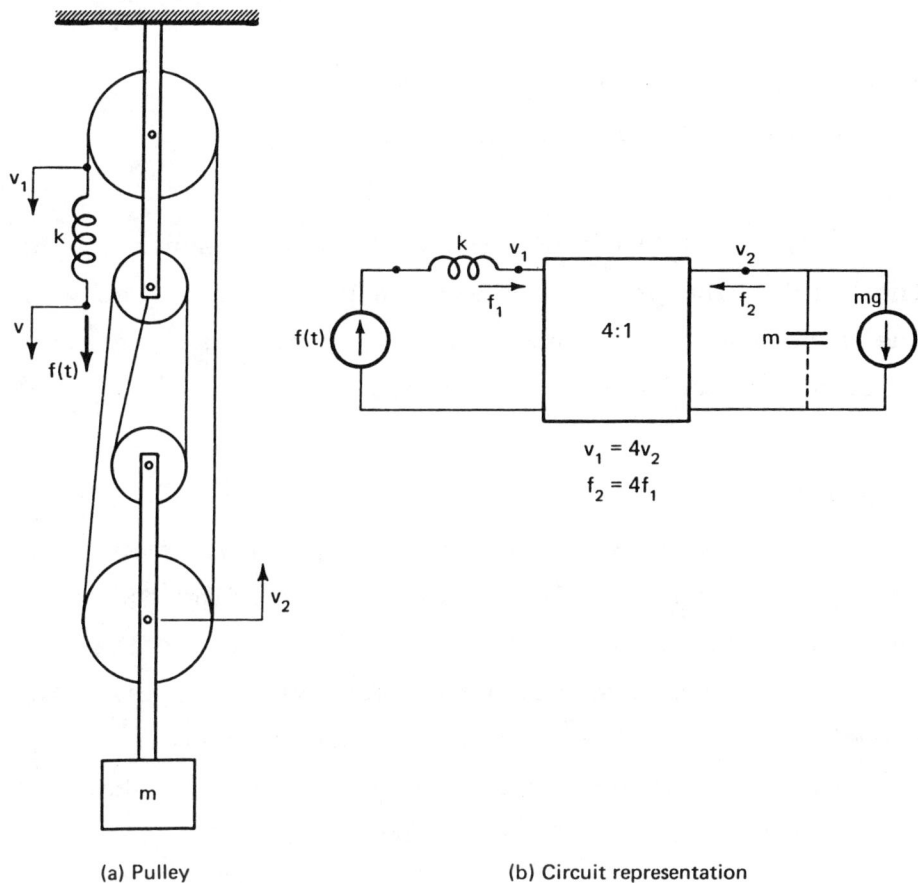

(a) Pulley   (b) Circuit representation

Figure 10.18 Pulley arrangement for example 10.7

f(t). The cable is looped in such a way that the resultant force, f_2, acting on the movable pulleys is 4 times the tensile force. Similarly, the motion of the free end of the cable is 4 times larger than that of the movable pulleys. Thus we may write

$$f_2 = 4f_1, \quad v_1 = 4v_2 \qquad \ldots (10.53)$$

These are the coupling equations of a device with a transformation ratio of 4.

Fig. 10.18(b) shows the circuit representation of the hoist. The primary side has a spring in series with the force source. The secondary side has a gravity force source and a mass to ground.

10.3.3 Gears

In translational mechanical systems, levers and pulleys are ideal couplers. The corresponding component in rotational mechanical systems is a set of gears. Fig. 10.19 shows a simple arrangement of two gears. The number of teeth on gear 1 is N_1 and on gear 2 is N_2. The relation between the angular velocities of shafts 1 and 2 and the number of gear teeth is:

$$\omega_1 = (N_2/N_1)\, \omega_2 \qquad \ldots (10.54)$$

The directions have been selected to avoid negatives. Comparison of (10.54) with the equivalent relation for levers [Eq. (10.49)] shows that the transformation ratio for the gear is N_2/N_1, whereas the ratio for the lever is l_2/l_1. To retain the convention that the larger number in the ideal coupling block be on

Two-port Networks and Mutually Coupled Components

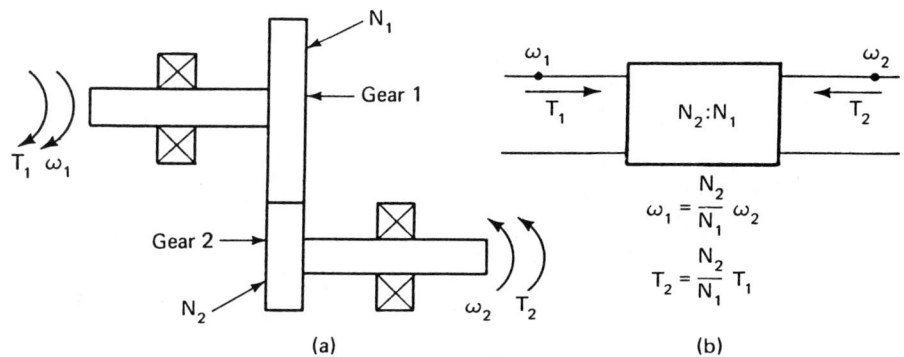

Figure 10.19 Representation of gears as an ideal coupler

the side with the larger across variable, we put N_2 on the primary side and N_1 on the secondary side. This is shown in Fig. 10.19(b). The relation between the torques associated with gears 1 and 2 is:

$$T_2 = (N_2/N_1) T_1 \qquad ..(10.55)$$

The gears, like the pivoted ideal lever, operate at all frequencies including zero frequency.

Example 10.8

In the gear arrangement shown in Fig. 10.20, a 40 tooth gear drives an 80 tooth gear through a 20-tooth idler gear. An input torque, $T_1(t)$, is applied to the driving gear. The driven gear is connected to a large flywheel, J, and a damper, B. Represent the gear arrangement by an analogous circuit and determine the transformation ratio.

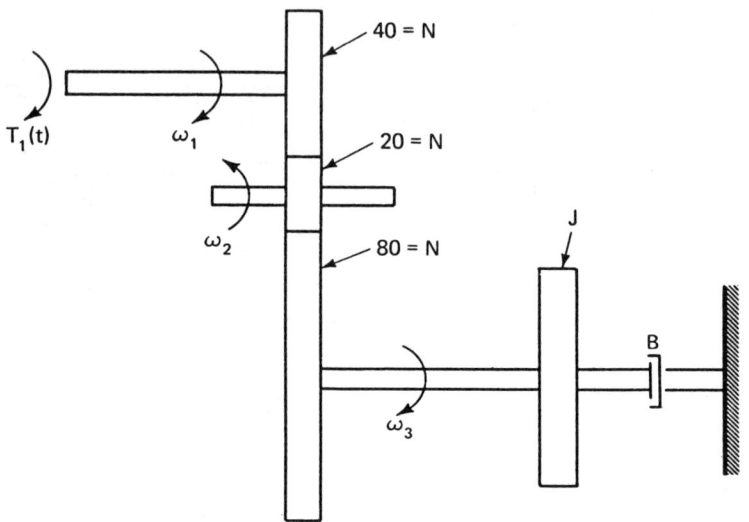

Figure 10.20 Gear arrangement for example 10.8

Let us designate the angular velocities of the driving gear, idler and driven gears as ω_1, ω_2 and ω_3 respectively. There is an ideal coupling action between each pair of gears (that is, 1 and 2, 2 and 3). Thus we may describe the rotational system with two ideal couplers, as shown in Fig. 10.21(a). The transformation ratio for coupler 1 (between gears 1 and 2) is 20 : 40 and so we may write that

$$\omega_1 = (1/2)\,\omega_2, \quad T_2 = (1/2)\,T_1 \qquad ..(10.66a)$$

The transformation ratio for ideal coupler 2 (between the idler and driven gear) is 80 : 20 and therefore:

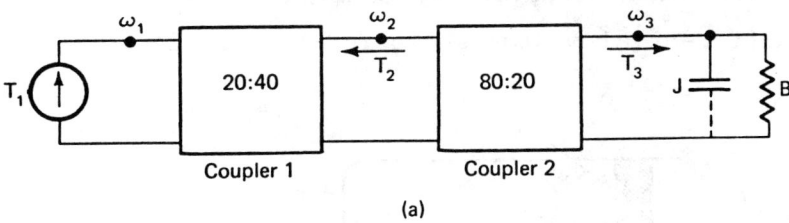

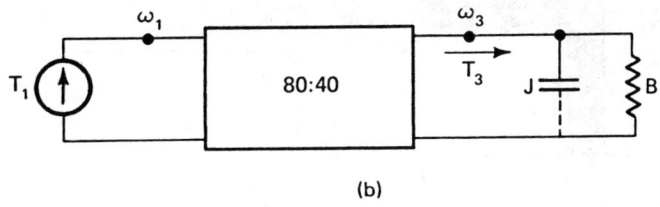

Figure 10.21 Circuit representation for example 10.8

$$\omega_2 = 4\omega_3, \quad T_3 = 4T_2 \qquad \ldots (10.66b)$$

From the above relations, we may express the variables associated with the driving and driven gears as:

$$\omega_1 = 2\omega_3, \quad T_3 = 2T_1 \qquad \ldots (10.66c)$$

Thus the transformation ratio that represents the entire gear arrangement is 2. This suggests that we may reduce the analogous circuit to one coupler as shown in Fig. 10.21(b). Of course, we could have done this immediately by recognizing that the idler gear merely changes the direction of rotation of the driven shaft and does not affect its angular velocity or torque.

10.3.4 Differential Piston

The ideal coupler component for fluid systems is the differential piston shown in Fig. 10.22(a).

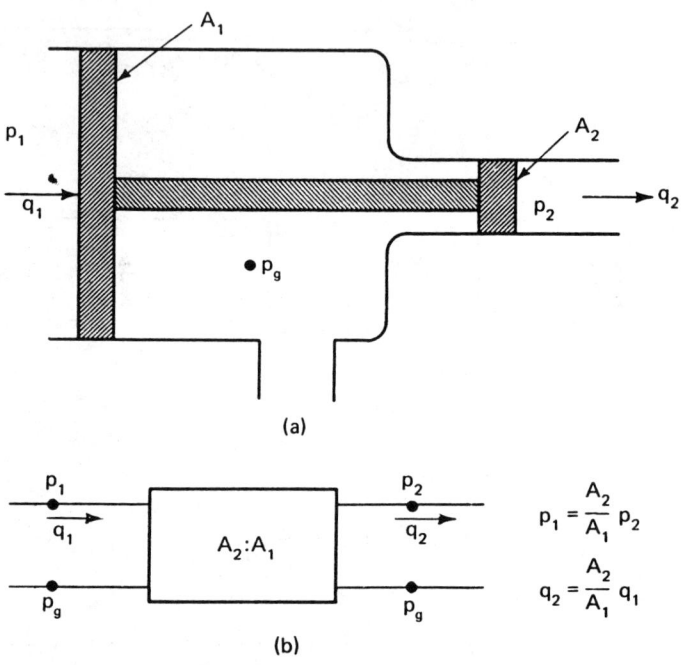

Figure 10.22 Differential piston

On the primary side, the pressure is p_1 and the flow is q_1. On the secondary side, the pressure is p_2 and the flow is q_2. Between the two sides is an enclosed region that is connected directly to a fixed reference pressure. The most convenient reference pressure and the one most easily maintained is the atmospheric reference at pressure, p_g. If the piston has negligible mass, the sum of the forces applied to it must equal zero and therefore:

$$(P_1 - p_g) = (A_2/A_1)(p_2 - p_g) \quad \quad ..(10.67)$$

where A_1 and A_2 are the cross-sectional areas on the primary and secondary sides respectively. Since there is only one piston velocity, we may relate the primary and secondary flows by:

$$q_2 = (A_2/A_1) q_1 \quad \quad ..(10.68)$$

Eqs. (10.67) and (10.68) show that the differential piston has the relationships required for an ideal coupler. The circuit representation is presented in Fig. 10.22(b). The transformation ratio is A_2/A_1. As in the gear and pulley coupling, the geometrical parameters are transposed within the coupling block. That is the area A_2 appears on the primary side and the area A_1 on the secondary side. In this case, it is the pressure which is always larger on the side of the larger number.

10.4 EQUIVALENT NETWORK FOR IDEAL COUPLERS

In general, the relations between the primary and secondary variables for ideal couplers may be expressed as:

$$(AV)_p = n (AV)_s \quad \quad ..(10.69a)$$

$$(TV)_s = n (TV)_p \quad \quad ..(10.69b)$$

where the subscripts p and s refer to primary and secondary respectively. We may consider the transformation ratio, n, as a scale factor between the sides. Thus, for example, it is possible to place scaled secondary variables into a single equivalent circuit with primary variables and to analyze the resulting circuit by the methods we have previously studied in Chapter VII. This procedure is sometimes called **reflecting** or **referring** to the primary. Conversely, we may also place scaled primary variables into a single equivalent circuit with secondary variables. This is **reflecting** or **referring** to the secondary.

When an ideally coupled circuit is put into the form of a single equivalent circuit, all the basic components, sources, and signal variables must be scaled. Eq. (10.69) will help us to determine the required scaling of sources and signal variables. Thus to reflect an across variable or across variable source from secondary to primary, we must multiply by the transformation ratio. A convenient way to remember this is to apply (10.69a) and to express it with the variable on the side we are reflecting to by itself. The coefficient of the other side is the scaling factor. To reflect from primary to secondary, then we would write (10.69b) as

$$(AV)_s = (1/n)(AV)_p \qquad \qquad ..(10.70a)$$

Two-port Networks and Mutually Coupled Components   681

and this would indicate division by the transformation ratio.

For the case of a through variable or through variable source, we would apply the same procedure to (10.69b). To shift the through variable from secondary to primary, we would write (10.69b) as

$$(TV)_p = (1/n)\,(TV)_s \qquad \ldots (10.70b)$$

and this would indicate division by the transformation ratio. On the other hand, a shifting to the secondary would mean using (10.69b) in its given form and then we would have to multiply by the transformation ratio.

The scaling factors for the reflection of the basic components are most conveniently treated by expressing the components in terms of impedance or admittance. We may define these in the primary and secondary circuits as:

$$Z_p = \frac{(AV)_p}{(TV)_p}, \qquad Y_p = \frac{(TV)_p}{(AV)_p} \qquad \ldots (10.71a)$$

$$Z_s = \frac{(AV)_s}{(TV)_s}, \qquad Y_s = \frac{(TV)_s}{(AV)_s} \qquad \ldots (10.71b)$$

We may now relate Z_p to Z_s and Y_p to Y_s by using (10.69) in conjunction with (10.71). The result is:

$$Z_p = n^2 Z_s \qquad \qquad ..(10.72a)$$

$$Y_p = (1/n^2) Y_s \qquad \qquad ..(10.72b)$$

Eq. (10.72a) may be interpreted in a similar manner to (10.69). When impedances are reflected from secondary to primary, we must multiply them by n^2. On the other hand, to reflect impedance from primary to secondary, divide the impedances by n^2. To reflect admittances, we would refer to (10.72b). Thus admittances reflected from secondary to primary should be divided by n^2 and those reflected from primary to secondary should be multiplied by n^2. The following examples will clarify the procedure.

Example 10.9

Fig. 10.23 shows the analogous circuit of a mechanical system that includes an ideal coupler. The primary circuit consists of a velocity source, V_1, and a series impedance, Z_1. The secondary circuit has a series impedance, Z_2, a shunt impedance, Z_3, and a force source, F_1. The transformation ratio is 3. Determine a single equivalent circuit: (a) in terms of primary variables and (b) in terms of secondary variables. For each circuit, express the velocity V_4 in terms of sources and impedances.

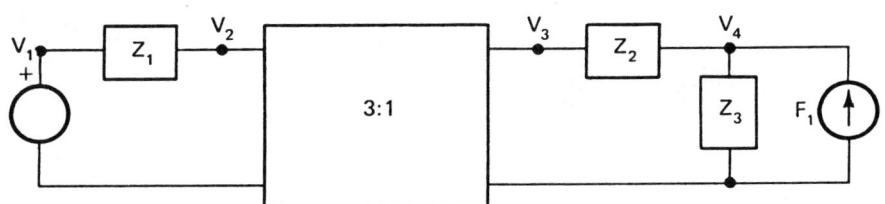

Figure 10.23 Analogous mechanical circuit (example 10.9)

(a) Primary Equivalent Circuit

To obtain the primary equivalent circuit, we must multiply all secondary impedances by 9 [Eq. (10.72a)]. The force source in the secondary must be divided by 3 [Eq. (10.69b)] and the velocity V_4 must be multiplied by 3 [Eq. (10.69a)].

If there had been a velocity source in the secondary, we would have treated it similar to a velocity and multiplied by 3. The equivalent circuit in terms of the primary variables is shown in Fig. 10.24. We may determine the variable, $3V_4$, by superposition.

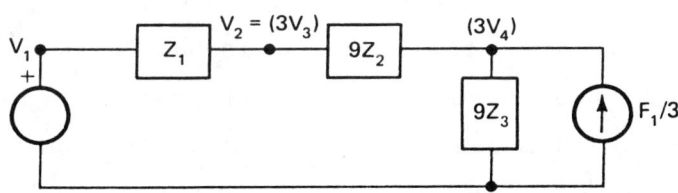

Figure 10.24 Equivalent network in terms of primary variables (example 10.9)

Thus:

$$\left[\frac{3V_4}{V_1}\right]_{F_1=0} = \frac{9Z_3}{Z_1 + 9Z_2 + 9Z_3} \quad \ldots(10.73a)$$

$$\left[\frac{3V_4}{F_1/3}\right]_{V_1=0} = \frac{(9Z_3)(Z_1 + 9Z_2)}{Z_1 + 9Z_2 + 9Z_3} \quad \ldots(10.73b)$$

The combined effect of both sources is then:

$$3V_4 = \frac{9Z_3(V_1)}{Z_1 + 9Z_2 + 9Z_3} + \frac{9Z_3(Z_1 + 9Z_2)(F_1/3)}{Z_1 + 9Z_2 + 9Z_3}$$

or $\quad V_4 = \dfrac{3Z_3 V_1 + Z_3(Z_1 + 9Z_2)F_1}{Z_1 + 9Z_2 + 9Z_3} \quad \ldots(10.74)$

(b) <u>Secondary Equivalent Circuit</u>

To obtain the secondary equivalent circuit, the impedance in the primary must be divided by 9 [Eq. (10.72a)]. The velocity source in the primary is divided by 3, when it is placed in the secondary [Eq. 10.69(a)]. Had there also been a force source in the primary, it would have required multiplication by 3 to bring it into the equivalent secondary circuit. Fig. 10.25 shows the secondary equivalent circuit. Once again, we must use superposition to determine V_4.

Two-port Networks and Mutually Coupled Components

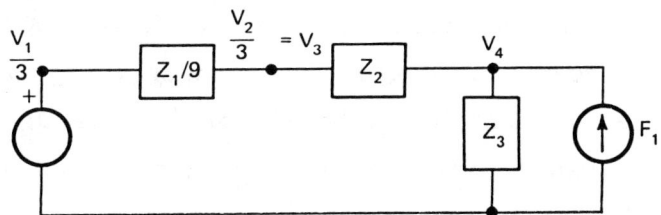

Figure 10.25 Equivalent network in terms of secondary variables (example 10.9)

We have:

$$\left[\frac{V_4}{V_1/3}\right]_{F_1=0} = \frac{9Z_3}{Z_1 + 9Z_2 + 9Z_3} \quad \ldots (10.75a)$$

$$\left[\frac{V_4}{F_1}\right]_{V_1=0} = \frac{Z_3(Z_1 + 9Z_2)}{Z_1 + 9Z_2 + 9Z_3} \quad \ldots (10.75b)$$

and the combined effect is:

$$V_4 = \frac{3Z_3 V_1 + Z_3(Z_1 + 9Z_2)F_1}{Z_1 + 9Z_2 + 9Z_3} \quad \ldots (10.76)$$

Note that the results for V_4 are the same in both cases. We may, therefore, always use either the primary or secondary equivalent circuit to determine the value of any signal variable.

Example 10.10

A motor drives a flywheel through a gearing arrangement as shown in Fig. 10.26. The motor has no friction and has a moment of inertia $J_1 = 1,000$ kg.m^2. The driving gear has 32 teeth and the driven gear has 16 teeth. The flywheel has a moment of inertia $J_2 = 2,250$ kg.m^2 and flywheel friction may be modelled as a damper with B = 250 N.m.s/rad. If the motor is suddenly turned on and then develops a torque $T(t) = 2,000\ U_s(t)$, N.m, find the angular velocity of the flywheel, $\omega_2(t)$. The flywheel is initially at rest.

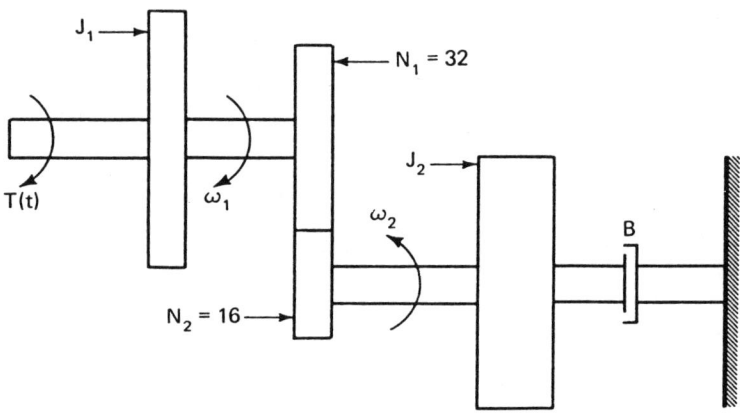

Figure 10.26 Mechanical circuit (example 10.10)

Fig. 10.27 shows the analogous circuit diagram. The gears are represented by the ideal coupler with the larger number on the side with larger angular velocity. Thus, for this coupler, the transformation ratio is 1/2. To derive the differential equation for this circuit, we shall use the admittances of the components. This requires the assumption of exponential signals throughout the circuit, namely:

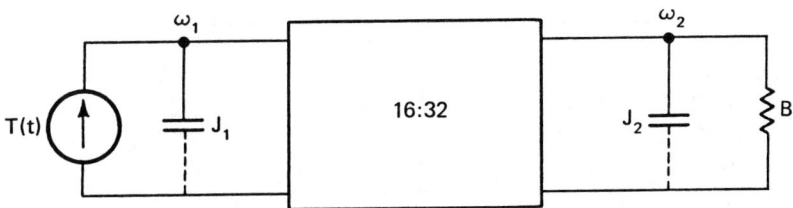

Figure 10.27 Analogous circuit (example 10.10)

$$T(t) = T \varepsilon^{st}, \quad \omega_2(t) = \Omega_2 \varepsilon^{st} \quad \quad ..(10.77)$$

We may proceed to obtain the governing equation for this system using these exponential functions even though the actual input torque is not an exponential function.

For this system, we express the components in terms of admittances and reflect the secondary components to the primary side as shown by the primary equivalent circuit in Fig. 10.28. When reflecting admittances from secondary to primary, we must divide by n^2 as indicated in (10.72b). In this case, $n = 1/2$ so that in effect, we must multiply the admittances by 4. Thus, for the flywheel and damper in primary terms

$$Y_j(s) = 4(2,250) \, s = 9,000 \, s,$$
$$Y_B = 4(250) = 1,000 \quad \quad ..(10.78)$$

The angular velocity of the flywheel is $(1/2)\Omega_2$ in

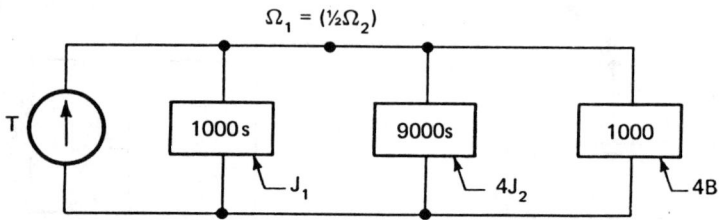

Figure 10.28 Equivalent circuit with primary variables and admittances (example 10.10)

accordance with (10.69a). From a circuit analysis of the equivalent primary circuit, the driving point admittance is:

$$Y_{dp} = \frac{T}{(1/2)\Omega_2} = 10,000\,s + 1,000 \qquad ..(10.79)$$

We may use this system function to find the differential equation in the time domain as:

$$10\,\frac{d[(1/2)\omega_2]}{dt} + (1/2)\omega_2 = \frac{T(t)}{1,000} = 2U_s(t) \qquad ..(10.80)$$

The solution of this equation is:

$$(1/2)\omega_2(t) = 2(1 - \varepsilon^{-t/10}) \qquad ..(10.81a)$$

so that:

$$\omega_2(t) = 4\,(1 - \varepsilon^{-t/10}) \qquad ..(10.81b)$$

10.5 SELECTION OF COUPLER FOR MAXIMUM POWER TRANSFER

We may transfer maximum power from a given source with fixed impedance to a specific load by inserting an ideal coupler between them. The procedure involves selecting the transformation ratio so that the equivalent circuit meets the maximum power condition given in (8.85).

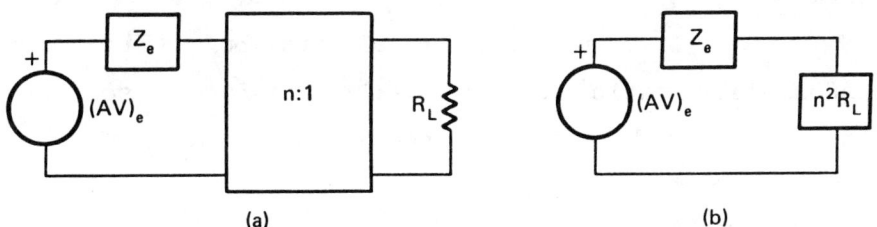

Figure 10.29 Transfer of power through an ideal coupler

Fig. 10.29(a) shows the Thevenin circuit with an ideal coupler placed between the source and the resistive load. If we reflect the load circuit into an equivalent source circuit [Fig. 10.29(b)], the value of the resistive load must be multiplied by n^2. If we now apply the condition for maximum power transfer, namely (8.85), the transformation ratio to achieve this is:

$$n_m^2 = \frac{\sqrt{R_e^2 + X_e^2}}{R_L} \qquad ..(10.82)$$

where n_m is the transformation ratio for maximum power transfer, R_e is the real part and X_e, the imaginary part of the impedance $Z_e(j\omega)$.

When the load is reactive, it will generally not be possible to select an ideal coupler for maximum power transfer. In this case, we would have to make some compromise between the two conditions given (8.87).

Example 10.11

An electric circuit containing an ideal transformer [Fig. (10.30)] has R_1 = 7.5 Ω, L = 5 H, C = 0.1 F, R_L = 1 Ω and e(t) = 10 sin 2t. Find the transformation ratio of the ideal transformer so that maximum power will be transferred to the 1 ohm resistor.

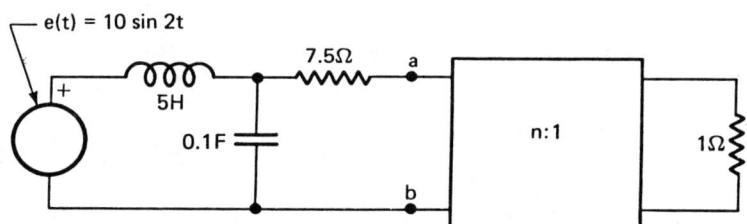

Figure 10.30 Electrical circuit (example 10.11)

From the terminals a-b on the primary side, the equivalent impedance looking toward the source is:

$$Z_e(s) = R_1 + \frac{(Ls)(1/Cs)}{Ls + (1/Cs)} = \frac{R_1 LCs + Ls^2 + R_1}{LCs^2 + 1} \quad ..(10.83)$$

$$Z_e(j\omega) = \frac{R_1(1 - \omega^2 LC) + j\omega L}{1 - \omega^2 LC} = 7.5 - j10 \quad ..(10.84)$$

Now, the application of (10.82) yields:

$$n_m^2 = \frac{\sqrt{(7.5)^2 + (10)^2}}{1} = 12.5 \qquad ..(10.85a)$$

or $\qquad n_m = 3.535 \qquad\qquad\qquad ..(10.85b)$

and the primary requires a coil with 3.535 times the number of turns on the secondary coil. A transformer with transformation ratio equal to 3.535 will permit the maximum power to be transferred to the load.

10.6 SUMMARY AND DISCUSSIONS

In this chapter, the well-known mutually coupled component known as the transformer, is examined. The mathematical representation of such a transformer is given. The conditions under which such a transformer approximates an ideal one are obtained. Some of the properties of the electrical and mechanical transformers are given. Also some of the ideal couplers like pivoted lever, the pulley arrangement, the gear and the differential piston are discussed. When ideal couplers are present, two types of equivalent networks, namely (i) the branches shifted to the primary side and (ii) the branches shifted to the secondary side, have been obtained. These equivalent networks can be analyzed using the techniques discussed earlier.

10.7 PROBLEMS

10.1 For the electrical circuit shown in Fig. P10.1, the ideal transformer has a primary-to-secondary turns ratio of 4. The primary has a capacitive reactance of 8 ohms, a resistance of 3 ohms and an inductive reactance of 4 ohms. The rms value of the voltage generator is 10 V. The secondary has a resistance of 1 ohm. Determine the power dissipated in the 1 ohm resistor. Also find the current through and voltage drop across the 1 ohm resistor.

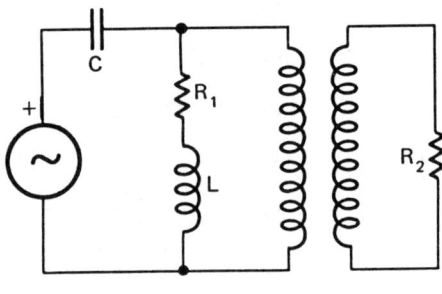

Figure P 10.1

10.2 Determine the output voltage, $e_o(t)$, for the electrical circuit shown in Fig. P10.2. The source $e(t) = 20 \cos 2,000t$, V and the turns ratio between primary and secondary is 2.

10.3 A source has the equivalent circuit shown in Fig. P10.3. The source supplies power to a load resistance of 8 ohms. To dissipate maximum power in this load resistance, it is proposed to use a transformer which may assumed to be ideal.

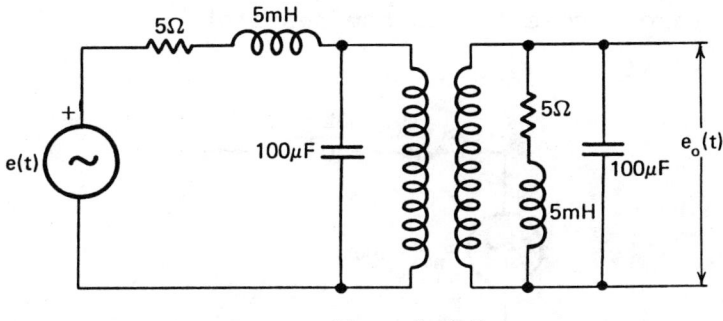

Figure P 10.2

Determine the turns ratio of the transformer which will result in maximum power being dissipated in the 8 ohm load. For this transformer, calculate the power transferred to the load, when the voltage source has an rms value of 20 V.

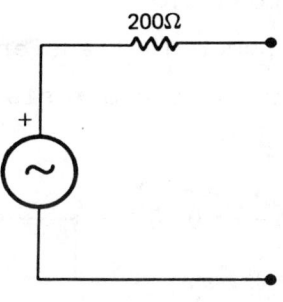

Figure P 10.3

10.4 An electric generator has a constant source voltage and frequency. The equivalent circuit of the generator is shown in Fig. P10.4. When the generator is connected to a 20 ohm resistive load through an ideal transformer, find the turns ratio that permits the maximum power to be transferred to the load. Determine also the maximum power and the voltage across the 20 ohm resistor.

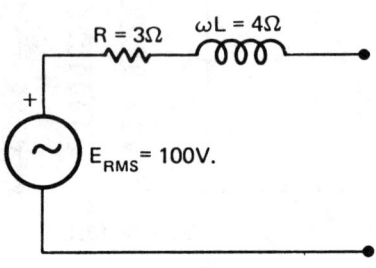

Figure P 10.4

10.5 A load resistance of 8 ohms is to be connected to the circuit shown in Fig. P10.5 so that maximum power is dissipated in the resistance.
(a) Find the turns ratio of an ideal transformer to achieve this condition.
(b) When maximum power is transferred, what is the voltage across the 8 ohm resistor?

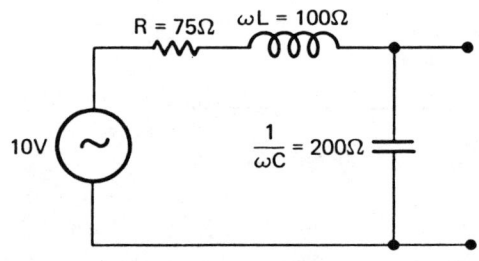

Figure P 10.5

10.6 A hydraulic press is represented in Fig. P10.6. The cross-sectional area of the smaller cylinder is A_1 and that of the larger cylinder is A_2. Assuming that the fluid is incompressible and its mass is negligible, represent the device as an ideal transformer.

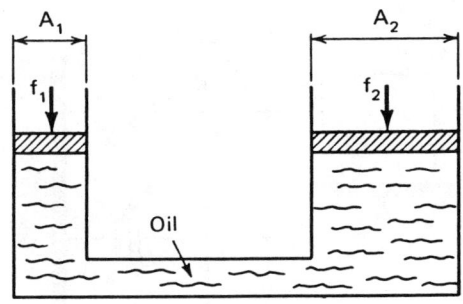

Figure P 10.6

10.7 A hydraulic press [Fig. P10.7] with $A_1 = 0.5$ m^2 is used to life an automobile of mass 1,000 kg.

(a) Find the force required at the input, f.

(b) Find the equivalent mass referred to the input.

(It is assumed that g = 9.81 m/s^2).

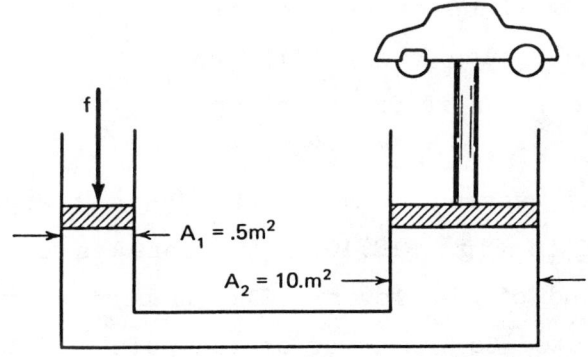

Figure P 10.7

10.8 A mechanical system contains a mass, spring, damper and lever as shown in Fig. P10.8. Determine the damping ratio of the system.

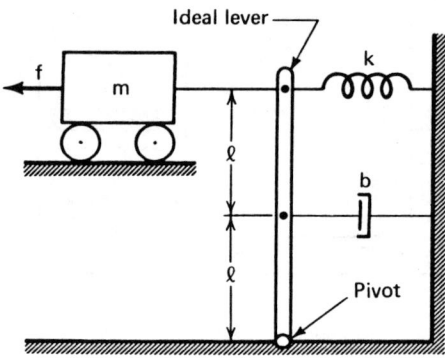

Figure P 10.8

10.9 The mechanical arrangement shown in Fig. P10.9 contains a mass, spring, damper and ideal lever. Find the position on the lever (x) at which the damper must be connected so that the resulting system will be critically damped. The mass m = 1 kg, the spring k = 4 N/m, the damper, b = 9 N.s/m and the lever is 0.5 m in length.

10.10 A mass, spring, damper and lever arrangement is shown in Fig. P10.10. The mass is connected to the end of the lever. The damper and spring are connected to the lever at the points shown.
   (a) Find the equivalent primary circuit for this arrangement.
   (b) Determine the damping ratio of the system.

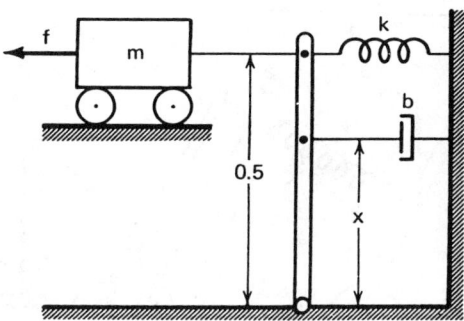

Figure P 10.9

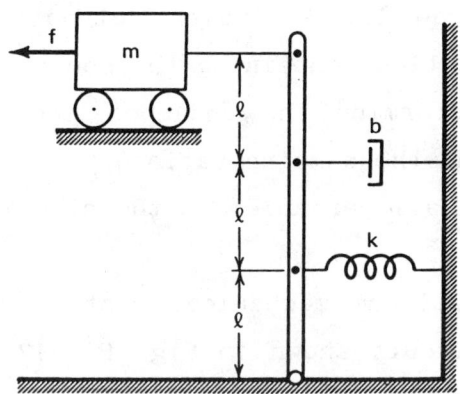

Figure P 10.10

10.11 In the system shown in Fig. P10.11, mass m slides over an oil film which provides viscous damping b_2 = 1.25 N.s/m. The other elemental values are: b_1 = 3 N.s/m and k = 2 N/m. A velocity input v(t) is applied at point A.

(a) Draw the circuit analog of the system showing the levers as ideal couplers.

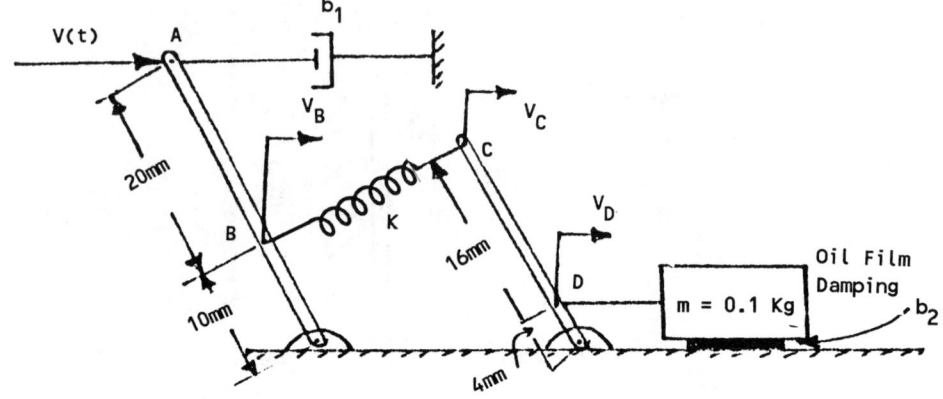

Figure P 10.11

(b) Draw the equivalent circuit referred to the section containing the source.

(c) Determine the undamped natural frequency ω_N and the damping ratio δ.

(Label all variables on the circuits).

10.12 A rotational mechanical system may be modeled by the circuit shown in Fig. P10.12. The system is used to drive a load which is equivalent to a damper of 1 N·s/rad.

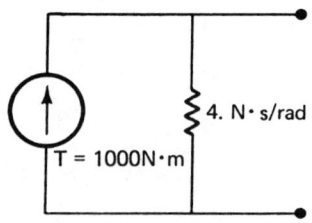

Figure P 10.12

(a) If the drive system is connected directly to the load, calculate the power transferred to the load.

(b) It is desired to transfer the maximum power possible from the system to the load. Find the gear ratio which must be used to accomplish this. In addition, find the value of the maximum power and the resulting angular speed of the load shaft.

10.13 In the rotational system shown in Fig. P10.13, the gears are considered as ideal couplers and the damping B_L = 62 N.m.s/rad is the load. The other elemental values are: B = 8 N.m.s/rad, J = 20 kg.m^2, K = 18 N.m/rad, T(t) = 10 cos 2t N.m. The gear N_1 contains 200 teeth and the gear N_2 contains 302 teeth.

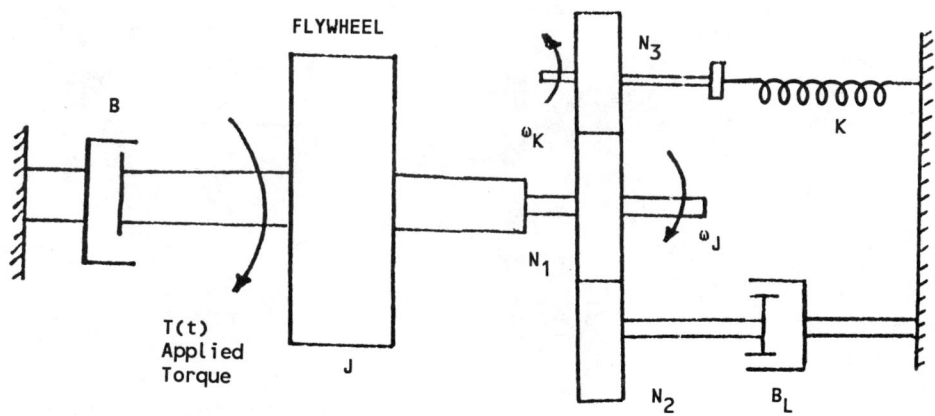

Figure P 10.13

(a) Draw the circuit analog of the system showing the gears as ideal couplers.

(b) Reduce the circuit to one in which there are no couplers, with all elements referred to the section containing the flywheel.

(c) Find the number of teeth N_3 on the gearwheel attached to the spring, if maximum power is to be transferred to the load B_L.

10.14 In a simplified sketch of a pressure-intensifier system shown in Fig. P10.14, the differential-area piston is assumed to be an ideal coupler and the fluid resistance R_2 is considered as the load.

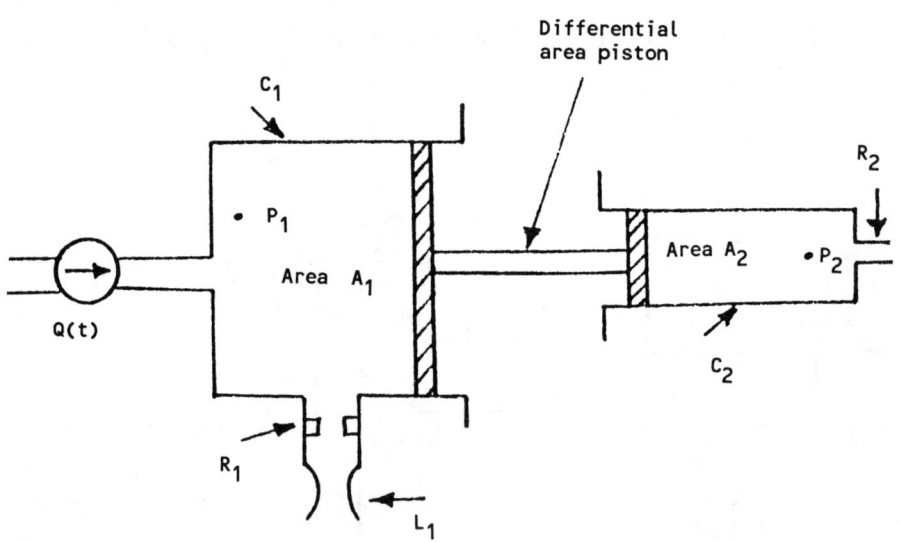

Figure P 10.14

The various elemental values are: $R = 10(10^6)$ $N.s/m^5$, $R_2 = 2(10^8)$ $N.s/m^5$, $L_1 = 3(10^1)$ $N.s^2/m^5$, $C_1 = 14(10^{-10})$ m^5/N, $C_2 = (10^{-10})$ m^5/N and $Q(t) = 2 \sin 10t$ kilopascals.

(a) Draw the analogous electrical circuit of the system, showing the ideal coupler.

(b) Redraw the circuit showing all elements referred to the primary side of the coupler.

(c) Determine the coupling ratio 'm' required, if maximum power is to be dissipated in R_2.

CHAPTER 11

NUMERICAL SOLUTIONS

11.1 INTRODUCTION

At this point, we have available powerful methods for the analysis of physical systems. However, it is clear that for anything beyond a second order system, the methods shown would be highly cumbersome. Although not obvious, this comment applies equally well to solutions based on the Laplace Transform where the problem becomes that of finding its inverse. A list of transform pairs extending to fourth order functions is surprisingly difficult to use due to the many forms such functions can take. A further difficulty is that of interpreting the final solutions and some form of computation becomes highly desirable.

However, if any form of computation becomes necessary, it seems more logical to seek a solution to the differential equation directly. Most large digital computers have at least one standard algorithm available and even small personal computers are able to provide solutions in a reasonable time, provided the order of the system function is not too large. A further advantage of the numerical solutions which are considered in this chapter is that they are readily extended to nonlinear equations where the Laplace transform is not applicable in any case.

The main problem is to formulate the differential equation in a form which is suitable for numerical solution. This chapter is not intended to be a detailed discussion of numerical methods and all that need be noted is that round-off errors in a computer make differentiation potentially inaccurate. Although small differences in the variables are used in numerical integration, the most reliable algorithms include checks to ensure that the errors do not exceed some maximum permissible level that is specified by the user. In this respect, a digital computer is quite similar to an analog computer where integration is much to be preferred. Probably, the simplest approach to rearranging a differential equation to make it suitable for integration is that which was found convenient for analog computation. It involves a graphical representation of the differential equation in the form of a block diagram which may be used an aid to obtain a set of first order differential equations that forms an alternate statement of the original equation. This is also known as the **state-space** form which will be discussed in Section 11.3.

11.2 BLOCK DIAGRAM OF DIFFERENTIAL EQUATION

In Chapter II, we considered mathematical operations on a function, the main ones being differentiation and integration. At that time, the interest was in extending the ability to differentiate and integrate algebraic expressions to that of differentiating and integrating functions in graphical terms. It is already shown that it is possible to think of each of the storage elements as being devices which can differentiate or integrate a variable. For example, the force acting through an ideal mass is proportional to the derivative of its velocity. Although there are limitations since the mass cannot be completely isolated, this forms the basis of an accelerometer.

We may, therefore, consider that it is possible to have a **device** which, on paper, acts as an ideal integrator. That is, whatever signal is applied to its input will be integrated and appear at its output terminal. To some extent, this picture is based on the integrator of an electronic analog computer which can come close to the ideal. It may be noted that the earliest analog computers were mechanical in nature, and were called **differential analyzers**. The mechanical integrator was somewhat similar to the common instrument used for measuring small areas known as a **planimeter**. The first step in the process of interconnecting the components of such computers to simulate a differential equation is to form a diagram in which each component is represented by a block. In the case of the electronic analog computer, it is also necessary to revise this diagram to account for a reversal of sign which is normally present in the amplifiers and to scale the values so that the signal voltages are within their range. Presently, the operational amplifier (OA) has become an off-the-shelf component like a resistor or a capacitor, it can be conveniently used as an integrator and its performance comes close to that of the ideal provided that the signal voltages are kept within the linear range of operation. Also, from the block diagram which forms the basis of programming, it is a very convenient method of converting an ordinary differential equation into a form suitable for numerical solution. Although it certainly is not the purpose of this chapter to consider analog computation in any detail, it seems fair to acknowledge that in many situations, the fact that the integral of a signal is available without the delay that is inevitable in a digital simulation still makes the technique of analog computation valuable, especially in instrumentation.

The block symbols corresponding to the mathematical operations that we require are shown in Fig. 11.1 where all are considered to be ideal. As a first step, let us consider a second order differential

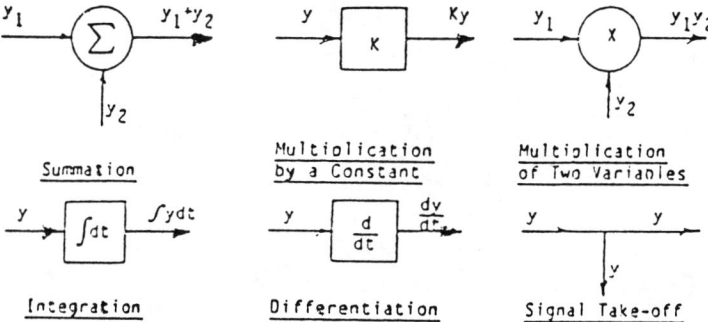

Figure 11.1 The Block Symbols

equation as an example to describe the process.

$$a_2 y'' + a_1 y' + a_0 y = f(t) \qquad ..(11.1)$$

where $y'' = \dfrac{d^2 y}{dt^2}$, and $y' = \dfrac{dy}{dt}$.

This equation is rearranged so that the highest order derivative term is equated to all the other terms. We get:

$$y'' = -\frac{a_1}{a_2} y' - \frac{a_0}{a_2} y + f(t) \qquad ..(11.2)$$

We now hypothesize that this highest order derivative will be available at the output of the summing device. If this is to be the case, then the signals which are applied to this summer must be $-(a_1/a_2)\, y'$, $-(a_0/a_2)\, y$ and $+f(t)$. Fig. 11.2 shows this first stage in the construction of the block diagram.

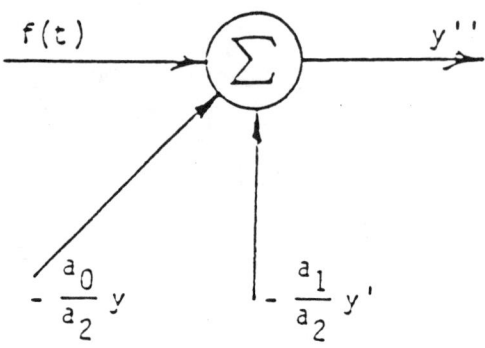

Figure 11.2 The signals at the Summer

If the signal at the output of the summer is the second derivative, it can be integrated to yield the first derivative which in turn can be integrated to yield the variable itself. The diagram is completed by taking the y' signal, multiplying it by $-(a_1/a_2)$, connecting this as an input to the summer, repeating this for the y signal (a_0 replacing a_1), and finally connecting the driving function, f(t), to the input of the summer. Fig. 11.3 shows the complete diagram. It must be emphasized that Fig. 11.3 contains exactly the same information as that in the original equation (11.1), the only difference is the graphical nature of showing it. The same procedure may be used to represent

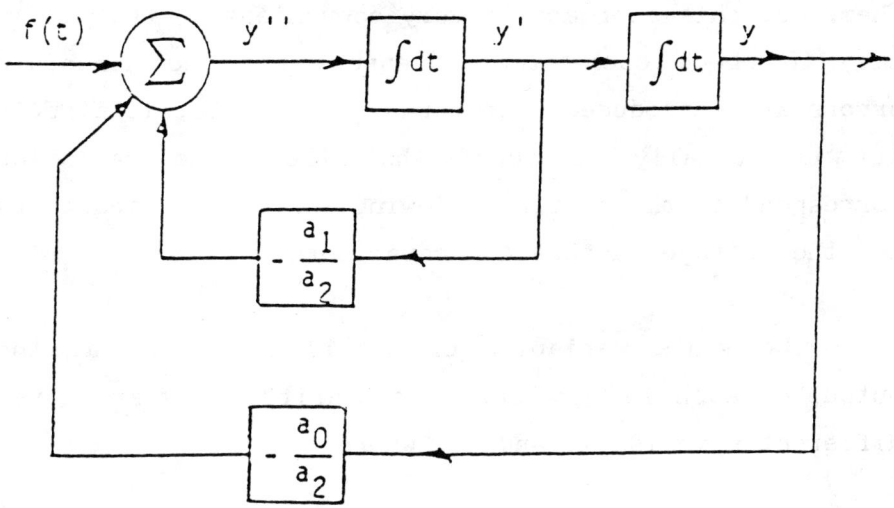

Figure 11.3 Block diagram of second order equation

equations of higher order.

11.3 STATE EQUATIONS

We may now observe that if the output of each integrator is known at any value of time, all other variables may be obtained using only simple arithmetic operations. These variables are sufficient to characterize the system or represent its **state**. By taking the output of each integrator as one of a new set of variables, we effectively convert a single differential equation of a second order into two simultaneous differential equations of first order. In general, it is possible to convert a differential equation of order **n** into **n** simultaneous first order differential equations. These are the **state equations** of the system. There are more elegant ways of obtaining them, but this approach is very convenient in that it is easy to check each step to ensure that no algebraic errors are introduced. In terms of the electric circuit it will usually be found that the state variables correspond to the currents flowing through the inductors and the voltages across the capacitors.

The state variables of (11.1) are those at the output of each integrator. For clarity, they are given different symbols, x_1 and x_2, where

$$x_1 = y \qquad \qquad ..(11.3)$$

$$x_2 = \frac{dy}{dt} = x_1' \qquad \qquad ..(11.4)$$

The second derivative of the original variable is now expressed in terms of the state variables in an alternative form of (11.2). That is:

$$x_2' = \frac{d^2y}{dt^2} = -\frac{a_1}{a_2} x_2 - \frac{a_0}{a_2} x_1 + \frac{1}{a_2} f(t) \qquad ..(11.5)$$

Eqs. (11.4) and (11.5) are expressions for the first derivative of each of the state variables. They may be conveniently rewritten in matrix form as

Numerical Solutions

$$\begin{bmatrix} x'_1 \\ x'_2 \end{bmatrix} = \begin{bmatrix} 0 & 1 \\ -\dfrac{a_0}{a_2} & -\dfrac{a_1}{a_2} \end{bmatrix} \begin{bmatrix} x_1 \\ x_2 \end{bmatrix} + \begin{bmatrix} 0 \\ \dfrac{1}{a_2} \end{bmatrix} u(t) \quad ..(11.6)$$

The general form of these equations for a system of order n is, therefore,

$$\mathbb{X}' = \mathbb{A}\mathbb{X} + \mathbb{B}\mathbb{U} \quad ..(11.7)$$

where $\mathbb{X}$ is the vector of state variables,

$\mathbb{U}$ is the vector of driving functions,

$\mathbb{A}$ is a square matrix of order n,

and $\mathbb{B}$ is a matrix having n rows and columns corresponding to $\mathbb{U}$.

If we recall that the original variable was y rather than x, and note that in general the output of a system may consist of more than one variable or a combination of variables, the general form of the output equations becomes

$$\mathbb{Y} = \mathbb{C}\mathbb{X} + \mathbb{D}\mathbb{U} \quad ..(11.8)$$

where $\mathbb{Y}$ is now a vector of output variables,

$\mathbb{C}$ is a matrix having the same number of rows as

Y and n columns,

and D is a matrix of appropriate dimensions.

In the case of the example of the second order equation (11.1), the output (11.3) can be expressed in this more general form as:

$$Y = \begin{bmatrix} 1 & 0 \end{bmatrix} \begin{bmatrix} x_1 \\ x_2 \end{bmatrix} + [0] f(t) \qquad ..(11.9)$$

It is useful to note that the state equations corresponding to a system function are readily obtained. The procedure simply requires that the corresponding differential equation first be written. If the numerator of the system functions has any terms in s, all that need be recalled is the procedure for obtaining the solution to composite driving functions, namely to obtain the response to a unit step first of all, and then evaluate the response as the appropriate combination of the step response and its derivatives. Again the procedure is most simply explained by considering the following system function.

$$G(s) = \frac{b_1 s + b_0}{a_2 s^2 + a_1 s + a_0} \qquad ..(11.10)$$

The corresponding differential equation is

$$a_2 y'' + a_1 y' + a_0 y = b_1 \frac{df(t)}{dt} + b_0 f(t) \qquad ..(11.11)$$

We have already seen that the solution is obtained by first getting the step response, that is the solution to

$$a_2 y'' + a_1 y' + a_0 y = U_s(t) \qquad ..(11.12)$$

This is exactly the same as (11.1) except for the driving function and therefore the first of the state equations is the same as before, namely

$$\begin{bmatrix} x_1' \\ x_2' \end{bmatrix} = \begin{bmatrix} 0 & 1 \\ -\dfrac{a_0}{a_2} & -\dfrac{a_1}{a_2} \end{bmatrix} \begin{bmatrix} x_1 \\ x_2 \end{bmatrix} + \begin{bmatrix} 0 \\ \dfrac{1}{a_2} \end{bmatrix} U_s(t)$$

$$..(11.13)$$

The solution for the actual driving function is obtained, as before, by multiplying this solution for x_1 by b_0 and adding it to the derivative of x_1 (namely x_2) multiplied by b_1 so that the second of the state equations take the form

$$[y] = \begin{bmatrix} b_0 & b_1 \end{bmatrix} \begin{bmatrix} x_1 \\ x_2 \end{bmatrix} \qquad ..(11.14)$$

Thus we are now in a position to obtain the state equations of any of the systems that have been considered up to this point, provided the system function is first derived. Any of the methods considered in Chapter VII may be used and the only remaining decision is that of finding a suitable algorithm for the solution of the state equations.

11.4 NUMERICAL INTEGRATION

In essence a numerical integration consists of choosing a small increment or step for the time variable, and then using the state equations to calculate the resulting change in each state variable, assuming that the rate of change is constant throughout the interval. This assumption is not normally correct, and it is necessary to limit the step size to keep the resulting error within acceptable limits. At first sight, the solution to this problem might appear to be that of choosing an extremely small increment, but this would result in excessive time of computation which would not be acceptable, even on the largest and fastest of digital computers. Practical methods such as those to be considered in Section 11.5 also estimate the curvature, usually indirectly in order to avoid computation of derivatives, and seek a better fit while still using a reasonable size of step.

Although any algorithm in which the step size is kept constant is unlikely to be efficient, it is useful to start by considering such an approach as providing an

Numerical Solutions

approximate solution where the first few points obtained may be reasonably close. Let us consider a first order differential equation to show the detials. Consider

$$a_1 x' + a_0 x = U_s(t) \qquad \qquad ..(11.15)$$

Since it is a first order differential equation, there is only one state equation

$$x' = -\frac{x_0}{x_1} x + \frac{1}{x_1} U_s(t)$$

$$= -ax + bu \qquad \qquad ..(11.16)$$

where $a = \dfrac{x_0}{x_1}$, $b = \dfrac{1}{x_1}$, and $u = U_s(t)$

If the increment in time is h, then the change in x during this first interval is

$$\dot{x}_{(0)} = (-ax_{(0)} + bu)h \qquad \qquad ..(11.17)$$

where the subscripts within brackets indicate the number of points that have been previously been computed. The value of x at the end of the interval is presumed to be

$$x_{(1)} = x_{(0)} + h\, \dot{x}_{(0)} \qquad \qquad ..(11.18)$$

This process is repeated until enough points have been obtained to define the solution.

To illustrate the process and demonstrate the inaccuracy of such a simplistic algorithm, the following numerical example will compare the computed values with the exact values obtained from an analytical solution.

Example 11.1

Calculate the first ten points of the solution to the differential equation given below using a constant step size of 0.2 second, if the initial condition $x(0) = 0$.

$$2x' + x = U_s(t) \qquad \qquad ..(11.19)$$

Solution
The first step is to rearrange the equation in state space form. That is:

$$x' = -0.5\,x + 0.5\,U_s(t) \qquad \qquad ..(11.20)$$

Because of the repetitive nature of the calculations, it will be found to be most convenient, if they are arranged in tabular form. The columns are numbered so that the probability of operating on the wrong value is reduced. The calculation of each point proceeds on one line from left to right with the value in column #7

Table 11.1

#1	#2	#2	#4
t	x	0.5x	0.5u
0	0	0	0.5
0.2	0.1	0.5	0.5
0.4	0.19	0.095	0.5
0.6	0.271	0.1355	0.5
0.8	0.3439	0.17195	0.5
1.0	0.40951	0.204755	0.5
1.2	0.468559	0.2342795	0.5
1.4	0.5217031	0.26085155	0.5
1.6	0.56953279	0.284766395	0.5
1.8	0.612579511	0.30628976	0.5
2.0			

#5	#6	#7
		#2 + #6
#4 − #3	0.2(#5)	
0.5	0.1	0.1
0.45	0.09	0.19
0.405	0.081	0.271
0.3645	0.0729	0.3439
0.38205	0.06561	0.40951
0.295245	0.05049	0.468559
0.2657205	0.0531441	0.5217031
0.23914845	0.04782969	0.56953279
0.215233605	0.043046721	0.612579511
0.193710244	0.0487420489	0.65132156

(The table has been put in two portions because of space considerations).

becoming the starting point for the calculation of the next point on the following line column #2. Referring to (11.16), the coefficients are a = 0.5 and b = 0.5 and it is usually convenient to include these values directly in the column headings. Table 11.1 shows the various calculations.

This should be sufficient to illustrate the process, but some comments or observations are appropriate. In column #3, we can see that in the last complete line, the value is not exactly one half of that in column #2, this being due to the rounding off error in the particular calculator used for these calculations. A similar error can be expected in column #6. Unless such calculations are repeated using the same model of calculator, there are likely to be differences. Students are therefore urged to perform these calculations in detail using their own calculator. Small differences are to be expected and no solution is considered to be **wrong**, although all have different errors. In many situations where the limitations on the accuracy of parameter values to three significant figures (for example) means that there is no point in pretending that any results are accurate to a greater number of significant figures. However, in numerical integration calculations, it is extremely important to use all of the available accuracy of a calculator during the integration.

From the analytical solution using the methods of

Chapter V, we can obtain the response as

$$x = (1 - \varepsilon^{-0.5t}) U_s(t) \qquad ..(11.21)$$

Substituting values of t from 0 to 2.0 gives the values shown in Table 11.2, which can be compared with those in Table 11.1.

Table 11.2

Time	Numerical	Analytical
0	0	0
0.2	0.1	0.095162582
0.4	0.19	0.181269247
0.6	0.271	0.259181779
0.8	0.3439	0.329679954
1.0	0.40951	0.393469340
1.2	0.468559	0.451188364
1.4	0.5217031	0.503414696
1.6	0.56953279	0.550671036
1.8	0.612579511	0.593430340
2.0	0.651321560	0.632120559

It is convenient to compare the values at a time

equal to the time constant, namely 2.0 seconds. Evidently, this simple algorithm has significant error and a smaller step size is necessary, if there is to be any chance of obtaining a reasonably accurate solution. It is not practical to do this manually, but the result obtained on a small computer when the step is reduced to 0.02 is 0.633967659 which still cannot be considered acceptable. When the interval is further reduced to 0.002 seconds, the value becomes 0.632304577 which now is in agreement as far as the third significant figure. However, the time required to get to this point in the solution is far greater than that using a method such as the one described in the next section. One surprising aspect of this approximate algorithm is that it does integrate to the correct final value. It is important to note that the final value does not constitute a valid test of the validity of an algorithm for integration.

Example 11.2

Calculate the first four points of the solution of the differential equation given below using a constant step size of 0.2 second, if the initial conditions are $y(0) = 0$ and $y'(0) = 0$.

$$8y'' + 6y' + y = U_s(t) \qquad ..(11.22)$$

Solution

Again the first step is to rearrange the equation in state space form. In this case, there are two state

Numerical Solutions

variables to be chosen. That is:

$$x_1 = y$$
and $$x_2 = y'$$ ..(11.23)

and the state equations are:

$$\begin{bmatrix} x'_1 \\ x'_2 \end{bmatrix} = \begin{bmatrix} 0 & 1 \\ -0.125 & -0.75 \end{bmatrix} \begin{bmatrix} x_1 \\ x_2 \end{bmatrix} + \begin{bmatrix} 0 \\ 0.125 \end{bmatrix} U_s(t)$$

..(11.24)

Because of the repetitive nature of the calculations, again it is found to be most convenient if they are arranged in a tabular form. However, we now have rather more values to calculate and the width of a single page is not sufficient. The calculations are therefore arranged in columns and are shown in Table 11.3.

Since y is equal to x_1, our solution is found in line #2 with the additional point at t = 1.0 second found in line #11 as 0.0916454688. For all other points, there are two values (rows #10 and #11) which must be transcribed from the bottom of each column to lines #2 and #3 respectively at the top of the next column.

Table 11.3

#			
#1	Time	0	0.2
#2	x_1	0	0
#3	$x_2 = x_1'$	0	0.25
#4	0.125#2	0	0
#5	0.75#3	0	0.01875
#6	$0.125 U_s(t)$	0.125	0.125
#7	x_2'	0.125	0.10625
#8	0.2#3	0	0.005
#9	0.2#7	0.025	0.02125
#10	x_1 = #2 + #8	0	0.005
#11	x_2 = #3 + #8	0.025	0.04625

0.4	0.6	0.8
0.005	0.01425	0.0270875
0.04625	0.641875	0.079203125
0.000625	0.00178125	0.0033859375
0.0346875	0.048140625	0.0594023438
0.125	0.125	0.125
0.0896875	0.075078125	0.0622117188
0.00925	0.0128375	0.015840625
0.0179375	0.015015625	0.0124423438
0.01425	0.0270875	0.042928125
0.0641875	0.079203125	0.0916454688

Numerical Solutions

Evidently as the order of the differential equation increases, the number of calculations increases dramatically and it is impractical to perform such a solution manually beyond the first few points. It is also impractical from the point of view of the errors involved. A comparison of this solution and the analytical solution shown in Table 11.3 shows that at best this algorithm can give only an approximation.

Table 11.4

Time	Numerical	Analytical
0	0	0
0.2	0	0.00237856874
0.4	0.005	0.00905591684
0.6	0.01425	0.0194022678
0.8	0.0270875	0.0328585399
1.0	0.042928125	0.0489290936

Just as with Example 11.1, it is necessary to reduce the step size in order to get a solution which is reasonably close to the correct value, and again the time required for such a computation is prohibitive. The need for a better algorithm which also includes a check of the error should now be quite apparent.

11.5 RUNGE-KUTTA METHODS

There are many useful methods available for the

numerical solution of differential equations, these being considered in texts on numerical analysis and methods. Since it is not the purpose of this chapter to examine the problem of numerical integration with a view to determining the **best** method, only the very common Runge-Kutta approach will be considered. This method has the merit that it is self-starting and most computer centers provide access to at least one form. For situations where the integration is lengthy, it may be beneficial to change to another algorithm requiring less computation after the first few points have been obtained. However, for the small systems that we are in a position to consider, there is little to be gained by introducing such complexity.

From our examination of the crude method described in Section 11.4, it should be evident that the main problem is related to computation of the derivatives of the state variables in a form which allows for the fact that they cannot be expected to be constant over a complete interval. The basic philosophy of the Runge-Kutta approach is to probe ahead of the current value of our time variable to estimate values of the derivatives at several points. A weighted average is then obtained and used to make the final calculation for a particular step. The result is that a fourth order Runge-Kutta method can be expected to require four calculations of the derivatives for each step. If a particular implementation also includes an accuracy check as part of control over the size of the step, it

Numerical Solutions

may require additional computation for some steps.

Before considering a set of state equations, let us consider a single equation having the form

$$x' = f(x, t) \qquad \qquad ..(11.25)$$

given the initial condition that

$$y = y_0 \text{ at } x = x_0 \qquad \qquad ..(11.26)$$

The derivative is calculated at three or more closely adjacent points in the x and t domains and the expression used for the next value of x takes the form

$$x_{i+1} = x_i + u_0 k_0 + u_1 k_1 + u_2 k_2 + u_3 k_3 + \ldots \qquad ..(11.27)$$

where the values of k_i are obtained by substitution in (11.25) and the values of u_i are the weighting factors mentioned above and arranged such that

$$u_0 + u_1 + u_2 + u_3 + \ldots = 1 \qquad \qquad ..(11.28)$$

If the step size is h, the values of k_1 are defined by means of the following set of equations.

$$k_0 = h\,f(x_i, t_i)$$

$$k_1 = h\,f(x_i + b_{1,0}k_0,\ t_i + a_1 h)$$

$$k_2 = h\,f(x_i + b_{2,0}k_0 + b_{2,1}k_1,\ t_i + a_2 h)$$

$$k_3 = h\,f(x_i + b_{3,0}k_0 + b_{3,1}k_1 + b_{3,2}k_2,\ t_i + a_3 h)$$

$$..(11.29)$$

The next step consists of finding values of the coefficients such that when the expressions in (11.28) are expanded about the operating point using a Taylor's series expansion, the two formulations are in agreement for a specified number of terms in the expansion. For a single equation such as that which we are considering, a widely used fourth-order Runge-Kutta formula has the form

$$x_{i+1} = x_i + \frac{1}{6}(k_0 + 2k_1 + 2k_2 + 2k_3) \quad ..(11.30)$$

where $k_0 = h\,f(x_i, t_i)$

$$k_1 = h\,f(x_i + \tfrac{1}{2}k_0,\ t_i + \tfrac{1}{2}h)$$

$$k_2 = h\,f(x_i + \tfrac{1}{2}k_1,\ t_i + \tfrac{1}{2}h)$$

$$k_3 = h\,f(x_i + k_2,\ t_i + h)$$

$$..(11.31)$$

Numerical Solutions

Because of the nature of the equilibrium equations used to model our physical circuits and systems, this set of equations is of little value. Indeed, the reader should now be able to obtain the analytical solution for a first order system far more quickly than by using this numerical solution. For a set of simultaneous differential equations in the form of state equations, a commonly used fourth order formula is:

$$x_{i+1} = x_i + \frac{1}{6} k_0 + 2(1 - \sqrt{\frac{1}{2}})k_1 + 2(1 + \sqrt{\frac{1}{2}})k_2 + k_3$$

..(11.32)

where $k_0 = h\, f(x_i, t_i)$

$$k_1 = h\, f(x_i + \frac{1}{2} k_0,\, t_i + \frac{1}{2} h)$$

$$k_2 = h\, f\left[x_i + (-\frac{1}{2} + \sqrt{\frac{1}{2}})\, k_0 + (1 - \sqrt{\frac{1}{2}})k_1,\, t_i + \frac{1}{2}h\right]$$

$$k_3 = h\, f\left[x_i + (-\sqrt{\frac{1}{2}})k_2 + (1 + \sqrt{\frac{1}{2}})k_2,\, t_i + h\right]$$

..(11.33)

Evidently, expressions such as those above have been developed for use in digital computer programs and are not expected to be used in manual calculations. They have been included so as to give some idea of the complexity which has the final result of their being simple to use.

Example 11.3

Use the fourth-order Runge-Kutta method to obtain the solution to the first order differential equation considered in Example 10.2, namely

$$8 y'' + 6y' + y = U_s(t) \qquad ..(11.34)$$

The step size is to be 0.2 seconds as before.

Solution

When rearranged into state variable form as before this corresponds exactly to (11.25) other than that now x is the vector set of variables and our task is simply that of determining the numerical values for the matrix

$$\mathbb{A} = \begin{bmatrix} 0 & 1 \\ -0.125 & -0.75 \end{bmatrix} \qquad ..(11.35a)$$

$$\mathbb{B} = \begin{bmatrix} 0 \\ 0.125 \end{bmatrix} \qquad ..(11.35b)$$

A typical Runge-Kutta subprogram on a computer will require the user to program the function

$$y' = \mathbb{A} x + \mathbb{B} u \qquad ..(11.36)$$

usually in the form of a subroutine. For a step response, u is a constant, but the nature of the steps in x and t in (11.32) is such that a time-varying driving function may be used. Although we shall not do so in this text, this is the point where any nonlinearity can be included in the model. Naturally the details of linking this part of the computation with the Runge-Kutta program will depend on the particular installation. The results obtained are given in Table 11.5.

Table 11.5

Time	Numerical	Analytical
0	0	0
0.2	0.00237857389	0.00237856874
0.4	0.00905592636	0.00905591684
0.6	0.01940228130	0.01940226780
0.8	0.03285855720	0.03285853990
1.0	0.04892911440	0.04892909360

The almost complete agreement between the numerical and analytical solutions is evident. Depending on the particular computer and the error criterion used, the computed values can be expected to vary somewhat. However, as long as the reader is capable of checking the initial response (including derivatives) using the methods described in Chapter IV, it should be possible to judge whether or not a numerical solution is likely to be valid. Again, it must be emphasized that the final value does not constitute a valid check.

Example 11.4

Obtain the step response corresponding to the system function

$$G(s) = \frac{1}{s^3 + 2s^2 + 2s + 2} \qquad ..(11.37)$$

Solution

This system function corresponds to the differential equation

$$y''' + 2y'' + 2y' + 2y = U_s(t) \qquad ..(11.38)$$

We therefore choose the state variables

$$x_1 = y$$

$$x_2 = x_1' = y'$$

$$x_3 = x_2' = y''$$

and thus $x_3' = y''' = -2y'' - 2y' - 2y + U_s(t)$
$\qquad\qquad\quad = -2x_3 - 2x_2 - 2x_1 + U_s(t)$

$$..(11.39)$$

These are the state equations which are expressed in matrix form as

$$\begin{bmatrix} x_1' \\ x_2' \\ x_3' \end{bmatrix} = \begin{bmatrix} 0 & 1 & 0 \\ 0 & 0 & 1 \\ -2 & -2 & -2 \end{bmatrix} \begin{bmatrix} x_1 \\ x_2 \\ x_3 \end{bmatrix} + \begin{bmatrix} 0 \\ 0 \\ 1 \end{bmatrix} U_s(t)$$

..(11.40)

We are now in a position to provide the matrices A and B as data for computation noting that the output variable required is

$$y(t) = x_1(t) \qquad ..(11.41)$$

and the result is plotted in Fig. 11.4.

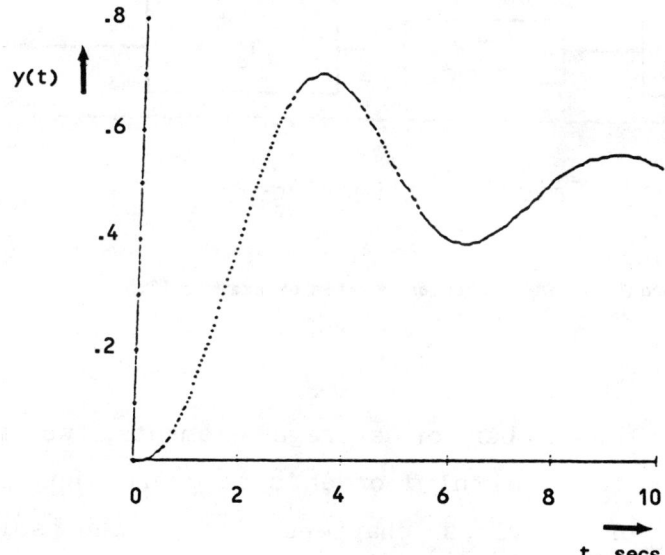

Figure 11.4 The step response of the transfer function of example 11.4

Example 11.5

For the mechanical system shown in Fig. 11.5, derive the equilibrium equations and use them to obtain the system function

$$G(s) = \frac{V_1}{F} \quad \quad ..(11.42)$$

Obtain the state equations and use them to obtain a numerical solution for $V_1(t)$, if the system is initially at rest and a force of 1.0 N is suddenly applied. The parameters have the following values: $m_1 = m_2 = 1.0$ kg, $k = 1.0$ N/m, $b = 0.1$ N.s.m.

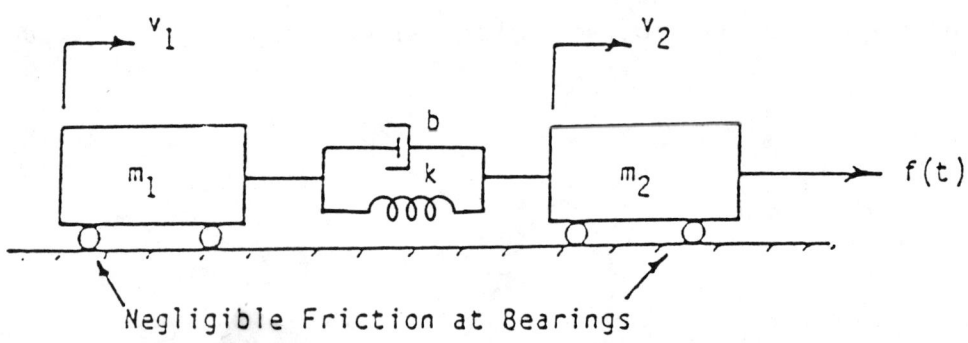

Figure 11.5 The mechanical system of example 11.5

Solution

From the number of storage elements, we should expect this to be a third order system and thus beyond the scope of previous chapters. But the solution proceeds by using all of the techniques previously described. First, the analogous circuit shown in Fig.

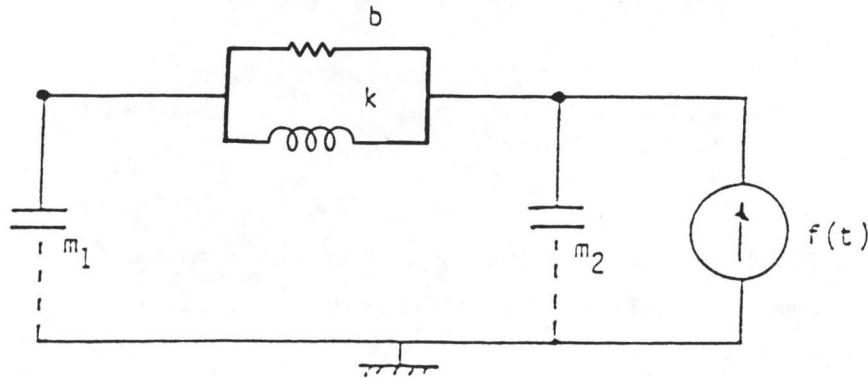

Figure 11.6 The electrical equivalent circuit of the mechanical system of figure 11.5

11.6 is obtained and from this the node-to-datum equations are written by inspection.

$$\begin{bmatrix} m_1 s + b + \dfrac{k}{s} & -(b + \dfrac{k}{s}) \\ -(b + \dfrac{k}{s}) & m_2 s + b + \dfrac{k}{s} \end{bmatrix} \begin{bmatrix} v_1 \\ v_2 \end{bmatrix} = \begin{bmatrix} 0 \\ F \end{bmatrix}$$

..(11.43)

The system function may be obtained by any suitable means. It will normally be better to continue with the symbols for each parameter so that changes in the numerical values can conveniently be investigated without repeating the complete solution. The result is:

$$\frac{V_1}{F} = \frac{k + bs}{sk(m_1+m_2) + b(m_1 + m_2)s^2 + m_1 m_2 s^3}$$

$$= \frac{1 + 0.1s}{2s + 0.2s^2 + s^3} \qquad ..(11.44)$$

using the numerical values given in the statement of the problem. As before, the differential equation is first taken as

$$y''' + 0.2\, y'' + 2y' = U_s(t) \qquad ..(11.45)$$

We may choose our state variables:

$$\begin{aligned}x_1 &= y(t)\\ x_2 &= x_1{'}\\ x_3 &= x_2{'}\end{aligned}$$

Thus $\quad x_3{'} = -2x_1 - 0.2x_2 - x_3 + U_s(t) \qquad ..(11.46)$

The matrix $\mathbb{A}$ is therefore given by

$$\mathbb{A} = \begin{bmatrix} 0 & 1 & 0 \\ 0 & 0 & 1 \\ 0 & -2 & -0.2 \end{bmatrix} \qquad ..(11.47)$$

and in this case the matrix $\mathbb{B}$ is a vector given by

Numerical Solutions

$$\mathbb{B} = \begin{bmatrix} 0 \\ 0 \\ 1 \end{bmatrix} \qquad \qquad ..(11.48)$$

These may be compared with (11.13) and (11.14). At this stage, we have state equations corresponding to a system function whose numerator is unity. In this example, the actual output variable required is a combination of x_1 and x_2, obtained from the numerator of the system function:

$$v_1 = x_1 + 0.1\, x_2 \qquad \qquad ..(11.49)$$

The first ten points of the resulting computation are shown in Table 11.5. They have been rounded out for clarity of presentation, but it must be emphasized that the complete values have been used throughout the computation. Also, the first ten seconds of the response are plotted in Fig. 11.7.

From the analogous circuit shown in Fig. 11.6, we can check that the initial velocity and acceleration are both zero, and examination of the plotted response shows that our numerical solution is in agreement.

11.6 SUMMARY AND DISCUSSIONS

In this chapter, state variable formulation of a given system is discussed. This is accomplished by forming the differential equation governing the system.

TABLE 11.5

Time	Velocity
0	0
0.1	0.000615
0.2	0.003275
0.3	0.00874
0.4	0.0179
0.5	0.0314
0.6	0.0500
0.7	0.0742
0.8	0.1043
0.9	0.1406
1.0	0.1834

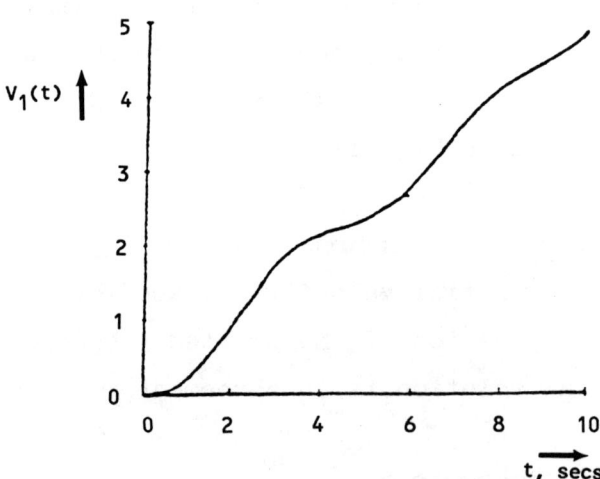

Figure 11.7 The response $V_1(t)$ of the mechanical system of example 11.5

It is also possible to obtain the state equations from the given system directly. This is not discussed here, as this is outside the scope of the present text. In addition, some numerical techniques for obtaining the time-domain solution of the state equations are discussed. It is to be emphasized that many such methods are available in the literature.

11.7 PROBLEMS

11.1 Obtain the state equations corresponding to the system function

$$G(s) = \frac{1}{s^2 + 5s + 2}$$

11.2 Obtain the state equations corresponding to the system function

$$G(s) = \frac{s + 3}{s^2 + 5s + 2}$$

11.3 Obtain the state equations corresponding to the system function

$$G(s) = \frac{0.01s^2 + 0.2s + 1}{s(s^4 + 0.4s^3 + 6.01s^2 + 0.6s + 3)}$$

11.4 Use the approximate method to calculate the point $y(0.2)$ of the solution to the differential equation

$$\frac{d^2y}{dt^2} + 3\frac{dy}{dt} + 2y = U_s(t)$$

(a) with a step size of 0.1 second.

(b) with a step size of 0.05 second.

The initial conditions are $y(0) = 0$ and $y'(0) = 0$.
From the analytical solution, determine the error in each case and express it as a percentage of the true value.

11.5 Use the approximate method to calculate the point $y(0.04)$ of the solution to the differential equation

$$\frac{d^2y}{dt^2} + 4\frac{dy}{dt} + y = 2 U_s(t) + U_i(t)$$

(a) with a step size of 0.02 second.

(b) with a step size of 0.01 second.

The initial conditions are $y(0) = 0$ and $y'(0) = 0$.
From the analytical solution, determine the error in each case and express it as a percentage of the true value.

11.6 A mechanical system has a system function relating output velocity to input force given by

$$G(s) = \frac{V_o}{F_i} = \frac{1}{s^2 + 12s + 100}$$

(a) Use the approximate method with a step size of 0.01 second to obtain the velocity at $t = 0.02$ second when $f_i(t) = U_s(t)$.

(b) Use a recognized method such as Runge-Kutta to obtain the response up to $t = 1.0$ second.

(c) From part (b), determine the maximum value of $v_o(t)$ and the time at which it occurs.

11.7 An electric circuit has the system function

$$G(s) = \frac{E_o}{E_i} = \frac{4(s+1)}{s^2 + 4s + 4}$$

If the input is a step of 1 volt, determine the output voltage using a Runge-Kutta algorithm for a period of 5 seconds.

Is this circuit critically damped? Note the shape of the response.

BIBLIOGRAPHY

1. D.E. Johnson, J.L. Hilburn and J.R. Johnson, Basic Electric Circuit Analysis, Fourth Edition, Prentice-Hall Inc., 1990.
2. J.W. Nilsson, Electric Circuits, Third Edition, Addison-Wesley Publishing Co., 1989.
3. J.D. Irwin, Basic Engineering Circuit Analysis, McMillan Publishing Co., 1989.
4. J.S. Shearer, A.T. Murphy and H.H. Richardson, Introduction to System Dynamics, Addison-Wesley Publishing Co., 1967.
5. W.T. Thomson, Theory of Vibration with Applications, Second Edition, Prentice-Hall Inc., 1981.
6. L. Meirovitch, Elements of Vibration Analysis, McGraw-Hill Book Co., 1975.

7. S. Timoshenko, Vibration Problems in Engineering, John Wiley and Sons., 1974.
8. J. L. Shearer and B. T. Kulakowski, Dynamic Modeling and Control of Engineering Systems, McMillan Publishing Co., 1990.
9. R. C. Dorf, Modern Control Systems, Fifth Edition, Addison-Wesley Publishing Co., 1989.
10. V. L. Streeter, Fluid Mechanics, McGraw-Hill Book Co., 1981.
11. J. M. Kirshner and S. Katz., Design Theory of Fluid Components, Academic Press, 1975.
12. F. Kreith, Principles of Heat Transfer, Intext Educational Publisher, N. Y., 1973.
13. J. P. Holman, Heat Transfer, McGraw-Hill Book Co., 1981.

ANSWERS TO SELECTED PROBLEMS

CHAPTER II

2.5(c) $y = \frac{2}{5} U_p(t) - \frac{4}{5} U_p(t - 5) + \frac{2}{5} U_p(t - 10)$

2.9(a) $g(t) = 5t^2 - 20t + 20, \quad t \geq 2$
$ = 0, \quad\quad\quad\quad\quad\quad\quad t \leq 2$

2.14(a) $g(t) = 2a\, U_s(t) - 3a\, U_s(t - T_1)$

$$+ \frac{a}{T_3 - T_2} U_r(t - T_2) - \frac{a}{T_3 - T_2} U_r(t - T_3)$$

2.19(a) $g(t) = U_r(t) - 1.5\, U_r(t - 1) - 1.5\, U_r(t - 2)$

$$+ 3\, U_r(t - 3)$$

(c) $h(t) = 2 U_s(t-1) - 2 U_s(t-3)$

2.25 (b) 2780 m

CHAPTER III

3.1 (a) Capacitance, $C = 0.005$ F

    (b) Inertance, $L = 5{,}000$ N·s^2/m^5

    (c) Capacitance, $C = 2.5(10^{-10})$ m^5/N

    (d) Resistance, $R = 250$ ohms

3.2 (a) Damper (Resistance) $B = 0.6$ N·m·s/rad

    (b) Spring (Inductance) $k = 800$ N/m

    (c) Capacitance $C = 40(10^{-10})$ m^5/N

    (d) Mass (Capacitance) $m = 2.25$ kg

3.3 $i(3.36) = 11.22$ A, $W(3.36) = 125.89$ joules

3.5 $W(5) = 0.045$ joules

3.7 $q_o = 5$ m^3/s, $h = 3$ m

3.8 (a) $l = 4.58$ m, $d = 0.0985$ m

    (b) $t = 19.53$ s

    (c) $W = 0.00733$ joules

Answers to Selected Problems 745

3.16 (a) $F = 50 (\sin 4t) U_s(t)$, N

 (b) $m = 5$ kg

3.18 25 percent decrease in current

CHAPTER IV

4.7 (a) 5 A, 0 V, 3.57 A, 7.14 V

 (b) 0 V/s, 100 A/s

4.10 (a) 0 rad/s, 100 rad/s, 500 N.m

 (b) 5 rad/s^2

4.11 (a) 6 N/s, 1.6 N/s

 (b) 6.67 N, 1.67 m/s

CHAPTER V

5.2 (a) circuit 1 → $e_1(t) = 4 \varepsilon^{-t/RC}$, $i(t) = \frac{4}{R} \varepsilon^{-t/RC}$

 circuit 2 → $e_1(t) = 4(1 - \varepsilon^{-t/RC})$,

 $i(t) = \frac{4}{R} \varepsilon^{-t/RC}$

 (b) circuit 1 → $e_1(t) = \frac{20}{C} U_r(t)$, $i(t) = 20 U_s(t)$

 circuit 2 → $e_1(t) = 20R\, U_s(t)$, $i(t) = 20 U_s(t)$

5.6 (a) $p(t) = 49,000\, \varepsilon^{-t/40}$ Pa, (b) $h = 2.32$ m

5.8 (a) 61,313 Pa,

(b) 61,313 Pa

5.11 (a) $v_m(0.25) = 0.6$ m/s

(b) $t = 1.89$ s

(c) no effect

5.12 $v(t) = 0.69 \, \varepsilon^{-0.31t}$ m/s,

$x(t) = 0.69(1 - \varepsilon^{-0.31t})$

5.13 (a) $t = 0.96$ s

(b) $h = 8.34$ m

5.14 $x = 4$ m

5.16 $v = 1.47$ m/s, $x = 2.72$ m

5.18 $\omega_1(t) = \left[5t - \frac{5}{2}(1 - \varepsilon^{-2t})\right] U_s(t)$

$- \left[5(t - 10) - \frac{5}{2}(1 - \varepsilon^{-2(t - 10)})\right] U_s(t - 10)$

rad/s

5.19 $\omega(t) = \frac{T}{B} - \left[\frac{T}{B} + \omega_o\right] \varepsilon^{-Bt/J}$

5.21 $J = 1.035(10^5)$ kg·m^2

Answers to Selected Problems   747

5.24 (a) 131.25 W

 (b) $\tau = 38.7$ hours

CHAPTER VI

6.1 $i_R(t) = \varepsilon^{-5,000\,t}(0.01 - 50t)$ A

 $e_L(t) = \varepsilon^{-5,000\,t}(1 - 2,500t)$ V

6.3 $Q(t) = \left[125\,\varepsilon^{-2,000t} - 110\,\varepsilon^{-4,000t}\right](10^{-6})$ coulombs

6.8 24 kPa

6.9 $P(t) = \left[2,000 - 8,000\,\varepsilon^{-15t} + 6,000\,\varepsilon^{-20t}\right]$

6.10 $v(t) = 0.5\cos 4t$ m/s

6.11 (a) $v(t) = (\varepsilon^{-t} - \varepsilon^{-2t})$ m/s

 (b) $v(t) = (\frac{1}{2} - \varepsilon^{-t} + \frac{1}{2}\varepsilon^{-2t})$ m/s

 (c) $v(t) = (-\varepsilon^{-t} + 2\,\varepsilon^{-2t})$ m/s

 (d) $v(t) = 2\left[\varepsilon^{-t} - \varepsilon^{-2t}\right]U_s(t)$

  $-\left[\varepsilon^{-(t-2)} - \varepsilon^{-2(t-2)}\right]U_s(t-2)$

6.13 $v(t) = \left[20\varepsilon^{-2t} - 20\varepsilon^{-4t}\right]U_s(t)$

$\qquad - \left[20\varepsilon^{-2(t-2)} - 20\varepsilon^{-4(t-2)}\right]U_s(t-2)$

$x(t) = \left[5 - 10\varepsilon^{-2t} + 5\varepsilon^{-4t}\right]U_s(t)$

$\qquad - \left[5 - 10\varepsilon^{-2(t-2)} + 5\varepsilon^{-4(t-2)}\right]U_s(t-2)$

6.16 $\omega(t) = 90 - 45\varepsilon^{-t}(t+2)$ rad/s

6.17 $\omega(t) = (50 - 150\varepsilon^{-2t} + 100\varepsilon^{-3t})$ rad/s

$T_B(t) = (1,125 - 3,375\varepsilon^{-2t} + 2,250\varepsilon^{-3t})$ N.m

6.18 $\theta(t) = \left[3.06t^2 - 6.89t + 3.05 - 2.5\varepsilon^{-t}\right.$

$\qquad\qquad\left. - 0.55\varepsilon^{-8t}\right]$

CHAPTER VII

7.1 (a)

$$Z_{dp} = \frac{Z_1[(Z_2+Z_5)(Z_3+Z_4)+Z_3Z_4]+Z_2[Z_3Z_4+Z_5(Z_3+Z_4)]}{(Z_2+Z_5)(Z_3+Z_4)+Z_3Z_4}$$

Answers to Selected Problems   749

(b) $Z_{dp} = \dfrac{Z_1 Z_5 (Z_3 + Z_4) + Z_1 Z_2 (Z_3 + Z_4 + Z_5)}{(Z_1 + Z_2)(Z_3 + Z_4 + Z_5) + Z_5 (Z_3 + Z_4)}$

7.3 (a) $Z_{dp} = \dfrac{RLC_1 C_2 s^3 + LC_2 s^2 + R(C_1 + C_2)s + 1}{L_1 C_1 C_2 s^3 + (C_1 + C_2)s}$

(b) $Y_{dp} = \dfrac{2.5 s^2 + 3.5 s + 1}{5 s^2 + 10 s + 2}$

7.5 (a) $\dfrac{E_o}{E_i} = \dfrac{R_2 (R_1 C_1 s + 1)}{R_2 (R_1 C_1 s + 1) + R_1 (R_2 C_2 s + 1)}$

(b) $\dfrac{E_o}{E_i} = \dfrac{R_2 C_1 s + 1}{(LC_2 s^2 + 1)(R_2 C_1 s + 1) + (R_2 + Ls) R_1 C_1 C_2 s^2 + R_1 (C_1 + C_2)s}$

7.6 $\dfrac{E_o}{E_i} = \dfrac{2s + 1}{5s + 1}$

7.7

$Y_{dp} = \dfrac{L_2 C_1 C_2 s^3 + R_2 C_1 C_2 s^2 + (C_1 + C_2)s}{R_1 L_2 C_1 C_2 s^3 + [R_1 R_2 C_1 C_2 + L_2 C_2] s^2 + [R_1 C_1 + R_1 C_2 + R_2 C_2] s + 1}$

7.9 $\dfrac{Q_o}{Q_i} = \dfrac{1}{L_2 Cs^2 + R_2 Cs + 1}$

7.11 $\dfrac{V_3}{F} = \dfrac{b \dfrac{k}{s}}{(m_1 s + b)[m_2 s (m_3 s + \dfrac{k}{s}) + m_3 k] + m_1 bs (m_3 s + \dfrac{k}{s})}$

7.14 $f(t) = 0.25 \, \varepsilon^{-6t} \sin 8t$

7.15 $p_o(t) = 3t/2 + 3(1-\varepsilon^{-2t})/4$

7.20 $e_o(t) = 0.5 - 0.173 \, \varepsilon^{-0.44t} + 0.673 \, \varepsilon^{-4.56t}$

CHAPTER VIII

8.4 $Q(t) = 35.4(10^{-6}) \cos(0.1t - 45°) \, m^3/s$

8.6 at 0.4 Hz $b_{eq} = 3.98 \, N.s/m$, $k_{eq} = 10.05 \, N/m$

8.7 2.58 N.s/m

8.9 $\omega(t) = \dfrac{T}{B[1+(\omega J/B)^2]^{1/2}} \cos(\omega t - \phi) - \dfrac{BT}{B^2+(\omega J)^2} \varepsilon^{-Bt/J}$

$M = \dfrac{1}{B[1+(\omega J/B)^2]^{1/2}}$, $\phi = \tan^{-1} \dfrac{\omega J}{B}$

8.10 $\omega(t) = 0.52 \sin(5t - 87°) \, rad/s$

CHAPTER X

10.1 $P = 2.84 \, W$. $\Delta e = 1.68 \, V$, $i = 1.68 \, A$

Answers to Selected Problems

10.3   n = 5,   P = 0.5 W

10.4   n = 1/2,   P = 625 W,   V = 111.9 V

10.7   (a) f = 490.5 N
       (b) m_e = 2.5 kg

10.8 $\delta = b/(8\sqrt{km})$

10.10 $\delta = 2b/(3\sqrt{km})$

INDEX

A-type (storage) element 116, 120, 418, 419, 628
A-type source, 162
Abscissa of convergence, 595
Acceleration, 140
Across variable, 5, 18, 103, 118, 140, 153
Active, 161
Admittance
 driving point, 427, 481
 equivalent, 422
 generalized, 419
 parameters, short circuit 643
 referred, 682
Amplitude
 exponential function, 418, 431
 impulse, 41
 parabola, 37
 ramp, 33
 sinusoid, 496
 step, 24
Analogy, 18, 141
 force-current, 138

Angular frequency, 49
Angular velocity, 141
Argument. 26

Bilateral, 108, 124, 132, 447
Bode plot, 561
  magnitude, 563
  phase, 574
Block diagram, 61
Branch, 444
Bulk modulus, 130

Capacitance
  electrical, 114
  fluid, 128
  mechanical, see mass
  thermal, 157
Capacitor, ideal, 114
Characteristic equation, 275, 347
  from system function, 437
  roots of, 347, 437
Charge, 106
Circuit, 1, 271, 339
  analogous, 19, 442
  first order type 1, 273
  first order type 2, 281
  formulation of, 196
  general first order, 269, 287
  general second order, 340
Compatibility, 345
  see also Kirchhoff's voltage law
Complex
  number, 47, 406
  root, 347
  s-plane, 438
  variable, 406, 593
Components, 3
  coupled, 640
  dissipative, 6, 111, 124 418, 419, 625
  electrical, 106
  fluid, 117
  mechanical, 137
  storage, 6, 113
  thermal, 152, 160
  two-terminal, 6, 104
Conductance, electrical 108
Conduction, thermal, 153
Conservation
  of electrical charge, 103
  of energy, 103
  of mass, 103

Index

Continuity, 103, 202
   see also Kirchhoff's current law
Convection, 153, 156
Convolution, 607
Coulomb friction, 141, 172
Coupler, 653, 666
   electrical, 654
   equivalent network, 679
   fluid, 678
   ideal, 653, 666
   mechanical (gears), 674
   mechanical (lever), 667
   mechanical (pulley), 670
   mechanical (unpivoted lever), 660
Critically damped system, 351
Crystal, quartz, 582
Current, electric, 106
Current source, 166
D'Alembert's principle, 209
Damped sine wave, 438
Damper
   rotational, 143
   translational, 142
Damping, 141
   constant, 348
   critical, 351
   ratio, 340
Decomposition of equations, 287, 355
Degrees of freedom, 196
Delay, see shifting
Delayed function, 27
Density, fluid, 118, 124
Determinant, 451
Differentiation, 41, 67
   graphical, 69
Direction, see polarity convention
Displacement. 140
Distributed model, 16
Doublet, see unit doublet
Dynamic, 10

Effective value, 528
Electromotive force, see voltage
Element, 105
   see also component
Elemental equation, 105
Elements
   analogous, 183
   storage, 216
Energy, 107
   dissipation, 110
   electrical, 110

fluid, 132
   mechanical, 148, 152
   storage, 113, 115
Equilibrium, 19, 103
   equations, 441
Equivalent circuit, 137, 679
   A-type element with
      initial charge, 245
   final step response, 230
   initial response, 228
   Norton, 480
   Thevenin, 474
   T-type element with
      initial flow, 246
Euler, 49, 349, 521
Exponential function, 46
Exponential input, 406

Faraday's law, 111
Filter, 536
   band-pass, 549
   bridged-T, 550
   high-pass, 543
   low-pass, 536
   Notch, 549
Final condition circuit,
   230
Final conditions, 224
Final value, 224, 276
Final value theorem, 611

First-order system, 271
   sinusoidal response,
      497
   step response, 273
Flow, 24, 118
   laminar, 121
   turbulent, 123
Flow divider, 433
Flow rate, thermal, 154
Flow source, 167
Fluid inductance, 125
Fluid inertance, 125
Flux linkage, 111
Force, 104, 138
   restoring, 148
Force source, 166
Forced response, 12, 274,
      357, 498, 509
Formulation of system
   equations, 340, 440
Fourier's law, 153, 154
Free-body diagram,
   104, 138
Frequency, 496
   resonant, 556
   response, 11, 496, 502.
      522
Friction, 141
   coulomb, 141, 172
   static, 141

Index

Viscous, 121, 141
Functions, 21
   advanced, 27, 34
   composite, 29, 34, 55
   decomposition of, 35
   delayed, 30
   driving, see input
   exponential order, 594
   multiplication of, 58
   singularity, 21
   system, 405, 431, 440
   transfer, see system function

Gears, 674
Generalized
   admittance, 419
   impedance, 405, 418
Graphical
   differentiation, 69
   integration, 66
Ground, 132, 196

Harmonic motion, 148
Head, 134
Heat flow, 153
Heat source, 167
Helmholtz, 558
Homogeneous equation, 273, 346

Homogeneous solution, 273, 346, 497, 509
Hooke's law, 129
Hybrid parameters, 652

Imaginary, 47, 406
Impedance, 287
   combination, 421
   driving point, 427, 474
   equivalent, 422, 690
   generalized, 405, 418
   mechanical, 420
   parameters, open-circuit, 640
   referred, 682
Impulse, see unit impulse
Impulse response 295, 371
Inductance
   electrical, 111
   fluid, see inertance
   mechanical, see spring
   mutual, 655
Inductor, 111
Inductor, ideal, 113
Inertance, 125
Initial condition
   circuit, 220, 223
Initial conditions, 217, 271

mathematical approach, 234
of T-type and A-type elements, 217
representation by circuit, 220, 223
Initial storage circuits, 243, 245, 246
Initial value theorem, 610
Input, 8, 273, 303
Insulation, thermal, 155
Integral,
  Contour, 597
  Convolution, 607
  definite, 62
Integration, 32, 61
  graphical, 66
  numerical, 714

Junction, 103, 198

Kirchhoff's current law, 208, 440
Kirchhoff's voltage law, 209, 441

L'Hopital, 413
Laplace transforms, 593
  definition, 593
  inverse, 596
  one-sided, 594
  properties, 597
Lever, 667
Linear, 14, 120
  equation, 356
  system, 303, 467
Load, 161, 170
Loop, 456
Lumped model, 16

Magnitude ratio, 503, 511, 522, 563
Mass, 16, 150
  mutual, 660
  rotational, 152
  thermal, 157
  translational, 151
  as two-terminal element, 151
Mathematical model, 3
Matrix
  admittance, 447, 643
  hybrid, 652
  impedance, 459, 642
  inverse, 469
Maximum power transfer, 531, 689
Mesh, 456

Mesh equations, 456
    impedance matrix, 459
Mobility, 420
Modeling, 2, 15
Moment of inertia, 152
Momentum, 125
Mutual
    inductance, 655
    mass, 660

Natural frequency
    damped, 349
    undamped, 340
Natural response, 12, 274
    of first order system, 274
    of second-order system, 346
Newton's law of cooling, 153, 156
Node, 196, 442
Node-to-datum equations, 444
    admittance matrix, 447
Norton, 480
Numerical Integration, 714
Numerical Solutions, 704

Ohm's law, 108

Open circuit voltage, 474
Operating point, 171
Operator, 175
Output. 8. 305
Overdamped, 347

Parabolic function, see unit parabola
Partial fraction expansion, 612
Particular solution, see forced response
Passive, 161
Pendulum, 11, 148
Permeability, 112
Permittivity, 115
Phase, 494, 503, 511. 522. 574
Piezoelectric phenomena, 582
Pipe, fluid resistance, 7
Piston, 136, 678
Polarity convention, 116, 201, 656
Pole, 437
Potential, 106
Potential divider, 433
Power, 103
    average, 528

electrical, 107
factor, 530
fluid, 119
mechanical, 140
sinusoidal steady-state, 526
Pressure, 119, 130
Pressure source, 163
Primary, 659
Pulley, 670
Pulse, 29
Pump, 125

Q-factor, 557, 586

Ramp function, see unit ramp
Reactance, 532
Real, 47, 347, 406
Reciprocity, 473, 642, 644
Rectangular pulse, 29, 311
Resistance
  electrical, 108
  fluid, 121
  mechanical, 141
  thermal, 154
Resistivity, electrical, 109
Resistor, ideal, 108
Resonance, 554
  parallel, 559, 584
  series, 555, 584
Response, 8
  complete, 276, 356, 504, 511
  composite function, 303, 371
  critically damped, 351
  exponential input, 407
  forced, 12, 274, 357. 498, 509
  impulse, 295, 371
  initial stored energy, 314, 381
  natural, 12, 274, 346
  overdamped, 347
  parabolic, 292
  ramp, 292, 371
  sinusoidal, 11, 496
  sinusoidal steady-state, 496, 502, 518
  step, 11, 273, 355
  underdamped, 348
Reynold's number, 123
Runge-Kutta methods, 723

Secondary, 659
Shifting, 27, 305
Short circuit current, 481

Sign convention, see polarity convention
Signal variable, 5, 17
Sine wave, 11, 50
Singularity functions, 24
Sinusoidal response, 495
Sinusoidal steady-state, 496, 502, 518
Solution, see response
Source, 161
   across variable, 162
   ideal, 162
   through variable, 166
Specific heat, 130, 157
Spring, 16, 145
   restoring force equivalent, 148
   stiffness, 145
   torsional, 146
   translational, 145
Steady-state, 11
Step function, see unit step
Step response,
   of first order systems, 273
   of second order systems, 355
Structure, 205
Superposition, 304, 315, 371, 471, 598
State equations, 705, 709
Symbol, compoennt, 6, 116, 127, 144, 147, 655, 707
System, 9
System function, 405, 431, 440

T-type storage element, 116, 217, 418, 419, 626
T-type source, 166
Tank, 131
Temperature, 154
Temperature source, 165
Terminal, fictitious, 131, 150, 152, 159
Thermal,
   capacitance, 157
   conducitivity, 154
   energy, 153
   power, 153
   radiation, 153
   resistance, 154
Thevenin, 474, 689
Through variable, 5, 18, 103, 118, 138, 153
Time constant, 272, 277
Torque, 138
Transfer function, see system function

Transformation ratio, 654, 689
Transformer, 654
Transient, 10, 502
Two-port networks, 639
Two-terminal-pair, 640

Underdamped, 348, 604
Unit
  doublet, 45
  impulse, 41
  parabola, 37
  ramp, 33
  step, 24

Valve, 24
Variables
  analogous, 182
  auxiliary, 33
  electrical, 106
  fluid, 118
  mechanical, 138
  thermal, 153
Velocity, 21, 140
Velocity source, 164
Viscosity, 122, 124
Voltage, 106
Voltage source, 162
Volume, 129, 134

Water hammer, 181
Weight, 174
Work, 106, 120

Zero, 438